Agricultural Development in Asia

Agricultural Development in Asia

Edited by

R. T. Shand

UNIVERSITY OF CALIFORNIA PRESS

BERKELEY AND LOS ANGELES

1969

630.95
S 528

First published in the United States by
University of California Press
Berkeley and Los Angeles, California

© Australian National University 1969

Printed and manufactured in Australia

SBN 520-01554-1

Library of Congress Catalog Card no. 72-75479

Foreword

IN most of the new Asian states which emerged from colonial rule after World War II, economic development was at first thought to be largely synonymous with industrialisation. Even among non-communists, the spectacular success of the Leninist strategy in the Soviet Union, with its emphasis on heavy industry in the framework of a socialist command economy, exerted a strong pull which was further reinforced by the initial achievements of the communist régime in China. Current Western thinking about economic development was at that time hardly less emphatic in stressing industrial development. To broad sections of politically articulate opinion in Asia, the very example of the West, whose wealth and power seemed so clearly to rest on industrial foundations, appeared to point in the same direction. So did the experience of the economic planners in the new countries who found it very much easier to plan for the modern industrial sector than for the large traditional sector of peasant agriculture.

In the past ten years the dangers of so unbalanced an approach to economic development have come to be increasingly recognised. A strategy of self-sufficient industrial development which was practicable for the Soviet Union with its immense area and natural resources has proved to be quite inappropriate to the smaller countries in Asia, as indeed in eastern Europe; they could not afford to neglect their export industries, and in most of the less developed Asian countries these were, and are bound to remain for some time, mainly agricultural. More important, with high and still rising rates of population growth, the problem of food supplies assumed increasing urgency, at any rate in the large overpopulated regions of India, Pakistan, and Indonesia. By the early 1960s, famine in India and widespread malnutrition in Java gave warning that these countries were rapidly approaching a Malthusian crisis of catastrophic proportions. Unlike the Soviet Union and Communist China, they were not nearly ready to cover large food deficits from imports financed by industrial exports. Gunnar Myrdal's monumental *Asian Drama* reflects the profound pessimism about the prospects for Asia inevitably induced by a study of India in the period before the new emphasis on agriculture in development strategy and the revolution in agricultural technology of the past five years.

The importance of Dr Shand's book lies in the contribution it makes, at a critical time, to a reassessment of these prospects. There are the striking technological advances in the form of new high-yielding varieties of crops such as rice, wheat, and rubber, and a new emphasis on agriculture in some of the main countries, especially India and Indonesia. There are also the unduly neglected success stories in agricultural development of some of the smaller countries of the region, such as Taiwan, Malaysia, and Thailand. By bringing together a group of ten specialists, each writing from intimate knowledge about the agricultural development in one country of the region, Dr Shand has achieved a total picture more balanced, authoritative, and up-to-date than could possibly be expected from any one pen.

No one who reads this book, especially the chapters by Sir John Crawford on India and Dr Penny on Indonesia, not to mention Mrs Richter on Burma, will come away confident that the immense problem of poverty in Asia, or even the problem of food supplies, will be easily solved. But one should at least see the possibility of progress—the main technical, economic, and socio-political ingredients of an effective approach to their solution.

As editor, Dr Shand set himself the task of giving the symposium a measure of cohesion by agreeing with his authors on a broad framework for each chapter including, first, a description of the major characteristics of the agricultural sector and of its role in the economy as a whole; second, an analysis of growth rates of output in agriculture and of their influence on the overall growth of the national economy; third, an attempt to identify factors which have promoted, and others which have inhibited, agricultural development; and fourth, some assessment of the potential for further growth in the agricultural sector, of the directions such growth might take, and of measures needed to promote it. If only because of differences from country to country in the availability of data and in stages of economic development, this framework was not to prevent each author from placing the emphasis and developing his subject as he thought best. As the reader will see, the authors have not failed to avail themselves of the latitude granted to them, but there is enough uniformity of approach to facilitate comparisons and Dr Shand has, in his concluding chapter, helped further to draw the threads together.

Inevitably, in a symposium of this kind, there are omissions. Particularly regrettable is the fact that it proved impossible to secure a chapter on Pakistan. On the other hand, to many readers outside Australia the inclusion of a chapter on the Australian Territory of Papua and New Guinea will be an unexpected bonus.

I commend this book as an outstanding contribution to the large
literature on economic development, outstanding both for its value to
students as an up-to-date analytical and descriptive account of the
main trends of agricultural development in non-communist Asia and
for its importance to policy-makers as a reassessment of the prospects
for agricultural development, and indeed general economic development, in one of the major regions of the world.

H. W. Arndt

Canberra
November 1968

Notes on Contributors

SIR JOHN CRAWFORD, C.B.E., Emeritus Professor of Economics, formerly Director of the Research School of Pacific Studies at the Australian National University, now Vice-Chancellor, led the Agricultural Section of the IBRD Mission to India in 1964-5 and again in 1967, and is recognised as an international expert on Indian economic problems.

E. S. CRAWCOUR, Professor and Head of the Department of Japanese at the Australian National University, is internationally recognised for his publications on the economic history of Japan.

E. K. FISK, Professorial Fellow in the Department of Economics, Research School of Pacific Studies, Australian National University, was recently a member of the Steering Committee for the Asian Agricultural Survey conducted by the Asian Development Bank. During fourteen years in the Malayan Civil Service, he was Chief Economist and first Head of the Economic Planning Division. He has since published extensively on the rural economy of the country.

RICHARD HOOLEY, currently with the National Planning Association and formerly Professor, School of Economics, University of the Philippines, is well known for his published works on the Philippine economy.

R. H. MYERS, now Professor of Economics, University of Miami, Coral Gables, and previously with the Department of Economics, Research School of Pacific Studies, Australian National University, is widely known for his historical work on the economies of China and Taiwan.

D. H. PENNY, formerly A.D.C. Visiting Professor of Economics in the Faculty of Agriculture, University of North Sumatra, is now a Fellow in the Department of Economics, Research School of Pacific Studies, Australian National University. He has been intimately concerned with the agricultural economy of Indonesia and has published extensively on the subject.

H. V. RICHTER, formerly with *The Economist* Intelligence Unit, now Research Fellow in the Department of Economics, Research School of Pacific Studies, Australian National University, has published on the economies of Malaya, Thailand, and Burma, with recent emphasis on Burmese agricultural problems.

VERNON W. RUTTAN, previously agricultural economist with the International Rice Research Institute and the University of the Philippines, now Professor and Head of the Department of Agricultural Economics, University of Minnesota, is a leading authority on the agricultural economy of the Philippines.

T. H. SILCOCK, Emeritus Professor of the University of Malaya, now with the Department of Economics, Research School of Pacific Studies, Australian National University, is internationally recognised for his work on the politics and economies of Southeast Asia, and has recently made a close study of Thai agriculture.

R. T. SHAND, Fellow in the Department of Economics, Research School of Pacific Studies, Australian National University, has published widely on the agricultural development problems of Papua-New Guinea.

Acknowledgments

MUCH advantage was gained by the circulation of individual chapters amongst the authors themselves, and in most cases, through presentation at seminars within the Department of Economics in the Research School of Pacific Studies at the Australian National University. A number of the authors also wish to record their gratitude to others as follows:

Japan: to Professor Rosovsky of Harvard University and Professor Ohkawa of Hitotsubashi University for helpful criticism and comment.

Taiwan: to the East Asian Research Center, Harvard University, for financial support to complete the latter stages of the study.

India: to colleagues in the IBRD Missions to India, particularly Dr W. D. Hopper of the Rockefeller Foundation, New Delhi, and Mr Wolf Ladejinsky, Consultant to the World Bank, for material used in the chapter, and to Mrs Elaine Treadgold, Research Assistant, Australian National University, in the course of writing.

Thailand: to Dr F. Nicholls and other officers of the Applied Science Research Corporation of Thailand, to Mr Chamnian Boonma of Kasetsart University, and to many agricultural officers, district officers, village schoolmasters, and farmers in different Thai villages.

Burma: to Mr L. J. Walinsky and to Burmese officials in Rangoon and Canberra, particularly U Kyaw Nyun.

Malaysia: to Tuan Haji Mohd. Ghazali bin Haji Jawi, Professor Ungku A. Aziz, Enche Thong Yaw Hong, Enche S. Selvadurai, Dr Agoes Salim, and many others.

Philippines: to Alice Mundo, Aida Recto, and John Sanders, for assistance in the preparation of tables and charts, and to L. M. Ilag, J. M. Lawas, A. J. Nyberg, G. L. Hicks, F. H. Golay, and R. E. Fonollera, for the use of previously unpublished materials.

Indonesia: to Professor H. W. Arndt, Mubyarto, Masri Singarimbun, Sulaeman Krisnandhi, Tan Hong Tong, Dahlan Thalib, and A. B. Lewis, for their helpful critical comments of earlier drafts.

Papua-New   to the Department of External Territories and in par-
Guinea: ticular to Dr V. Zmudski and Mr D. Lattin, for
                 providing useful unpublished data.

Asia: to Professor W. H. Nicholls of Vanderbilt University,
                 whose seminal article, 'An "agricultural surplus" as a
                 factor in economic development' (*Journal of Political
                 Economy*, Vol. 71, No. 1, 1963, pp. 1-29), influenced
                 the writing of this paper.

   Finally we wish to express our gratitude to Mr H. Gunther of the
Geography Department, A.N.U., for maps, and particularly to Mrs
Heather Harding, Secretary of the Department of Economics, Research School of Pacific Studies, Australian National University, and
to Mrs F. Johnson, Mrs D. Binnie, Mrs H. Michel, Mrs D. Love,
and Mrs E. Harriss of the Department, for their unfailing co-operation
in the preparation and typing of the manuscript. The editor also
wishes to add his own thanks to Mrs H. V. Richter for her counsel
and assistance during the editorial stage, and to Mrs R. P. Brown for
help in compiling the index.

                                                            R.T.S.

Canberra, 1969

Contents

Tables, Maps and Figures

Tables

Philippines

Indonesia

Maps

Figures

Currencies, Statistical Units, and Conversions

Currencies

Country	Unit	Rate per US dollar,[a] (selling) as at June 1968
Burma	Kyat (K)	4·80
India	Rupee (Rs)	7·58[b]
Indonesia	Rupiah (Rp)	320[c]
Japan	Yen (¥)	362
Malaysia	Malaysian dollar (M$)	3·09
Papua-New Guinea	Australian dollar (A$)	0·90
Philippines	Peso (₱)	3·91
Taiwan	New Taiwan dollar (NT$)	40·10
Thailand	Baht	20·80

[a] Multiple exchange rates operated in Taiwan from 1950 to 1958, Thailand post-war to 1954, the Philippines to 1961, and Indonesia to the present (1969). As members of the sterling area, Burma, India, Malaysia, and Australia (Papua-New Guinea) devalued with sterling by 30·5 per cent in September 1949. None of the countries listed devalued with sterling in November 1967.

[b] 1949-May 1960, around 4·78.

[c] Because of the devaluation over time of the exchange value of the currency and the complexity of the multiple exchange rate system, most economic statistics for Indonesia are given in US dollars. Rate cited is BE (Export Bonus) for October 1968.

Area

1 acre	= 0·40 hectare
1 hectare	= 2·47 acres
1 sq. mile	= 640 acres
	= 259 hectares
	= 2·59 sq. kilometres
1 sq. kilometre	= 100 hectares
	= 0·39 sq. mile
1 rai (Thailand)	= 0·40 acre
	= 0·16 hectare
1 acre	= 2·53 rai
1 hectare	= 6·25 rai

Area—continued

1 cho (Japan)	= 2·45 acres
	= 0·99 hectares
	= 10 tan
1 acre	= 0·41 cho
1 hectare	= 1·01 cho
1 chia (Taiwan)	= 2·40 acres
	= 0·97 hectare
1 acre	= 0·42 chia
1 hectare	= 1·03 chia

Weights

1 pound	= 0·45 kilogram
1 kilogram	= 2·20 pounds
1 long ton	= 2,240 pounds
	= 1·02 metric ton
1 short ton	= 2,000 pounds
	= 0·91 metric ton
1 metric ton	= 0·98 long ton
	= 1·10 short ton
	= 1,000 kilograms
	= 21·74 Spanish quintals
1 Spanish quintal (Philippines)	= 46·01 kilograms

Weight Equivalent of Volume Measures (approximate)

1 bushel of paddy (US)	= 45 pounds
	= 20·4 kilograms
1 bushel of paddy (Australian)	= 42 pounds
	= 19·1 kilograms
48·99 bushels of paddy (US)	= 1 metric ton
52·49 bushels of paddy (Australian)	= 1 metric ton
Burma: 1 basket of paddy	= 46 pounds
	= 20·9 kilograms
48 baskets of paddy	= 1 metric ton
1 basket of rice	= 76 pounds
	= 34·5 kilograms
Japan: 1 koku of rough rice (paddy)	= 106·3 kilograms
	= 234·4 pounds
1 koku of brown (husked) rice	= 150 kilograms
	= 330·7 pounds

Weight Equivalent of Volume Measures (approximate)—continued

Malaysia:	1 gantang of paddy	= 5·6 pounds
		= 2·5 kilograms
	394 gantangs of paddy	= 1 metric ton
	1 gantang of rice	= 8 pounds
		= 3·6 kilograms
	1 picul	= 133·3 pounds
		= 60·5 kilograms
	16·5 piculs	= 1 metric ton
	100 kati	= 1 picul
Philippines:	1 cavan of paddy (palay)	= 44 kilograms
		= 97 pounds
	22·7 cavans of paddy	= 1 metric ton
	1 cavan of milled rice	= 56 kilograms
		= 123·4 pounds
	1 cavan of shelled maize	= 57 kilograms
		= 125·7 pounds
	17·5 cavans of shelled maize	= 1 metric ton
Thailand:	1 picul	= 60 kilograms
		= 132·3 pounds
	16·7 piculs	= 1 metric ton

Yields

long tons per acre	× 2·51	= metric tons per hectare
pounds per acre	× 1·12	= kilograms per hectare
metric tons per hectare	× 0·40	= long tons per acre
kilograms per hectare	× 0·89	= pounds per acre

1

Japan, 1868-1920

E. S. Crawcour

THE case of Japan is famous as the most successful example, and perhaps still the only successful example of modern economic growth in Asia. The story of how she achieved this is now well-known and we need only remind ourselves of the broad outlines here. Although modern industrial development was insignificant before the 1880s, we take 1868 as our starting point because it was in that year that economic growth first became a national objective.

OUTLINE OF JAPANESE ECONOMIC DEVELOPMENT

Traditionally most Japanese economic historians have tried to fit the process of Japan's economic development to the stage theories of the European Historical school, and particularly to those of the Marxist school. They have been concerned with identifying these stages historically and with the process of evolution from one stage to the next. The feudal stage is generally supposed to have continued until the Restoration, but since the fifty years between 1868 and World War I is an awkwardly short time into which to fit the stages between feudalism and capitalism, there has been much discussion of the extent to which development can take place under feudalism. This approach seems to ignore the fact that the Industrial Revolution had already occurred in England, the United States, and some European countries, and was spreading rapidly by the mid-nineteenth century. There was no need for its spontaneous evolution in Japan. There is therefore now a tendency to move away from the idea of stages and pre-conditions and to concentrate on quantitative studies of the growth of the economy from the 1870s when quantitative data first became available.

There is no doubt that commercial developments of the kind described by Sheldon (1958) and agrarian developments of the kind described by Smith (1959) did occur in the century before 1868, but attempts to elevate them to the role assigned to the commercial and agrarian revolutions in English economic history have led to considerable confusion. The idea that Japanese economic growth was the

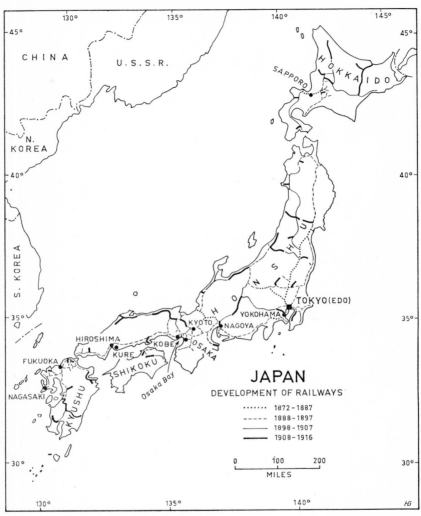

Map 1   Japan: development of railways, 1872-1916

Source: Tetsudōin [Railway Institute] (ed.), *Hompō tetsudō no shakai oyobi keizai ni oyoboseru eikyō* [The Social and Economic Influence of Railways in Japan], Hakubunkan, Tokyo, 1920, Map No. 2.

result of continuous and autonomous development can now scarcely be seriously maintained. Modern economic growth in Japan is clearly linked in some way with the impact of the West and Japan's response to it, and more specifically to the major changes of the mid-nineteenth century—the opening of foreign trade, the adoption of economic growth as a national objective, and the creation or adaptation of institutions favourable to it. Developments and trends before 1868 explain why Japan was able to respond as she did, but they contribute only indirectly to our understanding of how growth took place. In studying Japan's modern economic growth, therefore, we may start with the Japanese economy as it was in 1868 and omit the processes by which it reached this position, interesting though they may be.

Here we face our first major difficulty. Although it is only a hundred years ago, there is a remarkable lack of reliable factual information about the economic life of the time, and the impressions presented tend either to be coloured by the writer's theories of history or to suffer from a too uncritical reliance on what doubtful quantitative data we have. At one extreme we are shown a picture of a semi-serf rural population eking out a bare existence by subsistence agriculture and groaning under the weight of feudal oppression. At the other, we are told of prosperous farming communities with agricultural wage-labour, substantial rural industrial and commercial activity, and a relatively high degree of specialisation and division of labour—in fact a situation hardly differing from that of rural Japan thirty years ago. In part these caricatures reflect regional differences, but what we can say about the economy as a whole is rather limited.

The Japanese economy in 1868 was a traditional one with the usual heavy concentration of labour in rural pursuits. Some 80 per cent of the population of about 30 millions was classified as rural (agricultural, forestry, and fisheries) and only 4-5 per cent as engaged exclusively in secondary industry. A large proportion of the rural work force, however, spent some part—sometimes the whole—of its time in industrial pursuits, such as the processing of food and industrial crops, textiles, and other handicrafts. About 7 per cent of the population consisted of *samurai*—members of a ruling class military by origin and traditions, some of whom staffed the middle and upper echelons of the bureaucracy, and who as a class were heavy net consumers. Although a traditional economy, it was sophisticated and rather productive, at least in the more advanced regions, despite wide regional variation with more remote areas lagging far behind. Official regulation played a large, effective, and basically restrictive role. A well-developed system of taxation in kind provided the means both of extracting a surplus from the rural sector and of supplying food to quite large cities (Edo had a population of roughly one million). The opening of foreign trade in 1859 disturbed established patterns of production and distribution and initiated new economic opportunities. The Restoration of 1868 confirmed many of the economic and social changes that had been taking place unofficially for some time. The new government was firmly committed to bringing Japan into the modern world, and adopted modernisation and economic growth as national objectives (Crawcour 1965: 17-44).

The Meiji government's first step towards achieving the aims of economic growth was to create an institutional framework favourable to development in a capitalist free enterprise context. Between 1868 and 1875 existing legal social classes were abolished, restrictions on land use and on sale and mortgage of land were removed, mono-polistic guild-like privileges were withdrawn, and a national land tax was introduced, payable by the landowner in money and based on the estimated productivity of the land. A beginning was made with the establishment of a modern currency and financial institutions, stock exchanges, communications, and a national system of education. The budget, for which the land tax provided the largest single source of revenue, was used to build public utilities and other basic facilities and also to promote the development of industry through subsidies, provision of technical services, and the establishment of model factories. It was not until the 1880s, however, that these efforts began to bear fruit and not until around World War I that the modern sector began to set the pace of growth, either in terms of contribution to growth of output or as a source of investment funds. Meanwhile the development of agriculture and traditional industry, though possibly not an end in itself, played a major though steadily decreasing role. This was possible because the development of modern industry on the whole complemented rather than competed with existing methods of production. This was a key factor in achieving large increases in production with a limited amount of investment, but it was a strategy possible only because the traditional economy was capable of supplying most of the wants of the population and because those wants remained basically unchanged for a considerable time.

Until about 1900 only a relatively small part of the increase in secondary production was attributable to modern Western-style industry. Most of the increase was contributed by the development of small-scale factory production and 'manufacture' in the narrow sense. These are largely what Ohkawa and Rosovsky term 'hybrid' industries in which traditional methods of production were combined with some technical innovations such as chemical dyes or the appli-cation of electric power. It was around World War I that modern industry began to grow very rapidly, and it was at about the same time that the growth of agriculture appears to have faltered (Table 1.1). By 1920 the industrialisation of the Japanese economy was well under way, though features of a dual economy were to survive for a long time.

TABLE 1.1

SECTORAL CONTRIBUTIONS TO GROWTH:
QUINQUENNIAL INCREASE IN NATIONAL INCOME PRODUCED
(constant million yen)

Period	Primary	Secondary	Tertiary	Total
1872–82 to 1883–87	65 (17%)	115 (30%)	204 (53%)	384
1883–87 to 1888–92	176 (49%)	94 (26%)	92 (25%)	362
1888–92 to 1893–97	317 (44%)	172 (24%)	228 (32%)	717
1893–97 to 1898–1902	290 (34%)	265 (32%)	285 (34%)	840
1898–1902 to 1903–07	34 (15%)	10 (5%)	177 (80%)	221
1903–07 to 1908–12	249 (26%)	234 (25%)	466 (49%)	949
1908–12 to 1913–17	−15 (−2%)	442 (58%)	330 (44%)	757
1913–17 to 1918–22	384 (25%)	347 (22%)	827 (53%)	1,558

Source: Ohkawa *et al.* 1957: 17.

While the growth of industry as a whole has provided employment opportunities somewhat in advance of the growth of the working population, and has thus made possible a steady reduction in the proportion engaged in agriculture, the bulk of this increase in industrial employment has been in traditional industry. It is only under the impact of the very rapid growth of modern industry in the last few years that the proportion of workers in the 'traditional' sector as a whole, including both agriculture and 'traditional' industry, has begun to fall and the gaps in productivity and wages have begun to narrow significantly.

Agriculture in 1868

The Restoration had little immediate effect on the organisation of agrarian life, apart from raising the prestige of those more prosperous businessmen-farmers who had supported the Restoration movement. Even with the abolition of the semi-autonomous territorial governments in 1871, the village remained the most important unit of rural society. In all but the economically most advanced areas, the village headman and the senior members of the village, who by this time broadly corresponded with the most prosperous, continued to exercise a great deal of influence, and sometimes even detailed supervision, over the agricultural production of the village community. Legal changes in land tenure, confirming ownership for those previously registered as tax-paying cultivators (who were not necessarily the actual cultivators themselves), allowed a slow trend towards increased tenant-farming to proceed, but did nothing to change the small scale of farming. There were still wide regional differences in social organisation, land use, distribution of landholding and techniques, which make it impossible to speak of a typical Japanese farming community at this time (see Furushima 1963; Horie 1963).

At one extreme were the highly commercialised provinces around Osaka. In this region the impact of local 'feudal' administration was very light. Although some rice was grown, the major crops were cotton, oilseeds, vegetables, and other commercial crops. Double-cropping was the rule and the use of commercial fertilisers such as fish meal and oilcake was almost universal. Purchases of oilcake from nearby oil mills and of fish meal imported from the north of Japan often accounted for as much as two-thirds of current farm cash outlay.

In the cotton-growing districts of Izumi province along the southern coast of Osaka Bay, one of the economically most advanced areas, there was a relatively high degree of specialisation in the cotton and oil industries (Nakamura 1963: 37-119; Takebe 1960: Vol. VI, 127-65). The proportion of rice in total agricultural output was about the national average but much of it was grown for sale, and subsistence food production was very small. Paddy rice yields were quite high at about 3,000 kg per hectare. Although rice accounted for about two-thirds of the main crop, the remainder of the main crop and almost all of the second crop was devoted to cotton and oilseeds. The typical scale of farming was about 0·6 to 0·9 hectares, and since farm labour was expensive and often scarce, those who owned much more than this let the remainder either to efficient commercial tenant-farmers or in small plots to people whose main source of income was handicrafts. Some of these very small plots rented by persons engaged in industry or commerce were virtually housing sites, and the classification of their occupants as 'poor peasants' may therefore be misleading. About one-third of the land was let in one or other of these ways. Those in some 30-40 per cent of households, although classified as farmers, did not engage in agriculture at all but worked solely at handicrafts (mainly cotton ginning and spinning) or as artisans or retail traders. Of the remainder, most devoted not more than half of their time to agriculture. In some villages over half of the households worked solely at cotton-weaving for six months of the year. Many others worked at the textile trade for five or six hours every night, except during the dry summer months when they were fully occupied irrigating crops. Most villages had one or more oil-pressing mills employing wage labour.

Thus the majority of households bought the majority of their requirements, and every village contained anywhere from five to twenty shops
supplying all the necessities of life, including staple foods, vegetables,
and processed foods, as well as some luxuries (Andō 1958).

One landowner in this area (apparently representative of his class)
owning almost 10 hectares was able to engage enough labour to farm
3·4 hectares himself. The remainder was let to tenants. His main crop
was cotton, which was all sold, and contributed two-thirds of his
cash receipts. He grew sufficient staple food for his household and a
work force of eleven and sometimes had a small surplus of rice
and barley for sale. About 40 per cent of cash receipts were spent on
household expenses such as subsidiary foods, clothing, light, heat,
education, medical expenses, and entertainment. About 10 per cent
was saved and about 40 per cent was allocated to farming expenses,
of which half went for fertiliser (fish meal and urine) and one-quarter
for cash wages. Only about 13 per cent of his total cash income was
required for payment of taxes.

A substantial tenant-farmer farmed from one to two hectares and
employed from one to five wage labourers. He paid approximately
15 per cent of his gross output in rent and was therefore in a fairly
comfortable position. Below these were the so-called 'poor peasants',
in fact usually wage labourers or petty traders, who owned, or more
often rented, very small parcels of land big enough only for a house
and garden. In general the main characteristics of this type of agriculture were small-scale intensive commercial farming using some wage
labour, considerable part-time non-agricultural work, and wide opportunities for employment outside agriculture.

In the northern areas of the main island of Japan, climate, terrain,
and, up to a point, isolation from the chief centres of population
produced a rather different type of agriculture. Severe winters and
very heavy snowfalls on the Japan Sea side made double-cropping
impossible. Some 80 per cent of the cultivated land was planted to rice,
mainly as a commercial crop. As the region became more specialised
as a rice-exporting area, there was a tendency for rice fields to encroach on the dry-field area. In the rice-growing areas, therefore,
dry-fields were few and the subsistence crops grown on them also
tended to be comparatively unimportant. Most farms were of medium
size—0·7 to 1·0 hectares—but landholding was concentrated in a very
few hands and most land was tenant-farmed. Hill villages with few
rice fields tended to specialise in such commercial crops as rapeseed,
tea, flax, cocoons, or tobacco. Around the coast, fishing villages grew
some subsistence crops and supplemented their cash income from fish
sales by growing some rice for sale. Villages on the outskirts of cities
derived a cash income from sales of grain, vegetables, and processed

foods. Apart from the shipping industry, non-agricultural employment was limited and the region as a whole imported industrial goods in return for its exports of rice and some agricultural raw materials.

Some areas, such as parts of northeastern Honshu and large parts of Kyushu, were much less advanced. Techniques were a century or more behind the more advanced regions, productivity per hectare was lower, commercial production was more limited, and rural industry was far less common. In the Aizu district of northeast Honshu, for example, low productivity, very heavy taxation, and the absence of non-agricultural employment resulted in people deserting the land and leaving the area (Sakai 1963: 253-300). Population was further reduced by the disastrous famine of the 1780s, the effects of which were still apparent in the mid-nineteenth century. Since land was relatively abundant, tenantry did not develop. Instead, the extended family tended to survive, since, at least for those families who could achieve a certain minimum level of productivity, land was not scare and there was little incentive to push out family members. In fact the authorities of the Aizu fief parcelled out abandoned land to those assessed as capable of working it, in the interests of maintaining tax revenue. Sales of agricultural produce consisted of the marketing of whatever surpluses occurred in what was basically subsistence production with little specialisation in cash crops.

Although it is not easy to combine the results of local studies of Restoration agriculture into a clear picture at a national level, several features stand out. The scale of farming was small, averaging about one hectare or less, even when we exclude from the total the very small holdings of people not engaged in farming as an occupation. Levels of techniques and productivity per hectare varied widely. Where productivity was high it was associated with commercial production and labour-intensive techniques. The bulk of industrial output was produced in rural areas, and industrial activity tended to be concentrated in areas where agricultural productivity and commercial production were also at a comparatively high level. Given existing levels of agricultural practice, there does not appear to have been any general surplus of rural labour.[1] Throughout most of the country, national and local governments exercised more or less strict

[1] In the sense that peak seasonal demands and industrial employment kept the marginal productivity of rural labour well above zero. Despite restrictions, movements of rural labour often led to local shortages of agricultural labour. Scattered wage data and official attempts to peg agricultural wages in the face of labour shortages indicate that rural real wages rose steadily in the first half of the nineteenth century. I have seen no evidence that the practice of *mabiki* (disposing of unwanted children) was at all common at this time. On the theoretical implications of absence of surplus labour in the traditional economy see Jorgensen (1966: 45-60).

control over land use and land tenure. Although there was some official encouragement of cultivation of industrial crops, on the whole control tended to restrict the possibilities for innovation and to emphasise rice cultivation at the expense of possibly more profitable crops. As a national average, between one-fifth and one-quarter of cultivated land was tenant-farmed and tenant-farmers paid an average of about one-quarter of their crop in rent. Taxes, mainly in kind, averaged a little over one-third of agricultural output, but the incidence of taxation tended to become rather unequal, with the more progressive owner-cultivators probably paying at a somewhat lower rate. Of the remaining 60 per cent or so of output, about half was marketed and half consumed on the farm. In all these aspects there was wide regional variation.

AGRICULTURAL DEVELOPMENT, 1868-1920

The question of the rate of growth of output of Japanese agriculture in the period from the Restoration to World War I has been a subject of controversy in recent years. Official statistics (Table 1.2) show growth rates for primary industry (agriculture, forestry, and fisheries).

TABLE 1.2

AVERAGE ANNUAL GROWTH RATE OF PRIMARY INDUSTRY[a]

Period	Average annual growth rate (%)
1878–87 to 1883–92	3·0
1883–92 to 1888–97	2·6
1888–97 to 1893–1902	2·8
1893–1902 to 1898–1907	2·1
1898–1907 to 1903–12	2·0
1903–12 to 1908–17	2·1
1908–17 to 1913–22	1·8

[a] These percentages represent the average rates of growth of net agricultural output deflated by a price index of agricultural commodities, using 1928–32 prices.

Source: Ohkawa *et al.* 1957: 72–3.

The high rate of increase in the period before 1900—about 30 per cent from 1883-92 to 1893-1902—is difficult to explain, since there were no major changes in agricultural technique; agricultural prices did not rise relative to non-agricultural prices; and, according to official figures, the agricultural labour force remained practically constant. There was certainly a large increase in inputs of fertiliser and some element of catching up on the part of relatively backward regions through the adoption of improved practice from the more advanced areas. Nevertheless, such high rates of growth still seem rather implausible.

It is known that agricultural output statistics for the early years of the period are very inaccurate, and suggestions have been made that they should be raised substantially. Nakamura (1966) claims there is evidence that concealment of cultivated land and under-reporting of yields resulted in serious understatement of production in the early years and suggests almost doubling the estimates of agricultural production for the period 1878-82. This would have the effect of lowering the estimated average annual growth rate of agricultural output from 1878-82 to 1913-17 from 2·4 per cent to between 0·8 and 1·2 per cent. His demonstration that the jump in taxed land area between 1885 and 1890 was due to disclosure of previously concealed farmland is convincing and his extension of this to cultivated area seems legitimate. His assumptions about actual as opposed to reported yields in the early period are, however, much harder to accept and seem to be little more than an arbitrary figure within the range—in the middle of it, as it happens—between the 1680s figure (Jōkyō era) of 1,900 kg per hectare and the 1918-20 figure of about 2,800 kg per hectare. Some of his other assumptions, for example that calorie intake per head remained constant from the 1870s to almost 1920 despite a substantial rise in income per head, seem equally doubtful. On the other hand, the average daily per capita consumption of 1,351 calories indicated by the official output figures for the period 1878-82 is almost unbelievably low (Nakamura 1966: 97). Since this is supposed to have represented a rise of perhaps 50 per cent from the corresponding figure for the early Tokugawa period, this would give the thoroughly incredible level of 900 calories per head for the seventeenth-century Japanese, indicating that Nakamura's revision is in the right direction, even if it is unconvincing in its procedure and in the values he assigns.

A similar revision was adopted by Yamada (1963: 85-98), in order to show what the implications would be if, as he thought quite possible, the official figures grossly understated output in the early years. He reduced growth of gross output over the period 1880-4 to 1890-4 from 44 per cent to 22 per cent and the growth of net output over the same period from 37 per cent to 16 per cent. Although, like Nakamura, he seemed to believe that the revised figures produced a more plausible story, he was modest enough to allow that 'the question of which is closer to the facts requires further study' (Yamada 1963: 88). In fact it was generally accepted that the early official figures were too low, and they have now been reworked by Yamada in collaboration with Hayami and others[2] (Ohkawa, Shinohara, and Umemura 1966: Vol. IX). Their revision gives growth rates of farm output as shown in Table 1.3.

TABLE 1.3

RATE OF GROWTH OF FARM VALUE[a] OF GROSS AGRICULTURAL PRODUCTION[b]

Period	Average annual compound growth rate (%)
1875–84 to 1880–89	2·2
1880–89 to 1885–94	2·1
1885–94 to 1890–99	1·3
1890–99 to 1895–1904	1·7
1895–1904 to 1900–09	2·1
1900–09 to 1905–14	2·1
1905–14 to 1910–19	2·6

[a] 1904–6 average prices used.

[b] The source gives output figures in 1874–6 prices, 1904–6 prices, 1934–6 prices, and 1954–6 prices. I have chosen 1904–6 prices because they seem most relevant to this period. The 1934–5 prices understate the contribution made by the growth of sericulture.

Apart from higher output figures for the early period, the main feature of the new estimates is that whereas Ohkawa's figures showed a high rate of growth at the end of the nineteenth century followed by a fall around World War I, the new series shows just the opposite. Their revision is certainly in the right direction and may well be of the right order, but the level of agricultural output in the initial period can scarcely be regarded as finally determined. In explaining Japan's agricultural development, therefore, we run an embarrassing risk of explaining rather more or rather less than actually occurred. It is scarcely possible to provide an explanation that is adequate for both 4 and 2 per cent growth rates. If Ohkawa and Rosovsky's explanation is adequate for the higher growth rate, it is rather too adequate for the lower. It is worth while examining the main points of their explanation (see Ohkawa and Rosovsky 1960, 1965; Ohkawa *et al.* 1964) at this point. They assume that Japan entered the modern period with a markedly unfavourable man/land ratio and that this feature persisted throughout the period under view, since both farm population and cultivated area remained almost unchanged. A system of small-scale intensive farming was inherited from the pre-modern period and this also exhibited remarkable stability. There was some increase in inputs into agriculture, as Table 1.4 shows:

[2] The so-called LTES series.

B

TABLE 1.4

INDEX OF FERTILISER INPUT AND TOTAL CURRENT INPUTS
(1878–82 = 100)

Period	Fertiliser[a]	Total current inputs[b]
1878–82	100	100
1883–87	97	99
1888–92	94	106
1893–97	114	112
1898–1902	146	122
1903–07	203	123
1908–12	378	130
1913–17	503	133

[a] Index of nitrogen content.

[b] Index of farm value of current inputs deflated by index of prices paid by farmers.

Source: Ohkawa, Shinohara, and Umemura 1966: 185, 202.

Most of the increase in inputs of working capital was associated with greatly increased use of commercial fertilisers after the turn of the century. It is generally agreed that increases in fixed capital stock in agriculture were very slight, at least until the first decade of this century. According to the Ohkawa/Rosovsky account, apart from some limited increases in consumption and working capital, most of the increases in agricultural income were transferred to other sectors of the economy, where they enabled capital construction to proceed at a rate just fast enough to absorb the increase in working population. It has been estimated that net agricultural fixed capital stock (livestock, perennial plants, agricultural implements, and buildings) increased by no more than 6 per cent over the twenty years 1880-1900. If buildings (which formed a large proportion of the total and actually decreased over this period) are excluded, the increase over these two decades was 24 per cent. By contrast the corresponding figures (including buildings) for the following two decades, 1900-20, are much higher at 14 per cent and 39 per cent respectively (Ohkawa, Shinohara, and Umemura 1966: Vol. IX, 212).

The growth of inputs was therefore slight before 1900. Yamada (1963: 88, T.31.1) estimates the annual growth rate of total inputs from 1878-82 to 1918-22 at 0·27 per cent, with the bulk of this increase coming in the last ten years.[3] Technical progress must therefore bear the brunt of responsibility for growth of output.

[3] These input figures purport to be calculated independently of output.

Technical Progress

According to Ohkawa,

> . . . the technological progress of Japanese agriculture was characterized by the land and labour-saving type with a moderate increase of capital per unit of output . . . How to meet these severe triple requirements at the same time was a basic problem for the technological development of Japanese agriculture.

He believes (1964: 210) they were met by 'the progress of cultivating techniques in a broad sense, including seed improvement as a most important factor'. Elsewhere Ohkawa and Rosovsky (1960: 50) mention two major categories of improvements—the improvement of the land itself and improved methods of cultivation, including the use of better fertiliser.

Land improvement mainly took the form of rearrangement of fields for better irrigation and drainage. This process had been proceeding on a small scale since before the Restoration, but it gathered momentum in the 1890s with the introduction of improved underground drainage techniques. The Arable Land Consolidation Law of 1899 gave national backing to the movement, which made a considerable contribution to increasing land productivity after 1900. Before the Restoration, larger-scale flood control, irrigation, and land reclamation projects had been undertaken from time to time by the shogunate or by local administrations, sometimes on the initiative of merchant investors. After the Restoration, smaller-scale water utilisation programs, which had previously been handled by villages or groups of villages, were put under the control of local government bodies and Water Utilisation Associations. This does not seem to have been an improvement. The Meiji government sponsored very few large-scale works of this kind, and it was not until after the end of the period under review that they again made any significant contribution to the growth of agricultural output.

Seed improvement, which is singled out by Ohkawa as a major factor in technical progress, began to have an effect with the development of new varieties and their spread in the 1890s. It was at this time that the outstanding new varieties of *Shinriki* in central Japan and *Kameno-o* in the northeast began to raise yields. As early as the 1870s, however, associations of farmers were active in exchanging higher-yielding seeds, and their activities were responsible for a considerable improvement in average yields. Higher-yielding varieties had been known for some time, but their use had been restricted by official insistence on varieties of higher quality though lower yield. Increased yields in the second half of the nineteenth century were, therefore, accompanied by some fall in quality, and a system of inspection was instituted to prevent this fall going too far (Ogura 1963: 347). Nevertheless, it was not until the end of the century that new varieties, capable of responding with greatly increased applications of fertiliser, were widely used. In crops other than rice, such as wheat, barley, and vegetables, early experiments took the form of the introduction and acclimatisation of foreign varieties. These efforts achieved mixed results, and significant improvements do not seem to have been made until about 1890 (Ohkawa, Shinohara, and Umemura 1966: Vol. IX, 151).

The improvement of farm tools also began in the 1900s. Ploughing with draught-animals was practised in Japan from ancient times, but even in the 1890s its use was far from universal. Many farmers, we are told, still prepared their fields by hand with long-bladed mattocks (*kuwa*), possibly because of the inefficiency of existing animal-drawn ploughs. An 1874 survey of production shows an output of only 11,000 ploughs and 33,000 ploughshares compared with an output of 651,000 long-bladed mattocks (Chihōshi 1960: 123-4).[4] The population of draught animals was virtually unchanged between 1877 and 1907 (Umemura and Yamada 1962: 137). Modern short-bottomed ploughs which made deeper cultivation possible were developed after 1900. Fixed-comb threshers through which bundles of grain in the stalk were pulled by hand continued to be the rule until they began to be replaced by rotary threshers in the 1900s. Weeding implements showed little change from the traditional long-pronged hoes until an improved hand rotary hoe and standard spacing of rows made a substantial advance possible in the 1900s. One source (Ogura 1963: 411) sums up the situation as follows: 'Briefly, improved farm implements began to appear one after another in the 1900's, marking the dawn of modernization of farm implements'. In a more general way improved seed selection by the flotation method, cultivation of seedlings in seedbeds, standardisation of gaps between rows, better pest control, and crop rotation all helped to raise yields. Winter drying of paddy fields improved soil fertility and facilitated the planting of second crops, and this in turn was made possible by cold-resistant strains and improved means of ploughing hard, dry soil.

From the beginning of the Meiji period the government pursued a very active policy of promoting improved agricultural practices. Such policies followed naturally from the central place given to agriculture in traditional Japanese political and economic theory. As early as

4 Comparing quinquennial averages for 1877-81 and 1902-6, the draught oxen population rose only from 1,107,000 to 1,170,000 and the horse population actually fell from 1,515,000 to 1,466,000. In this period this certainly does not reflect any trend to mechanisation.

1870 the Bureau for the Encouragement of Agriculture was established. This was to be the forerunner of the agricultural side of the Ministry of Agriculture and Commerce set up in 1881. Early efforts by enthusiastic bureaucrats to introduce large-scale Western methods proved a failure but were not finally abandoned until the 1880s. At the same time, the government recognised that what a number of local associations of farmers had already been doing for some time offered better chances of success under Japanese conditions.

These local farmers' associations had been trying to improve their farming technique, taking as their starting point the best indigenous practices known to them. With more freedom to experiment after the Restoration these efforts seem to have succeeded in raising output rather rapidly in some areas, and as early as the 1870s the government began encouraging and facilitating this kind of exchange of information. In 1878 systematic communication was established between the central government and experienced local farmers in each area. This organisation acted both as a source of information for the government and as a channel for official directives. These local correspondents, being leading members of their own farming communities, were able to exercise considerable influence, and propagated progressive ideas through local discussion groups. The more famous or energetic of them toured larger districts or even the whole country and enjoyed a much better hearing than a city-based official could hope for. A start was made with agricultural education in the 1870s and efforts were made to include practical agricultural training in rural elementary schools. Vocational schools of agriculture, however, became important only after 1900. Experimental stations were established in the 1890s and tackled problems of adapting imported techniques and improving local practices on a scientific basis. Agricultural shows and prizes helped to encourage better farming. This kind of persuasion was supplemented by a stream of central and local laws and ordinances designed to compel the adoption of improved techniques, together with a system of control and inspection of many aspects of agriculture.

The effectiveness of official encouragement was amplified by a number of features of rural society. Dore (1960) has shown in an outstanding article how a tradition of experiment and gradual improvement had existed since pre-modern times and how the traditional respect paid to outstanding farmers helped them to disseminate their own or official ideas rapidly through tightly-knit village communities used to taking direction. He stresses also the innovating role played by landlords and farmer entrepreneurs and their close links with prefectural governments. The responsiveness of Japanese farmers to the operations of the price mechanism also was no new departure, although the range of responses possible was greatly widened by the institutional changes of the early years of the Meiji period.

Nevertheless, as late as 1903 it was still found necessary to issue regulations requiring compulsory adoption of such things as seed selection, disease and pest prevention measures, the management of seed beds, consolidation of holdings, and use of improved seeds and implements, including animal-drawn ploughs. Even in 1904 only half the rice seed was selected by the brine flotation method, and 'no rapid progress was made in disease and insect pest control techniques from 1868 to 1945' (Ogura 1963: 426).

Despite all these measures, therefore, technical change was not as rapid and spontaneous as it is sometimes represented. This should not surprise us, since experienced observers maintain that it takes something like thirty years for even a fairly minor change to become fully accepted by Japanese farmers of today. There seems to be no very good reason to believe that their grandfathers were much more progressive. Nevertheless, diffusion of better traditional practices from advanced to backward areas and from more progressive farmers to more conservative within the same area was undoubtedly an important factor in raising yields in the period before 1900. By World War I, all regions of Japan except Hokkaido had reached levels of paddy rice yields which had been attained only in the most advanced region in 1878. Also the relative gap between areas had narrowed considerably.[5] In crops other than rice, performance varied, with the most dramatic advances in sericulture and rather slow progress in upland crops.

Individual improvements in techniques of cultivation would not appear to have been capable of engendering a very rapid increase in output before the 1890s at the earliest. It is possible that, taken in combination, minor improvements reinforced one another to such an extent that their cumulative effect was quite large. Moreover, changes in demand, largely associated with the development of international trade, prompted changes in land use from less productive to more productive crops (e.g. the development of sericulture), as well as planting of existing crops on more favourable land (e.g. of rice on land previously planted to cotton or oilseeds). In these ways agriculture undoubtedly enjoyed some gains from trade over the period. Nevertheless, on balance, and in the absence of any significant increase in inputs of land, labour, and capital, it is hard to accept that 'the progress of cultivating techniques in a broad sense' could account for the very rapid growth of output which the official statistics show for the period 1878-1900. The LTES output figures

---

[5] For a quantitative discussion of the importance of the diffusion of existing best practice see Hayami and Yamada (1966).

Table 1.3) fit the picture very much better than the official figures do, since the former show a lower rate of growth in the 1880s and 1890s followed by a rise after about 1900. Even these figures indicate that the net value added per gainfully employed worker in agriculture in 1920 was $2\frac{1}{2}$ times the level in 1875, implying a somewhat high rate of increase.[6] It seems likely that the flow of labour input over this period may be understated. This point is elaborated in the following section.

Inputs Re-examined

During the Meiji period a transformation of traditional industry took place from a largely part-time cottage occupation to a largely full-time workshop or small factory basis. This may have led to a net increase in the agricultural labour force. Labour previously devoted to traditional industry on a part-time basis was now required full-time.[7] This transfer almost certainly led to a rise in the productivity of industrial labour in the sense that labour so transferred was able to produce, in addition to its former output, much of the output formerly produced by farmers in industrial side-occupations. Thus, in some regions at least, a release of some labour for agricultural work took place.

Those farmers and their families who were deprived of industrial by-employment, both by the physical movement of industry away from the farm and by the competition of such imports as textiles, cotton, and kerosene, would not only be available for agricultural work but would be actively seeking it to replace their lost or diminished subsidiary industrial incomes. Thus conditions might well have been created which prompted the adoption of more labour-using techniques than had been possible before. Although the Japanese were traditionally conscientious cultivators, they may well have become even more conscientious once opportunities for combining farming with other sources of income were reduced. Moreover, a farm worker who has enjoyed a well-earned night's repose might be expected to work more effectively next day than if he had been up spinning or weaving until the early hours of the morning.

At the same time, most of the improvements in agricultural practices in the Meiji period give the distinct impression of requiring

[6] Net value added from Ohkawa, Shinohara, and Umemura (1966: Vol. IX, 228). Gainfully employed population from the same source (1966: Vol. IX, 218). The use of 1934-6 prices probably understates the increase.

[7] It is quite clear that the population registered as agricultural in the early Meiji period, although on the whole fully employed, was not fully devoted to agriculture. The withdrawal of labour which was in any case not engaged in agriculture would, of course, not reduce that available for agriculture.

increased inputs of labour.[8] New varieties of seeds, new cultivation practices, and the development of artificial silkworm hatching were expanding the area of double-cropping and extending the growing season. These technical advances encouraged increased labour inputs by enabling a more even distribution of agricultural work through the year than had been possible with full employment under the old technology.

The fact that these advances were introduced contemporaneously with the reduction in opportunities for seasonal industrial work may well have been something more than a coincidence, since each of these factors complemented and reinforced the other.

On the side of land also, technical progress undoubtedly raised productivity by extending double-cropping and widening the geographical limits within which more profitable crops could be economically cultivated. After about 1890, increased use of commercial fertilisers, especially of soybean cake imported from Manchuria, made it possible to reduce the area of grassland (for fallowing and as a source of green manure) required to maintain soil fertility and correspondingly increased the area available for direct agricultural (including forestry) production. Finally, while Nakamura believed that there was considerable concealment of land at the beginning of the period and that this concealment decreased over time, producing statistical overstatement of the growth of land input, one factor may have worked in the opposite direction. Land once registered as taxable land was very difficult to conceal or to de-register. Some land formerly cultivated and registered as taxable had clearly gone out of cultivation through the effects of famine, the growth of rural industry, or from some other cause, before the Restoration. Whatever the effect of this on the official statistics, it would account for some increase in actual land input at little cost.

The possibility of some considerable increase in inputs, particularly of labour, would go some way towards making the official output statistics more credible. Nevertheless, just as technical progress took some time to raise output, so the movement of industry from cottage to workshop was gradual and was not in fact complete until well after the end of the period under review.

AGRICULTURE AND JAPANESE ECONOMIC DEVELOPMENT

Most explanations of the early stages of Japan's modern economic growth stress the importance of a spurt in the production of the tradi-

8 The only outstanding exception is the wider use of commercial fertiliser, which freed farmers from the extremely labour-consuming task of collecting vegetable compost.

tional economy—measured primarily by agriculture—coupled with rapid structural change from agriculture to secondary and tertiary industry. According to this view, Japan before the arrival of Perry was a closed economy in a virtually stationary state. Foreign trade was insignificant, population was stable, net investment was very low, and real income per head was at a low level and rising very slowly, although there may have been some secular tendency towards redistribution of income as between various groups. To move from this quasi-stationary state to a path of modern economic growth, a surplus had to be produced for investment in building a modern sector. In an economy such as Japan's at that time, which had no modern sector, this surplus could only have been produced (barring massive capital inflow from abroad) by a spurt in the output of traditional production, which was mainly agricultural. Rising agricultural productivity released labour to other sectors, and, as the output of the primary sector increased, the bulk of the increase was siphoned off, largely through the budget, to build the modern sector. This process is supposed to have continued until the eve of World War I, by which time the modern sector had become able to finance its own growth. The growth rate of agriculture could then be allowed to fall and deficits in the food budget could be made up by imports from newly-acquired overseas territories.

However, this is not an exclusive argument in favour of the spurt thesis, since a surplus for investment could conceivably have been available even if output had not been growing so rapidly. As we saw in the previous section, the nineteenth-century spurt in agricultural output was not as great as official figures show, and this would seem to make it advisable to consider alternatives to or at least modifications of the spurt thesis, at least as applied to agriculture.

Firstly, as Ohkawa says, almost any economy can be made to yield a surplus, although presumably this will be more difficult the lower the level of income per head. In pre-modern Japan, incomes were low by modern standards but rather high for a traditional economy. Moreover, as much as one-quarter of national income was taken in taxes, the bulk of which was raised from the agricultural sector. Since production did not depend significantly on the way these tax funds were spent, this constituted a kind of surplus. Most of this tax revenue was used to finance consumption by largely unproductive *samurai* and their dependants, while major public works or defence expenditures were usually financed by loans from the urban business classes. A redirection of resources from *samurai* consumption to industrial investment, therefore, could go some way towards providing finance for modernisation. Hence the rather attractive idea that it was not the peasants but the *samurai* who were exploited to build a new and strong Japan. There is some truth in this, since the compensation actually received by the *samurai* class as a whole, even at face value, amounted to no more than one-third of their former revenues (Smith 1955: 32). Although *daimyo* (feudal lords) did well out of the Meiji settlement, their compensation bonds were largely invested in banking or were otherwise made available for industrial construction. Moreover, ex-*samurai* under the necessity of earning their own living and unable to be absorbed in the new administration could have made a high-grade addition to the industrial labour force. In fact, however, many of them were directed towards agriculture, which lends some support to the idea that there was a shortage of agricultural labour at the time.

Secondly, the rate of transfer of resources from agriculture to industry is greatly exaggerated by the traditional account. According to official statistics, between 1868 and 1920 the proportion of population gainfully employed in agriculture fell from over 80 per cent to 52 per cent, and the proportion of national income produced in the agricultural sector fell from some 66 per cent to 30 per cent, while the relative importance of the secondary and tertiary sectors naturally rose in the same proportion.

While official statistics show real national income produced in secondary industry increasing something like sixfold between 1878 and 1908, an overwhelmingly high proportion of this growth (perhaps as much as 90 per cent) was in very small workshops employing fewer than ten workers. Thus industrial growth in this period was, generally speaking, the growth of traditional industry, and was probably financed to a large extent from within the traditional industrial sector. This idea derives some support from a study by Toya (1949: 503), who makes the surprising suggestion that farmers spent most of their increased income on luxury consumption, no doubt by eating rice instead of millet.[9] Where industrial investment was financed by landlords or wealthy farmers, these were almost always people who had already been involved for some time in traditional industry as well as in farming. Not all those farmers who became absentee landlords during the Meiji period went off to live as idle urban *rentiers*. Many, no doubt, followed their industrial interests as they moved from the cottage to the town, and continued to manage them actively in their new location.

Transfer of labour from agriculture to other sectors, too, was probably less massive than the official statistics imply. As we have seen, many of those who moved from agriculture to industry had in fact been engaged in industry, either wholly or in part, before the move

---

[9] Toya's evidence should not be taken too seriously, since it consists of only one case.

was recorded. This kind of statistical inaccuracy was made worse by the habit of shopkeepers, pedlars, industrial workers, and entrepreneurs of giving their occupation as farmers, partly from social snobbery (since, in theory at least, farming was the more honourable occupation), but partly also because in those days industry and commerce were thought to be ephemeral in comparison with traditional family links with the land. With the social and political as well as the economic changes of the Meiji period, this sort of misrepresentation of occupation declined, thus producing a bias in the statistics which exaggerated the true extent of labour transfer.

Thus, if we were able to separate the industrial from the agricultural activities of the rural population, it might well turn out that the scale of transfer of resources from agriculture was less than is commonly supposed. It seems likely that a large part of the spurt in traditional production was in traditional industry and that this was largely independent of resources from agriculture. This was possible in Japan because of an unusually low income elasticity of demand for food and other agricultural products. Little work has been done on the growth of traditional industry, presumably because the data are so poor, and a discussion of it here would in any case take us too far afield.

Unfortunately it is scarcely possible now to disentangle agriculture from traditional industry over this crucial period. Thus, while it is clear that resources were transferred at some stage from the traditional to the modern sector, several important questions must remain unsettled. Firstly, we do not yet know to what extent these resources depended on increases in agricultural productivity and to what extent the transfer was effected by reallocating existing resources. Clearly the two go together and rising productivity makes reallocation much easier. Secondly, we do not know how far industrial growth depended on resources from agriculture and how far it was financed by traditional industry. Either way the growth of traditional industrial output was just as important as that of agriculture. Thirdly, it is hard to distinguish 'traditional' from 'modern' in the industrial expansion of the Meiji period. If it were possible to produce output and employment series by 'traditional' and 'modern' sectors they would certainly give a far less dramatic, though no less interesting, picture of structural transformation than that reflected by the standard sectoral classification.[10]

Whatever the mechanisms, however, agriculture certainly met the demands of economic growth and Japan's development does not seem

[10] Rosovsky (1961: 16-19) estimated that two-thirds of construction investment was in the traditional sector in 1890, although the proportion fell rather rapidly in the twentieth century.

to have suffered from the kind of agricultural lag from which the U.S.S.R. and China are said to have suffered. Except for very occasional bad harvests, staple food production kept pace with population growth until about 1900 and may even have provided slight increases in consumption per head despite the rising proportion of non-agricultural population (Table 1.5). Even though, as we have argued above,

TABLE 1.5

AVERAGE ANNUAL PRODUCTION AND CONSUMPTION OF
CEREALS, 1877–1920

('000 metric tons of brown rice)

Period	Production[a]	Consumption	Consumption per head (kg)
1877–80	4,061	4,014	110
1881–85	4,494	4,466	119
1886–90	5,487	5,428	138
1891–95	6,053	6,055	148
1896–1900	5,890	6,082	141
1901–5	6,696	7,231	158
1906–10	7,137	7,511	155
1911–15	7,675	8,046	156
1916–20	8,534	9,187	168

[a] Umemura's production figures are higher for the early period, but nevertheless roughly double over the period 1875 to 1920.

Source: Nōrinshō 1955: 160–1, T. 48.

the real rate of transfer of population out of agriculture may not have been quite as great as it appears, this is still a great contribution to growth. The task set agriculture was lightened by a rather low income elasticity of demand for agricultural products and particularly for food products.[11]

As an earner of foreign exchange also, agriculture played a major role in the first phase of Japan's modern economic growth. Tea and raw silk formed the bulk of agricultural exports throughout this period. Although the contribution of agriculture to export earnings

[11] Estimates of income elasticity of demand for agricultural products and agricultural food products for the period 1878-1920 vary widely. For agricultural products they range from 0·82 according to Noda (1963) to 0·32 derived from Nakamura's output figures. For agricultural food products the range is from 0·63 given by Noda (1956: Vol. 1, 159-74) to 0·18 based on Nakamura. For the periods 1878-82 to 1918-22, Kaneda (1967: 10), using the LTES output data, calculated income elasticity of demand for agricultural products available for consumption to be 0·50 and that for agricultural food products available for consumption to be 0·39.

TABLE 1.6

AVERAGE ANNUAL AGRICULTURAL EXPORTS,[a] 1871–1920

(millions of current yen)

Period	Agricultural exports (A)	Total exports (B)	$\frac{A}{B}\%$
1871–75	12·4	18·9	66
1876–80	18·8	27·1	69
1881–85	23·8	35·2	68
1886–90	32·2	58·7	55
1891–95	54·1	102·5	53
1896–1900	66·7	188·0	35
1901–5	104·0	312·6	33
1906–10	146·5	452·9	32
1911–15	190·1	623·2	31
1916–20	462·6	1,863·2	25

[a] Includes rice, soybeans, tea, raw silk, and camphor.

Source: Tōyā Keizai Shimpō Sha 1927: 464, T. 484; 445, T. 471.

fell steadily (Table 1.6), its importance in the early decades while industry was being built up can hardly be exaggerated. Here, Japan was fortunate in having in its traditional economy two labour-intensive agricultural products for which the world market was expanding rapidly at this time and in which Western technology was not far, if at all, ahead of Japan's.

The role of Japanese agricultural development in providing a market for the products of a growing industrial sector is often cited. While the demands of the agricultural population should not necessarily determine the direction of industrial development, nevertheless the bulk of the population is bound to be rural in the early phases of economic development, and the ability of industry to supply their wants is as important as the ability of agriculture to supply the wants of the urban population. The fact that in Japan both traditional agriculture and industry were by and large capable of supplying the wants of the population, coupled with the fact that changes in traditional wants were gradual enough for changes in the composition of output to keep pace, was of crucial importance for the Japanese pattern of modern economic growth.

The relative contribution of agriculture to economic growth decreased steadily as the growth of other sectors proceeded. Its relative contribution to total output, budget, investment, exports, and even food supplies decreased steadily from around the 1880s and 1890s. Until about 1910 agriculture and secondary industry grew together and movements up or down in sectoral rates of growth tended to be in the same direction. By 1920 agriculture's positive role was over and it remained in a more or less depressed state until World War II. The further growth of agricultural productivity since the war has been achieved only with the help of massive protection, and although Japanese rice yields are now among the highest in the world, so are costs of production.

CONCLUSION

Nineteenth-century Japanese economic statistics are so much more comprehensive than those for most countries at a comparable stage of development that they tempt the economic historian into refinements of quantitative analysis which he would scarcely dare attempt elsewhere. Nevertheless their reliability is sufficiently doubtful to make sophisticated quantitative work a hazardous undertaking at best. This chapter introduces a note of scepticism and contents itself with identifying the main factors in Japanese agricultural development and in making some qualitative assessments of their importance. What is thereby lost in precision and certitude is partly offset by a gain in flexibility of explanation which seems necessary now that we are no longer quite sure how much we are called upon to explain. There is no doubt that Japan's modern economic growth and agriculture's part in it were remarkable achievements. If this analysis tends on the whole to make them seem less remarkable, it is only because it seeks to make them more understandable.

2

Taiwan

R. H. Myers

DURING the past seventy-five years Taiwan's rural economy has undergone a remarkable transformation. Farmers have turned from subsistence production to supplying the market with increasing quantities
of food and industrial crops. Traditional farming practices have
gradually been replaced by modern methods of soil preparation, crop
rotation, and plant care, for example. The introduction of chemical
fertilisers, improved seed varieties, and pesticides has increased the
productivity of land, and the adoption of labour-saving equipment,
such as pumps, harvesters, and rotary tiller ploughs, has raised labour
productivity.

Since 1900 Taiwan's population has risen fourfold, yet farm production has more than kept pace with demand. Surpluses have accumulated to permit an expansion of farm exports. During the last
quarter-century, when a substantial shift of rural labour to cities has
taken place, the supply of foodstuffs and fibres has continued to grow
to support this urban growth without an increasing dependence upon
food imports. In an age when agriculture has generally tended to
stagnate in Asia, Taiwan's agriculture has become increasingly prosperous and efficient. This chapter examines the ways in which the
rural economy has contributed to the island's long-run economic
development and analyses the factors which were responsible for the
dramatic progress in the countryside.

AGRICULTURE'S CONTRIBUTION
TO TAIWAN'S ECONOMIC DEVELOPMENT

Agriculture contributes to a country's economic growth in a number
of ways, but quantitative measurements of these in Asian countries,
with the possible exception of Japan, are few. The reason is simple.
As Simon Kuznets (1966: 236-56) has pointed out, there are conceptual
and empirical difficulties in measuring this industry's contribution
because of the complex interdependence between agriculture and
other sectors of the economy. There are no detailed studies of this
problem for Taiwan, but the following discussion should indicate

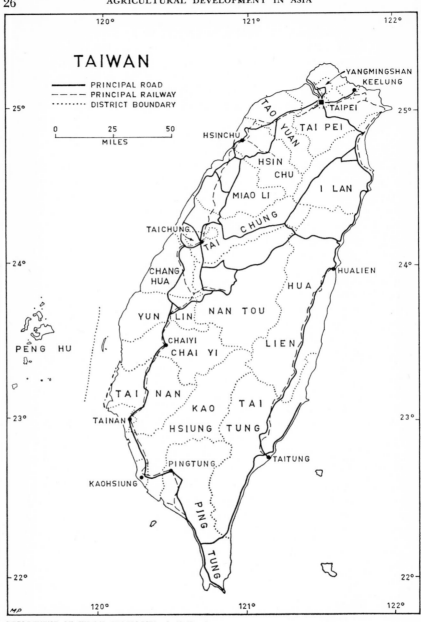

Map 2   Taiwan

how important agricultural development has been in the expansion
of Taiwan's foreign trade and national income.

The total value of exports grew tenfold between 1910 and 1940,
primarily as a result of expansion of agricultural exports. During
most of this period, the six leading agricultural exports provided
around four-fifths of total exports (Table 2.1). Rice and sugar were
the major contributors. In 1910 rice exports (to Japan) were 104,000
metric tons, or 17 per cent of the island's total rice production.

TABLE 2.1

AGRICULTURE IN TOTAL EXPORTS FOR SELECTED YEARS, 1910–65

Year	Relative importance of major agricultural commodities						
	Sugar	Rice	Tea	Bananas	Pineapples	Camphor	Total six commodities
	(%)	(%)	(%)	(%)	(%)	(%)	(%)
1910	58·6	11·7	10·7	0·6	0·1	6·6	88·3
1915	48·0	11·0	10·9	0·9	0·2	6·6	77·6
1920	65·8	7·9	3·1	0·8	0·4	3·5	81·5
1925	42·4	27·4	4·5	3·5	0·7	1·7	80·2
1930	58·8	16·0	3·7	3·6	1·5	1·0	84·6
1935	43·2	30·1	2·7	2·9	2·3	1·3	82·5
1940	39·3	15·5	3·7	5·0	2·7	0·8	67·0
1943	34·8	16·8	7·8	1·3	1·2	0·6	62·5
1950	79·8	3·1	2·9	1·4	0·1	1·3	88·6
1955	50·9	24·6	4·2	3·0	4·2	0·4	87·3
1960	43·8	2·5	3·7	4·0	5·0	0·2	59·2
1961	28·5	4·7	4·2	5·0	5·7	0·1	48·2
1962	20·8	3·1	3·3	3·6	4·5	0·2	35·5
1963	29·6	6·5	2·3	2·4	3·2	0·2	44·2
1964	29·2	3·9	1·8	7·2	3·0	0·3	45·4
1965	13·9	8·8	2·0	11·3	4·0	0·2	40·2

Source: Taiwan Provincial Civil Affairs Bureau 1946: 946–51; Bank of Taiwan 1966.

Twenty years later rice exports had trebled, and comprised 30 per
cent of the island's total output, while by 1937 the export volume had
reached 692,000 metric tons, which amounted to half the total rice
output (Bank of Taiwan 1958: Vol. I, 37-8). This was achieved without any reduction in the domestic level of per capita consumption,
despite a near doubling of population during the period. In the case
of sugar, total cane production trebled between 1910 and 1937 and
exports increased in the same proportion, while domestic per capita
consumption also rose slowly during the period.

World War II did not seriously affect the Taiwan economy until
1943, but from this point farm output and exports declined to low

levels. Economic recovery was not fully accomplished until the early 1950s. As the economy regained momentum, agricultural exports initially reasserted their pre-war dominance. However, over the last decade there has been a rapid decline in their relative importance despite a slowly rising trend in the export volume of major agricultural crops such as rice and sugar, and despite some recent export diversification (e.g. into bananas, canned pineapples, and canned mushrooms).[1]

During the 1910-44 period, the total value of imports trebled. Agricultural imports (food and fibres) averaged somewhat less than half of the total before 1930, and even less after that time (Table 2.2).

TABLE 2.2

RELATIVE IMPORTANCE OF FOOD AND FIBRES
IN TOTAL IMPORTS, 1910–65

Period	Food		Fibres		Total food and fibres (%)
	Wheat, flour, beans (%)	Total food (%)	Cotton (%)	Total fibres (%)	
1910–20	8	28	13	15	43
1920–30	10	30	13	15	45
1930–40	5	21	14	16	37
1940–44	6	19	15	17	36
1950–55	11	11	7	8	19
1955–60	11	14	9	11	25
1960–65	11	14	9	14	28

Source: Taiwan Provincial Civil Affairs Bureau 1946: 952–61; Bank of Taiwan 1966; Bureau of Accounting and Statistics 1962.

While imports of fibre maintained a fairly constant proportion of the total over most of the period, it is interesting to note that there was a growing trend towards self-sufficiency in food, particularly between 1910-20 and 1950-55.

Although it is clear that the agricultural sector was dominant in the Taiwan economy prior to World War II, precise estimates of its importance are not available. Since the war the economy has undergone a great structural transformation. The relative contribution of the agricultural sector in net national product has declined from 40 per

[1] Between 1956 and 1966 the export volume of bananas expanded from 20,000 to over 300,000 metric tons, canned pineapples from 20,000 to around 78,000 metric tons, and canned mushroom exports, which commenced in 1961, rose to around 33,000 metric tons by 1966 (Industry of Free China 1967: 124-5).

cent in 1953 to about 30 per cent in 1962,[2] and again to 25 per cent in 1964, owing to a relatively faster rate of growth in other sectors such as industry and mining (Directorate-General of Budgets, Accounts and Statistics, 1964: 19-31).

Before we examine agricultural growth performance in detail, one further perspective should be given. During the period under examination, the population of Taiwan underwent rapid change (Table 2.3).

TABLE 2.3

GROWTH OF AGRICULTURAL AND TOTAL POPULATION, 1905–65

Year	Agricultural population		Total population	
	(million)	Annual compound rate of increase (%)	(million)	Annual compound rate of increase (%)
1905	1·98	—	3·12	—
1921	2·23	0·74	3·84	1·30
1937	2·88	1·60	5·61	2·39
1950	4·00	2·56	7·55	2·32
1961	5·47	2·88	11·15	3·60
1965	5·74	1·22	12·63	3·16
1905–43	—	1·32	—	1·84
1905–65	—	1·79	—	2·36

Source: (1905 to 1921) Taiwan Provincial Civil Affairs Bureau 1946: 76, 513; (1937 to 1955) Joint Commission on Rural Reconstruction 1956: 7; (1961 and 1965) Department of Agriculture and Forestry 1967: 55.

Between 1905 and 1921 there was a slow rate of population increase, which was influenced considerably by a number of major epidemics. The rate accelerated during the twenties and thirties mainly because of a gradual decline in death rates. Since World War II the rate of increase accelerated further, under the continued influence of a declining death rate and because of the influx of nearly a million refugees from the Mainland during the late 1940s. The rate of increase since 1961 has slackened as immigration declined and as the birth rate fell. It is important to note that the agricultural population

[2] These shares were calculated with the Kuznets (1966: 240) formula:

$$\frac{P_a r_a}{\delta P} = 1 + \left(\frac{1}{\dfrac{P_b}{P_a} \cdot \dfrac{r_b}{r_a}} \right)$$

where P_a = product of agriculture
P_b = product of all other sectors
P = total product $(P_a + P_b)$
r_a = rate of growth of P_a
r_b = rate of growth of P_b

continued to increase over the whole period. The rate of increase to 1943 was below that of total population, but between 1937 and 1950 it was actually higher than that of total population. Since that time, it has again fallen well behind, as migration to urban areas has accelerated.[3]

Despite the problems imposed by such high rates of rural population growth, evidence, though limited, suggests that farm output increased fast enough to support the increasing numbers at rising standards of living.[4] Furthermore, this rising prosperity was achieved without serious inflation stemming from shortages of food or exportable crops. Before 1940 prices rose moderately, and while they increased more rapidly between 1950 and 1960, the increases have again been moderate since 1960 at a compound rate of about 2 per cent per annum (Industry of Free China 1967: 54).

Other relationships between agriculture and the rest of the economy, such as the expansion in rural demand for consumer goods and production inputs and the contribution of rural taxation in financing development, are relevant to the question of agriculture's role in Taiwan's economic development, but discussion of these topics must await further quantitative study.

THE GROWTH OF AGRICULTURE, 1910-1965

Before presenting the available estimates of the growth of Taiwan's farm output since 1910, some consideration should be given to the quality of Taiwan's crop statistics. The major difficulties in estimating farm output are to estimate accurately cultivated land and crop yield. In some Asian countries, notably pre-war China, the actual cultivated area may have exceeded the reported farm area by as much as 20 to 30 per cent, because farmers under-reported the amount of land they cultivated in order to evade higher taxes. Also, where traditional farming is practised, yield differences between areas are

[3] The rate of total population growth in Taiwan, 1905-65, was twice as rapid as that of Japan, 1879-1964 (Johnston 1966: 272, 305), and while Taiwan's rural population nearly trebled over the sixty years, Japan's rural population remained relatively constant from 1879 to 1955, after which it began to decline. Thus in Japan urban growth could absorb increases in the farming population from the outset, while in Taiwan the rural sector appeared to retain most of its own increase, at least until the 1960s.

[4] The Taiwan Nōjihō (1924: 46) reported in March 1924 that rice consumption per capita increased from 109 kg to 125 kg per annum between 1912 and 1922, taking into account total output, adjusted for exports and imports. A 1962 study showed that food expenditures, as a percentage of disposable household income, fell from a level of 64 per cent for the 1911-20 period to 43 per cent for the 1920-38 period, thus suggesting, indirectly, that a considerable expansion in disposable household income took place between periods (Li Teng-hui 1962: 173).

usually great, and a representative average yield can only be obtained by taking a large number of yield observations. For Taiwan it cannot be stated with certainty that these difficulties were completely surmounted. What can be said is that after 1895 Japanese colonial officials made a vigorous attempt to obtain accurate crop statistics, and that by 1910 administrative machinery had been in operation long enough to provide crop statistics which Japanese officials themselves regarded as a reasonably reliable basis for forecasts required for colonial policy.

After Taiwan was annexed by Japan in 1895, Japanese officials carefully studied the results of a land survey undertaken in the early 1880s by one of the last Chinese governors, Liu Ming-ch'uan.[5] Japanese officials recognised that Liu's survey was incomplete, because he had failed to convince the powerful absentee landlords that their land tax would not be increased after the survey was completed. These landlords successfully sabotaged Liu's attempts to survey their land. Japanese officials made every attempt to convince this same class that their survey would not augur a tax increase. They established a Land Commission to undertake an island-wide land survey, and between 1898 and 1902 district officials co-operated with the Land Commission to survey all land and make land maps. After the survey was completed, all landowners were required to report to district tax offices any exchange of land and any changes in size of holdings. Officials were ordered to provide annual estimates of cultivated land and output for all crops under their administration, and for this purpose the new land maps were used.

The very fact that another major land survey was never undertaken and that crop statistical reports, prepared by local officials, were used for estimating changes in total farm production, suggests that colonial officials were fairly satisfied with the quality of crop data.[6] The crop-reporting methods of the Department of Agriculture and Forestry in the Republic of China[7] are considered fairly accurate by agricultural experts in Nationalist China's top agricultural research agency, the Joint Commission on Rural Reconstruction.

How rapidly did agricultural output expand? Table 2.4 contains the results of four separate studies of the annual growth rate of farm

[5] The best account of Liu Ming-ch'uan's attempt to reform the Ch'ing land tax system between 1887 and 1893 can be found in Rinji Taiwan Tochi Chōsa Kyoku (1900: 1-29).

[6] This, of course, is still no assurance that such data were complete and unbiased, but it does somewhat strengthen the argument that they had reached an acceptable degree of reliability.

[7] These methods are continually under review so as to up-grade the quality and accuracy of statistical reporting of crops.

TABLE 2.4

ESTIMATES OF AVERAGE ANNUAL GROWTH RATES OF AGRICULTURAL OUTPUT, 1910-64[a]

Agricultural development phase	Period	Hsieh and Lee (1935-7 = 100) (%)	Myers and Ching[b] (1935-7 = 100) (%)	Ho (1901 = 100) (%)	Kao[c] (1953 = 100) (%)
Initial	1910-20	1·66	1·54	1·30	n.g.
Secondary	1920-39	4·19	4·48	4·31	n.g.
World War II	1939-45	-12·32	-12·97	-13·62	n.g.
Recovery	1945-52	12·93	n.g.	13·6	n.g.
Recent (1)	1952-60	3·98	n.g.	4·34	n.g.
Recent (2)	1953-64	n.g.	n.g.	n.g.	3·69
Pre-World War II	1910-39	3·31	3·45	3·46	n.g.
Post-World War II	1945-60	8·06	n.g.	8·56	n.g.
Overall period	1910-60	2·67	n.g.	2·64	n.g.

[a] Indices are calculated by aggregating the physical output of approximately the same range of agricultural products, each weighted by average prices of a chosen period. Hsieh and Lee, and Myers and Ching used average farm prices of 1935-7 as constant prices, Ho those of 1952-6, and Kao 1953. Ho compared his series with Hsieh and Lee (Ho 1966: 26-7), and showed a remarkable correlation between movements in the two series.

Rates are average rates of growth from a single year at the beginning of the period to a single year at the end of the period. In view of substantial year-to-year fluctuations, moving averages would have been preferable. Ho gives alternative estimates based on five-year moving averages which show a rate of growth of 1·95 per cent between 1908-12 and 1918-22, 3·63 per cent between 1918-22 and 1939-41, -7·11 per cent between 1939-41 and 1943-47, 8·07 per cent between 1943-47 and 1950-54, 5 per cent between 1950-54 and 1956-60.

[b] Total food production only.

[c] Crops only, this being the nearest in terms of coverage to the other authors' series. Kao shows 4·6 per cent average rate of change in each year compared to the previous year. The given 3·69 per cent is an average rate of growth from 1953 to 1964.

n.g. = not given.

Source: Hsieh and Lee 1966: 14; Myers and Ching 1964: 556; Ho 1966: 17-18; Kao 1967: 612.

output for various periods since 1910. These studies use the same official data and similar methods of index calculation, but vary slightly as to commodity coverage. All studies used more than fifty crops which included grain, fruit, fibres, special crops, but excluded fishery and forestry products. In this chapter, the period under review is divided into four development phases: an early growth phase; a phase of accelerated growth; wartime decline and recovery; recent development. It will be seen that the four studies show similar rates of growth for the various phases.

These studies show that agricultural output expanded slowly before the 1920s, accelerated markedly up to 1939, and then declined sharply during World War II. They show that post-war recovery was achieved roughly by 1953, and that since then the rate of growth has matched the achievement during the secondary or accelerated growth phase.

A comparison of Tables 2.3 and 2.4 shows that over the whole 1910-60 period the average annual growth rate of agricultural output exceeded that of total population by a rather narrow margin and that of agricultural population by a more substantial margin. The same relations held true within the important sub-periods (the initial, secondary, and recent growth phases). Growth in agricultural output exceeded that of agricultural population by a particularly wide margin in the secondary and recent phases.

The four principal crops over the period were rice, sugar cane, sweet potatoes, and peanuts. These occupied roughly 80 per cent of the total cultivated area throughout. The relative importance of the four changed somewhat within the period (Table 2.5). While rice remained the most important, there was some decline in its relative importance, particularly in the early years of the period. This was

TABLE 2.5

AREAS PLANTED TO FOUR MAJOR CROPS, 1901–5 TO 1961–5

Period	Average annual 4-crop total area ('000 hectares)	Relative importance in 4-crop total area			
		Rice (%)	Sweet potatoes (%)	Sugar cane (%)	Peanuts (%)
1901–5	507	78	15	4	3
1916–20	740	65	15	16	3
1936–40	952	68	14	15	3
1951–5	1,188	65	20	8	7
1961–5	1,206	64	19	8	8

Source: Taiwan Provincial Food Bureau 1967: 1, 3, 7; Joint Commission on Rural Reconstruction 1956: 20, 32, 36, 44; Department of Agriculture and Forestry 1967: 97.

the result of a rapid expansion in sugar cane acreage up to the early 1920s. The latter's prominence, however, was not sustained after World War II.

An analysis of these four crops shows in more detail what was taking place during the various development phases (Table 2.6).[8]

TABLE 2.6

GROWTH RATES FOR FOUR MAJOR CROPS, 1901–5 TO 1961–5

(annual averages, compound, in percentages)

Crop	1901–5 to 1916–20	1921–5 to 1936–40	1951–5 to 1961–5	1901–5 to 1961–5
Gross area[a]				
Brown rice	1·35	1·47	−0·04	1·13
Sugar cane	12·43	0·92	0·63	2·61
Sweet potatoes	2·71	0·68	−0·15	1·92
Peanuts	2·45	1·72	2·15	3·08
Total, 4 crops	2·56	1·28	0·15	1·46
Yield				
Brown rice	0·55	1·84	3·11	1·28
Sugar cane	0·22	4·82	1·79	1·58
Sweet potatoes	0·95	2·35	2·91	1·16
Peanuts	1·28	1·27	4·15	1·16
Production				
Brown rice	1·88	3·30	3·11	2·42
Sugar cane	12·63	6·04	2·35	4·19
Sweet potatoes	3·51	3·06	2·76	3·03
Peanuts	2·82	2·51	5·33	4·40

[a] Includes double-cropping.

Source: Taiwan Provincial Food Bureau 1967: 1, 3, 7; Joint Commission on Rural Reconstruction 1956: 20, 32, 36, 44; Department of Agriculture and Forestry 1967: 97.

Growth rates of output over the whole period were substantial for all four crops. The rates were more notable for sugar cane, sweet potato, and peanut production than for rice, but since rice production was such a large part of total production, its growth rate was the chief determining factor in the overall rate of growth of agricultural output. Both yield and area increases contributed to the output expansion of the four crops, though the relative importance of these two factors varied markedly within the fifty-year period.

8 Only growth rates are presented in the text. For more detail see Table 2.8, which gives indices of crop areas, yields, and output.

During the initial development phase (1900-20) the main source of output increase for the four crops was the expansion in cultivated area. The increase in area under sugar cane was particularly fast in this period. The contribution of increased productivity was more pronounced during the second phase. For rice and peanuts this roughly matched the contribution from area expansion, while the yield factor was considerably more important for sugar cane and sweet potatoes. During the recent period, the relative contributions were the reverse of those in the initial phase. Area increases made a marked contribution to output expansion only in the case of peanuts, while very high rates of yield increase were recorded for all four crops, particularly rice and peanuts. Broadly then, agricultural development was first of all largely extensive, then became a mixture of extensive and intensive elements, and recently became almost exclusively intensive.

The foregoing development performance is essentially the response of farmers to government agricultural policy, set within the framework and operation of a free market system.[9]

Over the whole period, government policy under Japanese and Nationalist administrations can be broadly divided into four major areas of endeavour. First, there was an attempt to provide farmers with a general environment conducive to economic development by establishing and maintaining law and order, by providing public health services, and by improving transportation and marketing facilities. Second, there was an attempt to give farmers greater knowledge and opportunity for more efficient farming by creating agricultural research units and a farm extension system, by making available new seed varieties and improved farming methods, and by instructing farmers in their use. Third, there was an attempt to provide, or to facilitate the supply of, necessary inputs which were beyond the normal capacity of farmers to acquire themselves (e.g. construction of irrigation facilities, supply of chemical fertilisers and pesticides, farm machinery and tools, and storage facilities). Finally, institutional reforms were introduced which significantly altered the land tenure system and provided a majority of farmers with a greater incentive to improve their farms (e.g. the land tax reform of 1892-1902, landlord tenant association in the late 1920s, and the land reform of 1959 to 1963).

The remainder of this chapter examines these policies in greater detail, within the context of the period, to evaluate their contribution to the rise of farm production under Japanese colonial rule and under Nationalist rule, since the recovery from World War II.

[9] The only period when substantial market intervention occurred was during World War II.

ESTABLISHING THE FOUNDATIONS FOR AGRICULTURAL
CHANGE, 1895-1920

The conditions the Japanese found in Taiwan were hardly conducive
for the rapid development of agriculture (Chang and Myers 1963;
Takegoshi 1905). Farmers had little peace and security to farm their
crops because of rampant banditry, feuding between Chinese clans,
and constant fighting between Chinese settlers and the indigenes.
Disease was widespread, and large numbers of people were afflicted
with malaria. Transportation was extremely poor; bridges and good
roads were virtually non-existent, thus preventing goods being shipped
any great distance. Commodity exchange was difficult and costly
for merchants, for weights and measures differed between markets,
and a number of currencies were in circulation. Farming techniques
were backward. New seeds and methods of making fertilisers were
slowly introduced by migrants from Fukien Province in China,
but considerable time elapsed before they were widely adopted.
Officials were merely concerned with collecting taxes and did nothing
to promote agricultural development. Even the responsibilities of
maintaining roads, providing village security, and recovering from
natural disasters were placed upon the villagers.

Japanese policy aimed at making Taiwan a rural appendage, and
for this purpose administrative policy was basically concerned with
eliminating obstacles that stood in the way of increasing the island's
farm production. Japanese officials were anxious to develop Taiwan's
resources to supply Japan with the food and industrial crops that she
could not afford to produce herself in sufficient quantity. They based
their agricultural policy on the premise that farmers would try to
improve farming and increase production if peace and security for
life and property prevailed, if the economic infrastructure was improved, and if new seeds and farming methods were made available to
farmers at low cost.

Between 1895 and 1899 the colonial administration concentrated
its efforts on solving the serious problem of unrest. With the use of
the armed forces, it isolated and eliminated guerilla units, at the same
time offering amnesty to any who laid down their arms. Despite considerable local opposition to the use of the army, these methods
quickly bore fruit, and by late 1899 all opposition to the Japanese
had been suppressed. A highly paid and efficient police force system
was then established in each district. Village headmen were required
to establish the *pao-chia* system. This involved forming every ten
households into a unit called a *chia*, then forming every ten *chia*
into another unit called a *pao*. To each unit a leader was assigned

from the village council, who was held accountable to the district police for the behaviour of households within the unit. This system was used to maintain peace, and later on to mobilise labour for road construction, and as a means of introducing new sugar cane seeds.

Between 1899 and 1905 the administration greatly improved the economy's infrastructure (Chang and Myers 1963: 436-42; Shigeno 1925: 303-4; Barclay 1954: 136-9). By August 1901 it had succeeded in constructing a railway line between Keelung and Hsinchu in northern Taiwan, and in June 1905 a 259-mile railway line was completed, connecting Hsinchu with the city of Kaohsiung in the southern extremity of the island, to give Taiwan its first trans-island transport line. Harbour construction in Keelung and Kaohsiung was completed in the same year to give the island two major deep water ports. By the end of 1905 the administration had successfully standardised weights and measures in all local markets. A central bank had also been established, and a new system of notes and coin, based on the gold standard, had been substituted for the old bi-metallic currency system of silver and copper. In the cities public health services were enforced: wells were dug and markets ordered to adopt sanitary methods of food handling and storage. Large-scale vaccination campaigns quickly followed, and by 1914 major diseases such as plague, cholera, and smallpox were under control. Swamps were drained, and districts with high malarial incidence isolated. The police gradually introduced these same public health policies into the rural areas, so that the countryside did not lag far behind the cities in improvements to public health conditions.

The colonial administration financed these programs by taxing agriculture and by establishing a number of monopoly bureaux which controlled the production and sale of tobacco, wine, camphor, and opium. The principal source of revenue from agriculture was the land tax. In order to increase land tax revenue, a land tax reform was enacted, which was based on a complete land survey of the island and a restructuring of land tenure relationships.

Taiwan's land tenure system was transplanted from Fukien Province of China some time in the early eighteenth century (Suzuki 1927; Ino 1965: 546-56; Wang 1966: 63-9). Early migrants to Taiwan reclaimed land, registered it with the local officials, and paid the land tax. Some of these landowners leased their land to other migrants, and in time many of these landowners moved to the walled cities (which were the administrative centres), and became merchants and money-lenders. These absentee landowning families were called *ta-tsu*. Their landholdings typically comprised a number of scattered plots, fragmented by a system of inheritance whereby family land was divided equally between the male progeny (Rinji Taiwan Kyūkan Chōsakai 1901: 547-9; Nakamura 1905: 1-142; Cheng 1963: 1-92). Despite this increasing fragmentation, rents were collected on these small plots.

The peasants who farmed these plots were called *hsiao-tsu*. They were free to manage their land as they saw fit, provided they paid their rents to the *ta-tsu*. As more migrants came to Taiwan, additional farm land was reclaimed and settled. Many *hsiao-tsu* families in turn leased some of their land to these new arrivals until they were firmly settled, so that by the end of the nineteenth century a complex land tenure system existed in which often three parties had claims on the same tract of land, and two rents were paid for it.

Probably close to half of the land reclaimed and farmed was not officially registered and was therefore not subject to tax. Farmers used informal arrangements to transfer land from one party to another in order to evade the land tax. Japanese officials were aware of this, and recognised that it could only be corrected by transferring ownership rights to the peasant class which actually farmed the land, the *hsiao-tsu*, and then forcing this class to pay a land tax on all the land it farmed or leased to other peasants. To achieve this, the administration decided to undertake a land survey, and to transfer ownership and reimburse the *ta-tsu* class so dispossessed. When in 1898 the administration published regulations stating that a land survey would be carried out (Taiwan Sōtokufu Minseibu Zaimu Kyoku Zeimuka 1918: 104-5), officials contacted the sons of the fifty-three most powerful landowners and explained to them the intentions of the new Land Commission and the purpose of the land survey. By 1902 the land survey had been completed, providing Japanese officials with an approximate measure of *ta-tsu* holdings by which to estimate land earnings. The *ta-tsu* were then given bonds in exchange for their rights to the land and to the rent from it. Many *ta-tsu* used their bonds to finance commercial enterprise in cities and market towns. It is not known how many failed in these endeavours or how many sold their bonds simply to maintain their grand living style, but the land tax reform did succeed in directing many of this class into productive activities, at the same time removing their authority from the countryside.

Ownership rights were then conferred on the *hsiao-tsu* families and they were made directly liable for land tax. Because the *hsiao-tsu* leased considerable areas to other households, there was in addition a large number of tenants, though their numbers were not precisely known at this time. The new landowning class was given deeds to the land which were protected under law, and land transactions could then be undertaken easily, without recourse to secrecy or informal agreement.

The Land Commission calculated new land tax rates and schedules for different classes of land. A new land tax was calculated by first estimating the value of the new owner's land, deducting the rent he formerly paid to the *ta-tsu*, and then fixing a tax percentage upon the land's new value. Generally the new land tax was much lower than the amount of rent paid by the *hsiao-tsu* to the *ta-tsu*. This was mainly because the old rent included both the land tax for which the *ta-tsu* was liable and an amount retained by the *ta-tsu* as his income. However, the new landowner was now compelled to pay a land tax on all the land he owned to the district tax office. In many instances the total land tax paid was greater than the total amount formerly paid to the *ta-tsu* because of the inclusion of much unregistered land. Tax evasion was now virtually impossible, and the state was guaranteed a larger fixed source of tax revenue.

Before the land tax reform only 350,575 hectares of land were taxed, but the land survey uncovered a taxable land area of 754,452 hectares. The reform made it possible for the administration to collect 50 per cent more revenue than before (Moyoshi 1933: 291), and was thus largely responsible for engendering the finance required for the improvements in the economy's infrastructure carried out during the early period of colonial rule. On the other hand farmers now had legal claim to their land, and thus a greater incentive to put their land to its most profitable use.

By 1898 the colonial administration had determined that the rice, sugar cane, and tea industries were to be developed as a source of supply for Japan (Mochiji 1912: 182-5). In that year the administration established an agricultural research station in Taipei. This station undertook research into the adaptation of foreign seed varieties to local conditions, into soil science, plant and animal diseases, animal breeding, and sericulture. It also taught new farming methods to a large number of students, who were sent home to pass on their new knowledge to other farmers. Perhaps the most significant technological development undertaken at the station in this early period was the adaptation of Hawaiian sugar cane seeds to Taiwan's southern soils. By 1901 these seeds were being distributed by police and *pao-chia* to farmers in the southern districts, where a rapid expansion of sugar cane cultivation commenced soon afterwards.

A network of agricultural research stations was gradually developed. By 1908 there were similar stations in the north, central, and southern sections of the island, which conducted research on problems specifically related to their locality. Besides these, institutes were established to study particular crops; for example Tainan city had a sugar cane experimental station, and An-p'ing city had a tea experimental station.[10]

The administration grafted an extension service on to these research stations, to channel research findings into the villages. Each district station was served by a district agricultural association comprising local officials, wealthy farmers, and landlords. In 1900 the first agricultural association was created near the Taipei research station (Joint Commission on Rural Reconstruction 1956: 11). Prominent landlords and wealthy farmers in the area were urged by officials to become due-paying members. The association developed an experimental garden and purchased seeds, fertilisers, and tools. It planted new high-yielding seeds, provided by the Taipei research station, in its experimental garden, and invited nearby villagers to observe the results and learn the methods necessary to give them similar yields in the field. Association members were encouraged to take these new seeds to their villages and instruct other farmers in their use. By late 1907 similar associations had been established in every district.

This extension and research system could not be expected to achieve widespread and fundamental change overnight. Time was needed to staff stations with experienced, trained personnel and to transmit results to the villages. Thus, although new seed varieties were tested and distributed to farmers (e.g. sugar cane seeds from Hawaii), their impact on crop yields seems to have still been minimal during the initial phase of agricultural development (Table 2.6).

As shown above, the major reason for the expansion in agricultural output during the initial phase was the increase in area of cultivated land. A number of factors contributed to this increase. First, as can be inferred from the objectives of colonial agricultural policy, market conditions were favourable for the island's farmers. Expanding Japanese demand caused commodity prices to rise steadily in Taiwan. There was, in addition, much more incentive for producing for the market. The establishment and maintenance of law and order in the countryside provided security of person and assets. The 1902 land reform gave the cultivator legal security on his farm. Producers undoubtedly benefited also from the growth and improvement of transport and market facilities.

As more land was reclaimed and brought under cultivation, there was a concurrent expansion of the area under irrigation. Indeed, despite the fast rate of increase in cultivated area between 1901 and 1920, the percentage which was classified as irrigated fell only from 56 to 49 per cent (Joint Commission on Rural Reconstruction 1956:

10 One of the few descriptions of these agricultural research stations can be found in a travel account by Liu Fan-cheng (1965: 49-53), who toured Taiwan and visited the Taipei station.

11). However, it must be kept in mind that irrigation facilities prior to 1920 were largely small and privately owned. Farmers typically built their own ponds and sluices to store rainwater for use in their fields. The colonial administration did little to develop local irrigation projects or to mitigate flood damage during this period, though district officials did in some instances provide local financial assistance to farmers.

Even though the overall rate of expansion of agricultural output was not as high in this as in the secondary phase which followed, the basic foundations had been established for an accelerated rate of progress by 1910, through the development of the rural infrastructure and the establishment of a research and extension system which worked in close association with the farmer. Nevertheless, at the end of the initial phase only a small beginning had been made in introducing modern technology to the countryside. There were still areas which were suitable for irrigation development, chemical fertilisers had barely been introduced, crop pest control was still in its infancy, and new farming tools had not yet made their appearance.

The longer-run achievements of Japanese colonial policy can be seen more clearly by examining the great spurt in farm production during the secondary phase from 1920-40.

THE ACCELERATION OF AGRICULTURAL DEVELOPMENT, 1920-40

Despite the fast rate of population growth after 1920 (Table 2.3) increases in area brought under cultivation were sufficient to prevent a decline in the average size of farm. In 1909 the average size of farm was 1·8 hectares. In 1930 this had risen to 1·9 hectares and reached a peak of 2·0 hectares in 1940, after which it gradually declined (Taiwan Provincial Civil Affairs Bureau 1946: 513, 516). There was, then, as in the initial phase, a contribution to growth from a larger rural labour force operating on a large cultivated area. However, as the foregoing analysis indicated (Table 2.6), there was a reduction in the rate of area expansion (*vis-à-vis* the initial phase) for all crops except rice, and the higher overall growth rates were thus primarily due to the contribution of increased productivity.

Once again the colonial administration played a leading role in promoting agricultural production, particularly in the area of irrigation construction, development and higher-yielding crop varieties, and supply of chemical fertilisers.

Certainly one of the more important changes in colonial farm policy after 1920 was the greater attention given to irrigation. After World War I, efforts were intensified to expand food exports to Japan, and for this purpose the administration began to invest more heavily than before in irrigation. Whereas total expenditures on irrigation came to only 28·5 million Taiwan dollars between 1911 and 1920, they rose to a total of 213 million dollars between 1921 and 1930, and still amounted to a total of 118 million dollars between 1931 and 1940 (Rada and Lee 1963: 27). Capital expenditures were highest during the 1921-30 period; for example the biggest single item of expenditure was the Chainan reservoir in the Tainan district, which was started in the twenties and completed in the early thirties. This serviced a total of 130,000 hectares. From 1931 to 1940 the majority of the expenditures were on maintenance work.

Overall, the increased spending on irrigation raised the area under irrigation from 267,177 hectares in 1920 to 529,279 hectares in 1940. According to Rada and Lee (1963: 44) more than 60 per cent of the increase in the output of paddy for the 1922-38 period could be attributed to the increase in irrigated area.

Progress was made in developing higher-yielding seed varieties. During the twenties a Javanese sugar cane variety was imported and, after a period of experimental crossing with local strains, new varieties were released during the thirties which replaced the existing Hawaiian varieties.

Probably the most outstanding success in crop research was the development of a new rice strain based on Japanese high-yielding varieties. An initial attempt to introduce Japanese rice was made as early as 1897, but failed. Research continued for another twenty-five years, until in 1922 a new high-yielding variety, *p'eng-lai*, was finally released to agricultural associations, which in turn distributed seed to farmers. It was first tried as the second rice crop, but once farmers found yields were higher than with native varieties, they began to use it for the first crop as well (Myers and Ching 1964: 566-7; Taiwan Nōjihō 1924: 439-50). By 1924 the area under *p'eng-lai* had expanded to over 25,000 hectares, and by 1940 the area was around 324,000 hectares, or 51 per cent of the total area planted to rice.

The yield potential of *p'eng-lai* rice could only be obtained with higher levels of labour, fertiliser, and water than were customarily used for native varieties. Seedlings could be planted more closely than with native varieties, thus giving more stalks per unit area. This meant additional weeding, more water, and heavier applications of fertiliser. Where farmers were able to apply these additional inputs, their income per unit area of *p'eng-lai* was well in excess of earnings from native varieties, or from sugar cane (Andō 1925: 27-8). In view of its apparent yield potential, it is surprising that *p'eng-lai* did not have a greater impact on yields during the period, for in 1940, although it was grown on 51 per cent of the total rice area, output from *p'eng-lai* was still only 55 per cent of total rice output. This may well be explained by difficulties farmers experienced in securing these additional high-cost inputs.

Evidence is very scanty concerning the new inputs being used by farmers during this period. The consumption of all fertilisers doubled between 1922 and 1938, and chemical fertilisers slowly replaced farm-produced fertilisers (Taiwan Provincial Civil Affairs Bureau 1946: 588-9). Gradual progress was also made in pest and plant disease control, and in the introduction of better tools and improved storage. The rate of acquisition of new tools and of building construction, however, was apparently slowed by the heavy competing commitments for seeds, fertiliser, and irrigation facilities.

An important factor, which perhaps explains the more widespread adoption of better farming methods, was a new development of land-lord-tenant relations in the late 1920s. Although the early land tax reform had been a distinct success, many Japanese officials believed the reformed land tenure system still hindered rural progress, because landlords did not encourage their tenants to develop the land and increase productivity as rapidly as they could. It was not until 1920-1 that an agency, the Production Promotion Bureau (Shokusan Kyoku), conducted its first survey of rural households to determine the number of tenants, owners, and part-owners in Taiwan. The results caused considerable alarm among top officials. They show that of 423,278 rural households, 42 per cent were pure tenants, 21 per cent were part-owners, and 37 per cent were owner-landlord households (Taiwan Sōtokufu Shokusan Kyoku 1926: 23-4). In the southern districts of Tainan and Kaohsiung the percentage of tenant households was lower than the island-wide average, whereas the percentage rose in the central and northern sections: Taichung (70 per cent), Hsinchu (53 per cent), and Taipei district (47 per cent). Tenancy appeared to be higher in those regions of the island which did not specialise in sugar cane.

There is some evidence that the percentage of tenant households declined slightly during the 1920s as more farmers bought additional land and achieved part-owner status (Taiwan Sōtokufu Shokusan Kyoku 1930: 3-4). This new trend might have resulted from increased farm income. In any case, it ceased during the 1930s. More investigation is required to explain both the general trends and regional variations in tenancy.

The prevailing land tenure system gave tenants little security to their land. Rents were very high, ranging between 40 and 60 per cent of the harvest (i.e. a very high, fixed proportion of output), and were paid in kind. Japanese surveys of land tenure practices revealed that nearly all landlord-tenant contracts rested upon oral agreement (Taiwan Sōtokufu Shokusan Kyoku 1931a: 2, 38-51; 1931b: 2-4, 34-53;

C

1931c: 3, 10, 19-27). Such arrangements could be broken easily. Landlords picked and discarded their tenants every other year. They
preferred tenants willing to pay higher rents, and would vigorously
eject one tenant family to make room for another willing to pay the
higher rent. If the harvest was good and farm prices rose, landlords
also pressed their tenants to pay higher rents than had been agreed
in the previous year. Thus, since so many contracts were based on oral
agreement, tenants had no security of tenure or permanency of rent
payment (Kajimoto 1932: 103, 109).

The Japanese appreciated that this large number of tenant households needed greater security to farm the land if they were going to
make any improvements (Taiwan Sōtokufu Shokusan Kyoku 1936:
1). In 1922 several Tainan officials formed a landlord-tenant association in which landlords had to agree to written contracts with their
tenants for a minimum term of five years. The association also arbi-
trated in disagreements between landlords and tenants. Despite
landlord resistance, district officials, assisted by the police, succeeded
in establishing a number of these associations throughout the Tainan
district. The colonial administration supported this policy, and in 1927
it ordered all district officials to introduce similar associations and
made funds available for this purpose. By 1935 there were 166
associations in operation, with 121 located at the village and township
level and 45 at the county level (Taiwan Sōtokufu Shokusan Kyoku
1936: 11). By that time about 70 per cent of all land farmed by
tenant households was under longer-term lease and written contract.
In Tainan district 80 per cent of tenant land was under such contracts, while in Taichung, which contained the highest percentage of
tenant households, 50 per cent of tenant land was under written
contract (Taiwan Sōtokufu Shokusan Kyoku 1936: 12).

WAR AND RECOVERY

In 1937 the Japanese realised that the island's economy would have
to be diversified if Taiwan was to participate successfully in Japan's
attempts to establish a new economic order in East Asia. By 1940
the administration had introduced a price policy which encouraged
the production of food and fibres at the expense of sugar cane (Taiwan
Keizai Nempō Kankokai 1941: 57-63). Increased administration spending on the development of new industries and the construction of
electric power plants, together with military mobilisation, called away
many farmers from their villages. As these policies were intensified
after 1940, farmers found it increasingly difficult to obtain necessary
inputs and to hire labour. Exports and imports declined to a mere
trickle. The Japanese were able to maintain the cultivated area and

output achieved in the late 1930s until around 1943 because of the highly organised system of farming which they had developed, but in 1944 and 1945 acute shortages of labour and materials took their toll, and farm output fell drastically.

The immediate post-war period (1946-9) was difficult for farmers. The departure of Japanese civil servants led to a dislocation of transport, marketing, rural research and extension services, and the severe shortage of consumer goods discouraged farmers from marketing their crops. With export trade disorganised and with little incentive to supply the cities, it was natural that farmers should revert to producing subsistence crops. There was in fact a great decline in the area devoted to sugar cane, and to some extent to rice, while the area under sweet potatoes and peanuts greatly expanded. Because of the drift of labour back to villages, a process set in motion by demobilisation and a decline in industrial production, farmers were able to revive production. Between 1946 and 1950 food output was higher (on a quinquennial average) than for the 1941-5 period.

It was not until 1950 that the Chinese Nationalist government began to take positive measures to control inflation, to rehabilitate trade, and to restore farm production to its pre-war levels. Had a smoother transition taken place from Japanese colonial to Chinese Nationalist rule, agricultural recovery might have been achieved much earlier. As it was, rural recovery did not really begin until early in 1950, and it was not until 1953 that crop yields and output levels regained the high levels of the late 1930s. These three years of recovery, however, were extremely important for post-war agricultural development. In this period the Nationalist government carried out an important land reform, and further modified and improved the organisational structure used by the Japanese to promote farm production.

Land reform was conceived and engineered by the official, Ch'en Cheng, later to become Vice-President of the Republic of China.[11] He was given the authority to launch a three-stage land reform scheme which was completed speedily and without bloodshed in 1954-5 (Ch'en 1951: 24-7, 81-2). The reform provided first, that tenant rents were to be reduced to 37 per cent of the harvest; second, that government land was to be sold to tenants and part-owners; and finally, that landlords were to relinquish all land over a prescribed maximum (ranging from 0·5 to 2 hectares for paddy land, and 1 to 4 hectares for non-paddy land, depending upon the class of land) at a fixed price to the government, which then sold this land to farmers. Farmers purchased the land with credit obtained through the newly-created

[11] Ch'en had introduced land reform in Hunan on Mainland China in the early 1940s.

land bank. They were given loans repayable over a period of up to
fifteen years at very low interest rates.

The program led to a transformation of the old land tenure system.
In 1950, 40 per cent of rural households were tenants, 33 per cent
owned their farms, and the remainder were part-owners. By 1956 only
17 per cent were tenants, 60 per cent were owners, and 23 per cent
were part-owners. By 1965 the proportions were 13, 67, and 20 per cent
respectively, indicating the continuing influence of the reforms. In
1965 virtually all households which had obtained long-term credit to
buy land in 1953-4 had made their final payments to the land bank,
and thus owned their land in full.

Several important administrative innovations were introduced as
the land reform was enacted. First, local and county government
budgets were integrated into regional and national budget accounts,
based on the revenue the government expected to collect, and the type
of economic development program it hoped to implement. Local
governments were ordered to spend on education, public health,
public works, and forms of rural improvement such as the distribution of new seeds, inoculation of livestock against disease, and the
construction of irrigation facilities and local roads (Hsieh 1963a: 3).
In this way, local government merged its resources with those of local
agricultural associations, to effect closer co-operation between the two.

Second, the old agricultural associations were revived, but their
membership was broadened to enlarge revenue and diversify activities.
They organised the young farming people into groups similar to the
United States 4-H clubs, developed agricultural extension education,
assisted local government in hog inoculation campaigns, and provided
minor banking facilities in towns (Bank of Taiwan 1962: 287-302).
Local irrigation associations were also revived to manage, improve,
and regulate local irrigation works (Hsieh 1963b: 32).

Finally, a new agency, the Joint Commission on Rural Reconstruction, was created to improve the efficiency of the organisational
system inherited from the Japanese. The new function of this agency
was to co-ordinate and implement agricultural policy and to undertake research over a wide range of problems. It was provided with
United States aid funds to finance many programs. Between 1950 and
1964 the Commission spent US$7·1 million. One-third of this amount
was allocated for water control projects and 10 per cent for crop
production. The remainder was distributed on programs such as crop
pest control, soil mapping, and promotion of special crops such as
mushrooms, asparagus, and tropical fruits. It also trained agricultural
specialists, financed capital projects, assisted farmers' organisations,
sponsored rural health programs, promoted fishery and forestry
development, and organised marketing and credit co-operatives (China

Publishing Co. 1952: 313-34). It served as a powerful catalytic agent in upgrading farming methods and developing new industries related to agriculture.

The Nationalist government has achieved a substantial development of both agriculture and industry since the post-war recovery. With the rapid progress of new industries, a transfer of resources has taken place out of agriculture, and a more complex linkage has developed between agriculture and the rest of the economy. These factors make it difficult to disentangle the effects of rapid general economic growth upon agriculture, from those produced by land reform and organisational change. Rather than pursue this complex problem further, attention is now turned to an examination of the nature of agricultural development since the post-war recovery and the types of changes taking place on farms today.

AGRICULTURAL DEVELOPMENT SINCE 1953

Between 1953 and 1965 farm output, comprising crops, livestock, forestry, and fishery products, has increased at the average annual compound rate of 5·5 per cent. The annual average growth rate of farm crops has been around 4·6 per cent, which exceeded the country's annual rate of population expansion by about 1·0 per cent per annum.

It should be strongly emphasised that this performance was achieved under circumstances which represented a real squeeze on the agricultural sector. On the one hand, there was a decline in the average size of the family farm, from 1·3 hectares in 1952 to 1·0 hectares in 1965. This was due to a growth rate in agricultural population which exceeded that of cultivated area over the period. Although urban population expanded at a particularly fast rate—it exceeded 50 per cent of the total population in 1960, and reached 54 per cent in 1965—it was not fast enough to prevent a continuous absolute increase in population within the rural sector. On the other hand, the Nationalist government spent nearly four-fifths of its annual budget in maintaining a military establishment of around a million men.

It was shown above (Table 2.6) that the sizeable growth rates recorded for the main crops in this period were due principally to increases in yields. Indeed, between 1951-5 and 1961-5, there was a decline in the area for rice and sweet potatoes, and only peanuts showed a substantial increase.

An analysis of the 1951-5 to 1961-5 period (Table 2.7) shows that a majority of the output expansion occurred in the first half of the period, from 1951-5 to 1956-60, and that growth actually decelerated in the second half. With the exception of rice, the major contributions to increased productivity were concentrated within the first

TABLE 2.7

GROWTH RATES FOR FOUR MAJOR CROPS BASED ON
QUINQUENNIAL AVERAGES, 1951–5 TO 1961–5

	1951–5 to 1956–60	1956–60 to 1961–5
Brown rice		
Area	0·10	−0·10
Yield	2·96	3·26
Production	3·01	3·14
Sugar cane		
Area	0·85	0·38
Yield	3·33	0·24
Production	4·16	0·57
Sweet potatoes		
Area	−0·85	0·51
Yield	5·24	0·61
Production	4·33	1·19
Peanuts		
Area	4·76	−0·04
Yield	3·45	2·94
Production	8·31	2·43

Source: Taiwan Provincial Food Bureau 1967: 1, 3, 7; Joint Commission on Rural Reconstruction 1956: 20, 32, 36, 44; Department of Agriculture and Forestry 1967: 97.

period, and additions to area under cultivation were virtually negligible thereafter.

Farming has recently begun to diversify significantly with the appearance of new crops such as pineapples, bananas, and mushrooms. Farmers have reacted to changes in market demand by switching to more profitable crops. These have been introduced into existing rotation systems, with the result that production has become a more continuous process. This in turn has been made possible and profitable by the use of capital inputs and improvements in technology, which have raised labour productivity.

In the early 1950s the government established six nitrogenous fertiliser plants, with the initial aim of self-sufficiency in fertiliser production. The operation of these plants raised output of chemical fertiliser from a level of around 59,000 metric tons in 1950 to about 522,000 tons in 1962. By 1964 the plants were supplying all domestic requirements of nitrogenous fertiliser with an additional surplus for export, which has rapidly increased since and has earned, for Taiwan, several million United States dollars annually (Yuan 1964: 10). Meanwhile, domestic fertiliser consumption has risen from a level of 300,000 metric tons in 1953, which matched the pre-war peak, to about 767,000 tons in 1965 (Dept of Agriculture and Forestry 1967: 284).

The fact that farming in Taiwan has become a more continuous process through the four seasons has intensified output per unit area, particularly over the last fifteen years (Table 2.8). Improved seed varieties have been introduced, for example, which have broadened the conditions under which production can take place, for example through greater drought resistance and a greater ability to withstand severe fluctuations in temperature.

Purchases of farm implements, with direct or indirect effects on productive efficiency, have become significant since 1960. Between 1961 and 1965 the number of power-driven sprayers and dusters rose from 966 to 4,489, water pumps from 10,114 to 32,107, rice dehuskers from 181,693 to 205,784, spacing gauges from 78,500 to 115,881, and rotary-tiller ploughs from 4,450 to 8,728 (Dept of Agriculture and Forestry 1967: 294-6). The sprayers, dusters, and spacing gauges contribute directly through improved conservation and planting of crops. Water pumps, dehuskers, and the rotary-tiller ploughs reduce labour inputs during the harvest period, a saving which has become important with Taiwan's multi-cropping system. A recent study of the plough indicates that, if adopted more widely, it will enable farmers to dispense with animal power and so eliminate the costs of land and other inputs required for their care (Joint Commission on Rural Reconstruction 1966: 7). In the spring months, timing of operations is particularly important for deriving maximum benefit from rainfall and temperature conditions. By the use of the plough, household labour is released for other farming tasks or to earn additional income off the farm.

Industrial development has provided new opportunities for non-farm employment. Bernard Gallin's 1957-8 study (1966: 109) of Hsin Hsing village near Taichung emphasises the importance of non-farm income for villages, and suggests that it was principally these increased opportunities which explains why villagers were able to improve their living standards during the 1950s. Prior to 1945 these villagers saved only to buy land, but since 1950 they have saved to buy consumer goods.

A considerable increase in non-farm earnings took place between 1952 and 1957 (Tsui 1959: 1; Bank of Taiwan 1962: 129). In 1957 farm families earned 63 per cent of annual income from farming as against 78 per cent in 1952. In 1957, for farms of less than 0·5 hectares, 41 per cent of household receipts were from non-agricultural income; the proportion declined to 26 per cent for the range 0·5-1·0 hectares, and to 14 per cent for farms larger than 2·0 hectares (Shen 1964: 294). Finally, a 1964 report on labour mobility showed that an

TABLE 2.8

INDICES OF AREA, YIELD, AND PRODUCTION OF FOUR MAJOR CROPS, 1901–5 TO 1961–5

(1901–5 = 100)

Period	Brown rice			Sugar cane			Sweet potatoes			Peanuts		
	A	Y	P	A	Y	P	A	Y	P	A	Y	P
1906–10	119	103	122	181	107	196	140	112	152	125	115	140
1911–15	124	104	128	362	79	319	148	109	156	119	98	116
1916–20	122	109	132	581	103	595	149	115	168	144	122	175
1921–25	131	117	154	586	117	670	161	133	208	150	147	233
1926–30	148	127	187	519	186	951	165	160	255	163	160	256
1931–35	168	142	238	467	231	1,064	179	177	307	188	169	314
1936–40	164	154	251	671	238	1,544	179	189	327	194	177	337
1941–45	152	130	198	690	193	1,321	201	155	300	131	120	151
1946–50	176	123	216	376	145	542	289	138	388	438	132	581
1951–55	196	159	310	443	214	932	319	151	466	500	133	733
1956–60	197	184	360	462	252	1,143	305	195	576	631	173	1,093
1961–65	196	216	420	471	255	1,176	313	201	611	619	200	1,233

Source: Taiwan Provincial Food Bureau 1967: 1, 3, 7; Joint Commission on Rural Reconstruction 1956: 20, 32, 36, 44; Department of Agriculture and Forestry 1967: 97.

average of eight out of ten farm households dispatched some farm labour to work off the farm (Tsui and Lin 1964: 1).

Urbanisation and rising living standards have increased the demand for vegetables, meat, and poultry, and have encouraged diversification in the rural sector. It is possible that smaller farms have been the most responsive to these new demands. A good example is a small family rice farm in Lung Hsi village in the Taichung district (Taiwan Provincial Dept of Agriculture and Forestry 1966: 1-30). Here, despite a declining size of farm and the necessity to borrow capital, the family were able to introduce mushroom production and hog raising, to repay debts, and to raise farm income and living standards, all within the space of a few years.

In the realm of agricultural policy there is one important area which hitherto has not received sufficient attention: the fertiliser-rice barter system. This was established to provide a stable rice supply for the government and has been administered by the Provincial Food Bureau. The Bureau determines relative prices twice a year and barters fertiliser mostly for unhulled rice, or 'paddy', which is delivered at the end of the crop season in question. About one-quarter of the rice crop is collected this way by the Bureau.

One study has seriously questioned the efficacy of the present system (Lewis 1967: 127-79). It showed that a sample of twenty-one rice growers in Taichung used a very large part of the fertiliser, acquired from the Bureau, for other crops. It was not possible, however, to measure the effect of this diversion accurately, owing to the difficulty in separating the yield effects of chemical from organic fertilisers used on the Taichung farms. The same problem was encountered when nitrogen, phosphate, and potash inputs were correlated against rice yield for the island as a whole over the 1951 to 1964 period. Correlation analysis did reveal a considerable yield variation which was not associated with the fertilisers acquired by barter. This variation had to be attributed to other inputs.

A comparison of fertiliser prices in Taiwan and Japan showed that a higher level existed in Taiwan. An estimation was then made of the likely response in rice production in Taiwan, assuming the price of fertiliser at the lower 1964 level in Japan. It was found that total output (and yield) would have been 44·5 per cent above the actual 1964 level of 2·2 million metric tons (Lewis 1967: 175). It was then concluded that:

A more generous distribution of fertilisers for paddy, at lower prices in paddy, along with increased open market supplies at market prices correspondingly lower than at present, would probably assist in increasing the production of the many other Taiwan crops on which fertiliser is needed.

CONCLUSION

The substantial accomplishments reviewed in this brief survey indicate what can be achieved by a combination of enlightened government policy and a responsive farming community.

Both the Japanese and Nationalist Chinese governments were prepared to introduce reforms where institutional obstacles lay in the path of agricultural development. They were also prepared to stimulate production more directly by making available to farmers improved seeds, new irrigation facilities, chemical fertilisers, and new farming techniques.

Farmers, for their part, showed they could respond to market incentives in raising output of traditional crops and in switching to newer, profitable alternatives, by working harder, by adopting new technology when it was to their advantage, and by improving management skills as the organisation of farming became more complex and demanding.

There seems every indication that a fast pace of agricultural development can be maintained in the future. The future demand for food and industrial crops can be expected to remain high as the urban population continues to rise. If urban conditions remain attractive, the present declining rate of rural population growth may continue and may begin to relieve the pressure of manpower in the rural sector. The efficient system of research and extension now available gives promise of advancing new techniques speedily to the farmer. Finally, the growth of capital inputs is only beginning on the farm and further potential exists for increases in consumption of certain types of fertiliser[12] and in labour-saving mechanisation. Considerable scope remains for improvements in disease and pest control[13] and in the local adaptation and introduction of improved farming practices. The responsiveness of farmers and their rising levels of education should ensure that the benefits of these improvements are quickly secured.

[12] One study of fertiliser prospects, however, points out that the use of nitrogenous fertilisers is fast approaching the desired limit for some soils (Society of Soil Scientists and Fertiliser Technologists of Taiwan 1963: 4-5). It proposed that yields should be increased by the use of more disease-resistant seeds, more organic manures, and the planting of green manure crops. The study of fertiliser-yield responses for different soils throughout Taiwan remains an urgent task.

[13] It is estimated, for example, that rodents reduce the harvest by as much as 10 to 20 per cent annually in some districts (see Lo 1961: 1; Horng 1959: 38-9).

3

India

J. G. Crawford

THIS chapter will attempt to show the major facts (in aggregate terms only) about Indian agriculture. More important, it will tell a story of major problems not adequately assessed in the Third Five-Year Plan (1960-1 to 1965-6) and certainly not fully appreciated by the government authorities. These problems were basically those inherent in population pressure on land resources and in agricultural policies shown to be inadequate for dealing with this pressure. The chapter contrasts past targets with performance, somewhat unfavourably, but does end with the promise of a closer match between need and probable achievement over the next few years. This promise arises from the increasing availability of a more profitable technology; from a more comprehensive and realistic set of components in agricultural policy; from a more adequate recognition of the interaction of agriculture with the industrial economy and with the external economic situation; and from a more determined and vigorous administration. The situation is still far short of ideal, but the improvement is sufficient to justify a change from pessimism to hopeful confidence that in the next few years India will adequately feed itself, independently of other than marginal external assistance through trade and aid.

The chapter draws extensively from reports by the author and his colleagues as members of the Agricultural Section of the International Bank for Reconstruction and Development's (World Bank's) Economic Missions to India in 1964-5 and again in 1966-7.[1] It is not a complete story. For example, since the emphasis is on the aggregate picture, the variations from village to village, district to district, and state to state only occasionally appear. Again, the chapter is centred largely on foodgrains, so omitting the livestock problem and the more general (and extremely important) problem of protein deficiency.

[1] The 1964-5 World Bank Economic Mission to India (Agricultural Section) included the following: Sir John Crawford (group leader), Mr Wolf Ladejinsky, Mr L. Garnier, Dr Louis Goreux and Dr W. D. Hopper. In 1966-7 Dr W. D. Hopper again assisted Sir John Crawford, joined this time by Mr L. Sonley and Mr A. Van Nimmen. As yet the reports of these Missions have not been published.

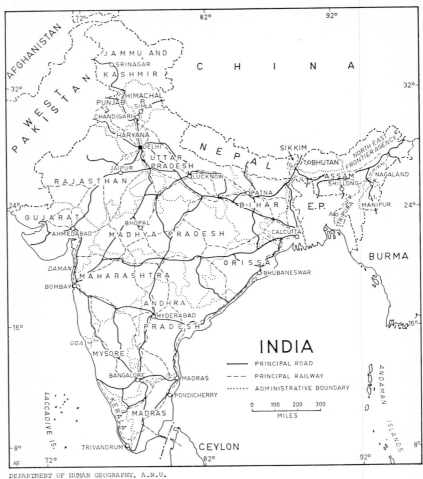

DEPARTMENT OF HUMAN GEOGRAPHY, A.N.U.

Map 3   India

The emphasis tends to be on immediately essential factors such as
price policy, new varieties, and input supplies, rather than on longer-
run but ultimately no less essential factors such as agrarian reform
(land tenure, consolidation of fragmented holdings), the task of
building up adequate extension services, and, not least, the social
tensions already growing between farmers with relatively large holdings of four hectares or more, assured water supply, and some capital,
and those poorly situated in all three respects. This imbalance may be

forgiven if the more limited objective of presentation is accepted: a statement of the major problems of food supply and related agricultural policies together with an indication of the hopeful change that has begun to appear in the situation.

Even this limited objective calls for an initial statement of the main background facts. This will be followed by an examination of agricultural targets and policies expressed in the Third Plan, and marked by substantial failure, and finally by a review of changes marking the period since the 'good year' of 1964-5.

THE RURAL SECTOR IN THE ECONOMY

The basic facts about Indian agriculture are readily enough available,[2] although the precision of use sometimes made of much of the data is open to considerable doubt.[3] In the first place the rural sector currently contributes around 40 per cent of net domestic product (Table 3.1). From 1948-9 to 1962-3 its relative importance remained stable, aside from short-term fluctuations, but more recently a declining trend has become apparent. Within the sector, agricultural production has been clearly dominant. Six major cereals provide about 85 per cent of foodgrain production (rice, sorghum (*jowar*), millet (*bajra*), maize, wheat, and barley).

Around 70 per cent of the total work force is dependent on agriculture. Approximately 22 per cent of commercial imports (excluding non-dollar PL480 aid) are of agricultural origin, while agricultural exports, including processed products, contribute more than three-quarters of total merchandise export income (Table 3.2). Over half of this contribution is derived from tea, jute manufactures, and cotton fabrics, though of these three, only jute manufactures showed important growth during 1960-1 to 1965-6.

Linkage between the rural and other sectors of the economy takes three main forms. First, there are inputs purchased by agriculture, such as fertilisers, pesticides, machinery, tools, electricity, fuel, and oil. It was estimated in 1958-9 that these amounted to hardly 1 per

[2] Where sources for tables in this chapter are not specifically given they are drawn from reports by the 1964-5 World Bank Mission. It will only be where important opinions have been specifically expressed and developed by other members in their particular reports that direct acknowledgment will be made to the individuals concerned. My own concurrence with such views will be clear from the context.

[3] Thus, for example, attempts to state per capita availability of foodgrains for consumption compound the error undoubtedly present still in estimates of production by highly problematical estimates of wastage in the fields and in storage. Again, the data on marketable proportion of production remains an unsafe basis for any very precise or detailed analyses.

Table 3.1

NATIONAL INCOME BY INDUSTRIAL ORIGIN: PERCENTAGE DISTRIBUTION

Sectors of origin	Average 5 years 1948–9 to 1952–3	Average 5 years 1953–4 to 1957–8	Average 5 years 1958–9 to 1962–3	1963–4	1964–5	1965–6[a]	1966–7[a]
Agriculture, animal husbandry and ancillary activities	48·8	46·1	46·4	n.a.	n.a.	n.a.	n.a.
Forestry	0·7	0·7	0·8	n.a.	n.a.	n.a.	n.a.
Fishery	0·4	0·5	0·5	n.a.	n.a.	n.a.	n.a.
Total agricultural	49·9	47·3	47·7	45·1	44·8	40·1	39·9
Mining, manufacturing and small enterprises	16·8	18·1	18·6	19·5	19·2	21·0 ⎫	
Commerce, transport and communication	18·1	18·1	17·0	17·1	17·4	18·3 ⎬	60·8
Other services	15·3	16·5	17·0	18·9	19·2	21·4 ⎭	
Net domestic product at factor cost	100·2	100·0	100·3	100·6	100·6	100·8	100·7
Net income earned from abroad	−0·2	0·0	−0·3	−0·6	−0·6	−0·8	−0·7
Net national output at factor cost (national income)	100·0	100·0	100·0	100·0	100·0	100·0	100·0

[a] Estimate.

Source: Central Statistical Organisation 1964; Planning Commission 1966a, 1967.

TABLE 3.2

EXPORTS OF PRINCIPAL COMMODITIES: PER CENT OF TOTAL VALUE

Commodity	1960–1	1961–2	1962–3	1963–4	1964–5	1965–6
Fish and fish preparations	0·8	0·6	0·6	0·7	0·9	0·9
Cashew kernels	3·0	2·7	2·8	2·6	3·6	3·5
Other fruits and vegetables	1·1	1·0	1·0	1·1	1·0	0·9
Spices	2·7	2·7	2·0	2·0	2·1	2·9
Sugar	0·3	2·2	2·5	3·2	2·6	1·4
Coffee	1·1	1·4	1·2	1·0	1·6	1·7
Tea	19·7	18·3	18·9	15·3	15·6	14·7
Tobacco	2·5	2·2	2·8	2·9	3·2	2·7
Total food, drink, and tobacco[a]	31·2	31·1	31·8	28·9	30·6	28·6
Oilcakes	2·2	2·6	4·5	4·4	5·0	4·5
Vegetable oils[b]	2·1	1·6	2·5	2·9	1·2	0·8
Hides and skins	1·4	1·2	1·6	1·2	1·1	1·3
Raw cotton and waste	1·9	3·0	2·5	2·1	1·7	1·5
Raw wool and hair	1·3	1·4	1·0	0·9	1·1	0·9
Other textile fibres	0·3	0·6	0·7	0·9	1·0	n.a.
Gums, resins, and lac	1·4	1·0	1·0	1·0	0·9	0·9
Total other agricultural products[a]	10·6	11·4	13·9	13·3	12·1	9·8
Leather and manufactures	4·0	3·7	3·4	3·2	3·4	3·6
Footwear	0·5	0·3	0·4	0·5	0·5	0·6
Cotton fabrics	9·2	7·2	7·0	6·7	7·2	7·0
Artsilk fabrics	0·5	1·0	1·0	1·4	0·9	0·6
Woollen manufactures	1·0	0·7	0·7	1·1	0·9	0·9
Jute manufactures	21·4	21·9	22·0	19·3	20·9	23·4
Coir manufactures	1·4	1·6	1·8	1·5	1·4	1·4
Total leather and fibre manufactures[a]	37·9	36·5	36·4	33·8	35·1	37·6
Total agricultural	79·7	79·0	82·1	76·0	77·8	76·0
Total other exports and re-exports	20·3	21·0	17·9	24·0	22·2	24·0
Total exports	100·0	100·0	100·0	100·0	100·0	100·0
Total exports (Rs crores)[c]	631	668	682	802	803	782[d]

[a] Excludes certain items not identified separately and included in 'other exports'.

[b] Includes 'essential oils'.

[c] 1 crore = 10 million.

[d] Export data in balance of payments for 1965–6 are on the 'negotiated document basis' beginning with October 1965 instead of the earlier shipment basis.

Source: Ministry of Finance, Government of India.

cent of gross agricultural output, although important changes are likely in the future and, indeed, are already occurring.

Second, there is an important relationship in agriculture's dependence upon industry for its consumer needs, and here the linkage has been somewhat stronger in the past, since rural incomes account for the bulk of expenditures on clothing, and are becoming increasingly important for other consumer and light capital goods. The recent depression associated with the droughts has shown the working of another significant aspect of this two-way relationship. Textile industries lost ground not only directly, because farm income fell, but also indirectly, because high prices for food cut into the urban incomes which provided the alternative outlet for industrial products.

Finally, large sections of India's manufacturing industry use agricultural raw materials. The two most important categories are textiles (cotton, wool, and jute) and food processing (wheat flour, edible oils, sugar, tea, etc.).

It will thus be readily appreciated that the total impact on national income of variations in the level of output in the agricultural sector is larger than the change in its direct contribution, because of these activities dependent upon it. This is qualified by the existence of a very large subsistence sector marked by little contact with the market, but despite this, the multiplier effect of the sector as a whole is significant—a point too often overlooked in some of India's past policies.

India has a high proportion of its land area under cultivation. In 1958-9 a net area of 131 million hectares was cropped, of which 20 million hectares were sown more than once, thus giving a total or gross cropped area of 151 million hectares (374 million acres). It was anticipated that in 1965-6 the net area would be around 134 million hectares, of which 26 million would produce an additional crop, giving a gross area of 160 million hectares (395 million acres). The net area cropped represented about 40 to 45 per cent of the total geographical area. If fallow land, land in use for other rural pursuits, and cultivable waste land were added, the proportion was probably in the range of 60 to 66 per cent. Some 23 million of the 131 million hectares (18 per cent) were irrigated, and of these, 3·5 million hectares were used for multiple-cropping. Forest land accounted for 53 million hectares in 1958-9, roughly 16 to 18 per cent of estimated total geographical area.

The key role of foodgrains in the rural sector is shown by the fact that they were grown on 76 per cent of the gross cropped area in 1958-9 (Table 3.3). Rice alone accounted for about 22 per cent of the gross acreage. About 80 per cent of the gross irrigated area was used for foodgrains and the majority of this was used either for rice or

TABLE 3.3

GROSS CROPPED AREA AND GROSS IRRIGATED AREA BY CROPS, 1958-9

| | Gross cropped area | | Gross irrigated area | | Percentage of crop area irrigated |
	million hectares	percentage	million hectares	percentage	
Rice	37·78 (81·00)	21·7	12·04 (29·77)	44·7	36·7
Wheat	12·61 (31·16)	8·4	4·02 (9·92)	14·9	31·8
Barley	3·31 (8·19)	2·2	1·37 (3·40)	5·0	41·5
Maize	4·26 (10·53)	2·8	0·45 (1·10)	1·7	10·5
Sorghum	17·98 (44·44)	11·9	0·62 (1·54)	2·3	3·5
Millet	11·41 (28·19)	7·6	0·34 (0·85)	1·3	3·0
Winter cereals	2·55 (6·31)	1·7	0·43 (1·07)	1·6	17·0
Other cereals and millets	5·32 (13·14)	3·5	0·12 (0·29)	0·4	2·2
Gram	10·05 (24·83)	6·7	1·21 (2·98)	4·5	12·0
Other pulses	14·31 (35·37)	9·5	0·86 (2·11)	3·2	6·0
Total foodgrains	114·59 (283·17)	76·0	21·46 (53·03)	79·6	18·7
Sugar cane	1·95 (4·82)	1·4	1·34 (3·31)	5·0	68·6
Other food crops[a]	4·22 (10·44)	2·8	1·26 (3·11)	4·7	27·1
Oilseeds[b]	13·07 (32·30)	8·8	0·42 (1·04)	1·6	3·2
Cotton	7·96 (19·68)	5·3	1·00 (2·47)	3·7	12·5
Jute and mesta	1·06 (2·62)	0·7	i.o.c.		
Other fibres	0·31 (0·76)	0·2	i.o.c.		
Tobacco	0·39 (0·97)	0·2	0·08 (0·20)	0·3	20·4
Fodder crops	5·42 (13·40)	3·6	1·03 (2·56)	3·8	19·1
Tea	0·32 (0·78)	0·2	i.o.c.		
Coffee	0·12 (0·30)	0·1	i.o.c.		
Rubber	0·12 (0·30)	0·1	i.o.c.		
Other non-food crops	0·89 (2·21)	0·6	0·36 (0·88)	1·3	39·7[c]
Total	150·85 (372·76)	100·0	26·95 (66·59)	100·0	17·9

Note: Bracketed figures are million acres, as in source.

i.o.c. = included in other non-food crops.

[a] Includes, in '000 hectares: chillies 590 (1,458), black pepper 93 (231), all fruits 994 (2,457), potatoes 337 (832), and other vegetables 1,187 (2,932).

[b] Includes, in '000 hectares: groundnuts 6,253 (15,451), castor 490 (1,212), sesame 1,888 (4,666), rape and mustard 1,238 (3,059), linseed 1,362 (3,365), and coconut 693 (1,713).

[c] For all other non-food crops marked i.o.c.

Source: Fertiliser Association of India 1965, 1967.

wheat. The proportion of total crop area which was irrigated was high for four crops—sugar cane, barley, rice, and wheat. It is noticeable that areas under tea, jute and mesta, and cotton, crops which figure prominently in Indian exports, accounted for only about 6 per cent of gross cropped area in 1958-9.

India has a large, although unevenly distributed, surface water (river) supply and has extensive groundwater resources, although, owing to lack of adequate survey work, these are not known with any accuracy. Both sources are concentrated in the north, and, although the limits are not accurately known, it is clear that both are far from fully used. Indeed, 'India still has immense potential resources for irrigation and its development is not likely to be hampered over a long-term period by the problem of water availability',[4] given, of course, the resources of capital and skill to exploit them. About two-thirds of all irrigated land is under minor schemes, including shallow wells, tubewells, tanks, or systems using river or spring water. The rest is under major and medium works comprising canal systems based on diversion barrage, designed to give 'drought insurance to large areas'. There are also some large storage dams, and, finally, there are schemes to control and rehabilitate water-logged and saline areas. As we will observe later, the development of groundwater resources through tube-well development is rapidly assuming a more vital importance.

Before examining the recent growth performance of the rural sector, a brief perspective is necessary of some of its chief social and economic characteristics.

Indian agriculture can be fairly described as traditional, in the sense that until recently changes in farming methods have been few and unspectacular, and in the related sense that farming is a way of life, centred on the village and conforming to the social structure of the community.

A three-tiered structure characterises rural organisation, comprising landowners, tenants (with varying degrees of security, but often with none), and landless labourers. Among landowners, there is an unequal distribution of ownership. In 1959 it was estimated (Planning Commission 1959: T. 4) that 75 per cent of rural households owned only 16 per cent of total area of holdings, while at the opposite end of the scale 7 per cent of households owned 52 per cent of the land. Similar disparities are revealed in terms of size of holdings. In 1960-1, 75 per cent of operational holdings were 3·2 hectares (8 acres) or less but accounted for only 30 per cent of total area operated, while 6 per cent of holdings were 8 hectares or more but accounted for around 35 per cent of the total area operated. About 22 per cent of

---

[4] L. Garnier in the 1964-5 Report of the World Bank Mission to India.

rural households owned no land at all. In 1961, 24 per cent of all workers in the agricultural sector were classed as labourers, as against owner and tenant cultivators (Ministry of Food and Agriculture 1963: T. 1.7).

Accurate data on tenancy are scarce and data from the various sources are by no means wholly comparable. According to the 1961 Census, 8 per cent of farm households had pure tenancy arrangements and 15 per cent had mixed tenancies (Sharma 1965: T.7). It is generally believed, however, that these figures underestimated the true position. Nevertheless, the census figures do indicate that in about half the states of India land cultivated under pure and mixed tenancy ranges from 31 per cent (Bihar) to 67 per cent (Kerala). This question of tenancy will, for reasons given later, become more serious with time.

Hardly less important is the continued existence of fragmented holdings. It is not possible to give an adequate discussion here, but it needs little imagination to appreciate that under any level of technology the cultivation of a holding in small fragments is likely to be less efficient than that of a consolidated holding. Since progress towards consolidation has been slow, the phenomenon must be noted as one of the constraints on progress in Indian agriculture. Agrarian reform was one of the great planks of the post-war independence movement. Progress in some states has been significant, but far short of longer-run necessity if India's agricultural expansion is to be at the rate required.[5]

To this story of an almost feudal structure of land use must be added other characteristics, to be discussed later, including low fertiliser usage, inappropriate water development policies (until recently), costly and inadequate credit systems, poorly organised extension services, and inadequate research.

The picture adds up to a traditional technology in which, as is true of any level of technology, some farmers are efficient within the existing constraints and others are decidedly inefficient. As will be shown, there has been growth within the traditional, social, and occupational structure: indeed, there is always room for some such expansion of output. Moreover, the constraints can be eased—even dramatically so—by appropriate policies, and much can be achieved even before substantial agrarian reform in relation to land consolidation and land tenure has been carried out. Additional fertiliser, more assured water supplies, high-yielding varieties of the various grains, cheaper and assured credit supplies, price guarantees, and other

[5] The author particularly regrets the non-publication as yet of W. Ladejinsky's contribution to an understanding of all these elements in the continuing problem of agrarian reform. The few words here do less than justice to his contribution to the Bank Reports.

elements in a sensible agricultural policy constituting in all a quite different level of technology, can be added to the existing farm structure with highly significant results. Indeed, they must be added in significant measure if the initial impact required to launch Indian agriculture on a higher level of growth is to be given. Again, that higher level of growth has to be achieved if the basic problem of population pressure is to be resolved.

THE MALTHUSIAN PRESSURE

The analysis in this chapter will be confined to foodgrain production, for two good reasons. First, foodgrains are the most important commodities in the agricultural sector and most of the problems of policy related to them are also relevant to other agricultural products. Second, and more immediately important for India, there has been a steadily developing food crisis, dramatically emphasised by the two recent droughts.

The trend in the relation of population growth to agricultural expansion has been moving in Malthusian fashion with no significant relief possible in terms of emigration or trade.[6] The growth rate in population has increased from about 0·8 per cent per annum (1901-31), to 1·3 per cent (1931-51), to 2·0 per cent (1951-61), and now is about 2·6 per cent per annum, with no certainty that it will not increase to, say, 2·8 per cent, before family planning becomes significantly effective.[7] With important regional exceptions in drought periods, food supply had been more or less adequate till 1961, ranging between an average of 386 to 476 grams (14 to 17 oz) of foodgrains per head per day (Table 3.4), though implying, of course, that great numbers managed with considerably less. The protein and vitamin supply situation was even less satisfactory. More recently, however, the rising annual growth rate of population has exceeded the average annual growth rate of foodgrain production (Fig. I). In the early 1950s population growth remained at about 2·0 per cent per annum. The then apparent trend rate in foodgrain production matched population growth and offered some improvement in food consumption standards as money incomes rose. By the 1960s population growth had risen to 2·5 per cent, while the trend in foodgrain output had fallen significantly below that figure. Regardless of drought, tell-tale signs appeared: rising food prices, increasing dependence on food imports, higher land rents, costly credit, and general farm dissatisfaction.

This chapter cannot deal further with the population story. We have to accept a situation in which, even if success attends the Indian

[6] For a more detailed application of the Malthusian argument see Crawford (1968).

[7] See Crawford (1968) and Chandrasekhar (1967).

TABLE 3.4

NET AVAILABILITY OF CEREALS AND PULSES

Year	Population (million)	Cereals (million tons) Production Gross	Net^b	Net imports	Withdrawals (+) (−) from government stocks	Net availability^c	Pulses (million tons) Net availability^c	Per capita net availability Cereals (gm)	Pulses (gm)	Total (gm) (oz)
1951	363·4	45·74	40·02	4·80	(+) 0·59	44·23	8·03	334·5	59·5	394 (13·9)
1952	369·6	46·40	40·60	3·93	(+) 0·62	43·91	7·97	326·0	59·5	386 (13·6)
1953	376·1	51·35	45·37	2·04	(−) 0·48	47·89	8·59	348·7	62·4	411 (14·5)
1954	382·9	61·08	53·44	0·83	(+) 0·20	54·07	9·72	385·6	70·9	456 (16·1)
1955	390·2	58·97	51·60	0·60	(−) 0·75	52·95	10·10	371·4	70·9	442 (15·6)
1956	397·8	57·53	50·34	1·40	(−) 0·60	52·34	10·21	360·0	70·9	422^d (14·9)
1957	405·8	60·20	52·68	3·63	(+) 0·86	55·45	10·61	374·2	70·9	445 (15·7)
1958	414·3	56·41	49·36	3·22	(−) 0·27	52·85	8·82	348·7	59·5	408 (14·4)
1959	423·3	65·49	57·30	3·86	(+) 0·49	60·67	11·54	391·2	73·7	465 (16·4)
1960	432·7	64·88	56·77	5·13	(+) 1·40	60·50	10·32	382·7	65·2	448 (15·8)
1961	442·7	69·31	60·65	3·49	(−) 0·17	64·31	11·11	396·9	68·0	465 (16·4)
1962	453·5	70·95	62·08	3·64	(−) 0·36	66·08	10·28	399·7	62·4	462 (16·3)
1963	464·3^e	67·01	58·63	4·55	(−) 0·02	63·20	9·99	371·4	59·5	431 (15·2)
1964	475·5^e	70·19	61·41	6·26	(−) 1·24	68·91	8·79	396·9	51·0	448 (15·8)
1965	487·0^e	76·56	66·99	7·45	(+) 1·06	73·38	10·88	413·9	62·4	476 (16·8)
1966	498·9^e	62·25	54·47	10·34	(+) 0·14	64·67	8·76	354·3	48·2	403 (14·2)

^a Production figures relate to agricultural year July-June; 1951 figure corresponds to the production of 1950–1, and so on for subsequent years. These estimates up to the year 1959–60 are adjusted with 1960–1 revised production index as the base. Figures for 1960–1 and 1961–2 are based on revised estimates, for 1962–3, 1963–4, and 1964–5 on partially revised estimates, and for 1965–6 on final estimates of production.

^b Net production has been taken as 87·5 per cent of the gross production, 12·5 per cent being provided for feed, seed requirements, and wastage.

^c Net availability = net production + net imports + decline in government stocks. Figures in respect of change in stocks with traders and producers over a year are not known. The estimates of net availability should not, therefore, be taken to be strictly equivalent to consumption.

^d Data do not add in original.

^e Provisional population figures relate to mid-year revised estimates. These estimates have been prepared by the office of the Registrar-General of India.

Source: Ministry of Finance 1967b: T. 1.5.

government's measures for population control, there will be little relief until the mid-1970s. Even with a falling birth rate absolute numbers will continue to grow; at the present time (1968) the growth rate is equivalent to 13 million additional mouths each year. It thus becomes important to know what has been happening to agriculture and its rate of expansion.

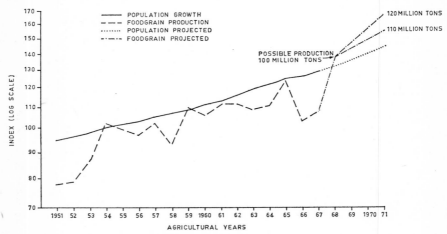

Fig. I   Indices of population and foodgrain production[a]

a Base: average of 1954 and 1955 crop years = 100 (index) = 386 million population = 71 million tons metric production.

Source: Planning Commission 1961; Ministry of Finance 1967a.

GROWTH RATES IN AGRICULTURE

Estimates of growth for the 1950-1 to 1964-5 period show that for the twenty-eight most important crops, a compound rate of expansion of 3·2 per cent was achieved over the period (Table 3.5). The growth rate for foodgrains was somewhat below that for non-foodgrains. The highest growth rates were recorded for crops relatively less important in terms of their contribution to total production—cotton, sugar cane, and groundnuts.

Foodgrains

In view of their singular importance, a special comment is called for on the growth rate of foodgrains, bringing into account the 'good' crop for 1964-5. The compound growth rate for the fifteen-year period was 3·0 per cent per annum. In fact, a somewhat better fit is obtained by fitting a linear trend to the data, which implies a decelerating

TABLE 3.5

GROWTH RATES OF AGRICULTURAL PRODUCTION, 1950–1 TO 1964–5

Commodity	Weight assigned in total production	Compound annual growth rate (%)
All agricultural crops[a]	100	3·2
Foodgrains	66·9	3·0
Non-foodgrains	33·1	3·6
Summer cereals	47·8	3·1
Rice	35·3	3·4
Winter cereals	10·5	3·3
Wheat	8·5	4·0
Pulses	8·6	1·6
Oilseeds	9·9	3·2
Groundnuts	5·7	4·2
Sugar cane	8·7	4·6
Cotton	2·8	4·4
Jute	1·4	3·5
Plantation crops	3·6	2·7

[a] Covers the 28 principal crops grown.

Source: Ministry of Food and Agriculture, Government of India. These data are based on official sources made available to the author in 1966 by way of a revision of his material which up until then did not include 1964–5.

percentage rate of growth.[8] If projected forward to 1970-1, the rate of growth from 1969-70 to 1970-1 would drop to 2·1 per cent, as against 3·6 per cent at the beginning of the period. This suggestion of decelerating growth receives strong support if trends are fitted to the period 1955-6 to 1964-5. The compound rate for this shorter period is 2·5 as against 3·0 per cent for the full period. The fit for the shorter period is not as good as for the longer period, but the evidence leaves a strong impression that, far from accelerating as planned, the growth rate in food production had actually been declining in the later years of the period since 1950-1. This is of particular importance in view of the high target rates of growth to be discussed later. The conclusion is all the more tenable if the two factors in growth are examined: changes in land area devoted to crops and productivity per hectare (Table 3.6).

For foodgrains there was a substantial contribution to increased production from growth of total area under crop during the First Plan period, but this dropped sharply during the Second and Third Plan periods. The contribution from increased agricultural productivity was somewhat lower relative to area expansion in the First

[8] Calculated by W. D. Hopper in the 1964-5 Report of the World Bank Mission to India.

TABLE 3.6

ANNUAL COMPOUND GROWTH RATES OF AGRICULTURAL
PRODUCTION, AREA, AND PRODUCTIVITY, 1951–2 TO 1964–5[a]

Plan periods	Crops	Area (%)	Productivity (%)	Production (%)
First Plan				
(1951–2 to 1955–6)	Foodgrains	3·28	2·79	6·18
	Non-foodgrains	2·44	0·74	3·21
	All crops	3·15	1·90	5·11
Second Plan				
(1956–7 to 1960–1)	Foodgrains	1·34	2·81	4·20
	Non-foodgrains	1·27	2·16	3·45
	All crops	1·33	2·57	3·93
Third Plan				
(1961–2 to 1964–5)	Foodgrains	0·03	2·23	2·26
	Non-foodgrains	0·52	3·66	4·21
	All crops	0·13	2·81	2·95

[a] These rates have been determined by a least squares fit of $I_t = ab^t$ to the years involved. The calculated rates have been carried to two decimal places, but at best the two decimal places are indicative only. Because of the short periods embraced by each rate calculation, each must be regarded as suggestive and no claim is made that the evidence is conclusive. It should be noted that 1965–6, the final year of the Third Plan, is excluded. This was the first of the two drought years.

Source: Table prepared for the World Bank Mission, 1966–7, by W. David Hopper, Rockefeller Foundation; included in the March 1967 Report on Agricultural Policy (Vol. II of Mission Report).

Plan period, but since it was at least maintained during the following two Plans, its contribution had become relatively greater by the Third Plan period. The net effect of the two influences was to produce the decelerating rate of expansion in output of foodgrains, and of all crops, noted above. For non-foodgrains the area under cultivation has grown steadily since 1950-1, but a significant proportion of this has been at the expense of the area devoted to foodcrops. Even with these transfers, however, there was a diminution in the contribution to expansion in output of non-foodgrains from crop area increases over the three Plan periods. In contrast, there was an accelerating contribution from productivity over the whole period, which was sufficient to produce an accelerating trend in output of non-foodgrain crops during the three Plans.

The declining importance of additions to cultivated area is particularly significant for the future of Indian agriculture. Some further additions can be expected but evidence strongly suggests that this method of adding to agricultural production will never regain its former importance. This prospect emphasises the need for ensuring that future area extensions should be as productive as possible, through provision of irrigation to enable multiple cropping, for example. It also shows that additionally heavy reliance must be placed upon productivity increases per unit area (including extra or multiple crops).

Three further observations on Table 3.5 need to be made. First, it belies the notion that no growth in agricultural productivity has occurred. The growth of productivity in foodgrain production is significant by world standards, though it happens to have been inadequate to offset declining growth in area under production. Second, the growth in productivity disproves the notion of utter stagnation, a term which can only be used in relation to the overall problem of food supply and economic growth including external economic viability (Crawford 1966). Third, a table of national aggregates conceals a picture of wide variation between states. There have been many encouraging performances, for example in the states of Gujarat, Madras, Mysore, and Himachal Pradesh, where the higher than all-India average results in agricultural growth are due predominantly to increases in productivity per unit area (Ministry of Food and Agriculture 1966).[9] Increasing labour supply, additional fertiliser, extra water, improved seed quality (not to be confused with the new high-yielding varieties now available), improved land practices, and, in some cases, measures of agrarian reform have played a part in varying degrees in each of these states, and also in Punjab.

Nevertheless, for India as a whole the results have been inadequate in relation to need, and have certainly fallen far short of targets set.

Planning and Achievement Contrasted

Ever since 1950-1 the agricultural sector has come under the influence of economic planning from the central government. Production targets have been set for the sector and for individual commodities within the general framework of three Five-Year Plans implemented between 1950-1 and 1965-6. These targets provide useful points of reference against which the agricultural performance within the period can be judged.

In relation to foodgrains, it appears that the targets set in the First and Second Plans were achieved (Table 3.7). The 100 million long ton target originally set for the Third Plan (102 million metric tons)

[9] Material covering the states for the period 1951-4 to 1958-61 was supplied to the 1964-5 Mission by B. S. Minhas and A. Vaidyanathan.

in 1965-6 was fixed with an expectation of 76 million long tons (77 million metric tons) as the actual starting base (1960-1), which would have given a target compound growth rate of 5·6 per cent per annum. Had it been known that production in 1960-1 was actually

TABLE 3.7

TARGETS AND PERFORMANCES—FOODGRAIN PRODUCTION
UNDER THREE PLANS

(million metric tons)

	First Plan (1950–1 to 1955–6)	Second Plan (1955–6 to 1960–1)	Third Plan (1960–1 to 1965–6)
Planned[a]			
1. Base level production	54·9[b]	66·0[b]	77·2[b]
2. Original target for last year of Plan	62·6[b]	76·2	101·6
Revised		[81·8][c]	[92·0][d]
3. Ratio 2 : 1	114	115	132
4. Average annual compound growth rate (%)	2·7	2·9	5·6
Actual			
5. Base level production	54·9	69·2	82·0
6. Production for last year of Plan	69·2	82·0	89·0[e] [to 1964–5]
7. Ratio 6 : 5	126	118	109
8. Average annual compound growth rate (%)	4·7	3·45	2·07[e]
Actual to planned			
9. Average annual compound growth rate (%), 6 : 1	4·7	4·4	3·6[e]

[a] Data converted from long tons.

[b] Figures are taken from Second and Third Plans.

[c] This figure was revised upwards during the course of the Second Plan (see Planning Commission 1961: 302).

[d] The expectation for 1965–6 as revised in 1963 was 92 million metric tons (90·5 million long tons). It came to be regarded as the revised target for the Third Plan. This figure, too, served as the base for discussion of Fourth Plan targets. However, the actual production figure was, because of drought, nearer 72 million metric tons. The figure 92 could be regarded as a possible 'norm' for 1965–6 had all the planned inputs, including water, been achieved. In this sense 89 million metric tons achieved in 1964–5 was possibly a little better than was reasonably to be expected, given the shortfall in that year.

[e] Growth rate has been taken over a four-year period to 1964–5, the following year being regarded as 'abnormal' because of drought.

Source: Planning Commission 1956, 1961; Ministry of Finance 1967a.

to be 80·7 million long tons (82 million metric tons), the apparent postulated growth rate would have been 4·4 per cent, which is still high, but clearly not as ambitious as the target actually set. Nevertheless, it was the higher growth rate that was set, for when the 1960-1 crop was harvested, it was recognised as a 'good' year (i.e. it was better than 'normal' seasonally), and it could not be described as the forecast fulfilment of the Second Plan. Indeed, fertiliser consumption (nitrogen and superphosphate) in that year did not rise, so it could well have been expected that yields would not have risen at all over the 1959-60 level, assuming the season to be normal. It seems proper, therefore, to discuss the Third Plan in terms of a base below the actual performance in 1960-1, that is, using the expected base of 77 million metric tons (metric measure to be used exclusively from here on).

The desired growth rate can then be taken as 5·6 per cent, which, in relation to the 4·1 per cent growth rates achieved in the previous Plans, was a very high target. Indeed, the foodgrain output target had to be revised downwards in mid-term to 92 million tons, representing an implicit compound growth rate of 3·5 per cent per annum.

Using the planned 77 million tons as the base, the achievement of 89 million tons in 1964-5 (Table 3.8) represents growth at a compound rate of 3·6 per cent per annum, which is close to the long-term growth for foodgrains over the past fifteen years (Table 3.5). In actual terms there appears to have been comparative stagnation between 1960-1 and 1964-5. This is, however, partly unreal since, as explained above, 1960-1 was a better-than-average year. The following year was on or better than the measured trend, and only 1962-3 and 1963-4 were below trend, probably because seasonal factors were adverse for India as a whole. There were in fact some input increases in both years, and although much lower than 'planned', these could have been expected to raise output somewhat in those years, although to levels well below the trend required for reaching the unrealistic output targets originally set for 1965-6, the final year of the Third Plan. Indeed, the adverse seasonal factors would have been more restrictive but for these increased inputs.

The 1965-6 and 1966-7 production figures show the effects of the severe drought experienced in these two years. The drought in 1965-6 produced the serious shortfall in actual output as against projected output (both original and revised), and the continuation of the drought into 1966-7 exacerbated the situation. Total foodgrain output in these two years was well below the actual figure recorded at the commencement of the Third Plan in 1960-1. Once again the situation might well have been worse but for the increase in inputs like fertiliser and water supplies which did occur in those years.

TABLE 3.8

TOTAL FOODGRAIN OUTPUT,[a] SELECTED YEARS 1949–50 TO 1965–6

(million tons)

Year[b]	Rice	Wheat	Sorghum	Millet	Other cereals	Pulses	Total foodgrains[c]
1949–50	25·1	6·8	7·0	3·1	8·6	10·0	60·7
1950–1	22·1	6·8	6·2	2·7	7·9	9·2	54·9
1955–6	28·7	8·9	6·7	3·5	9·8	11·7	69·2
1956–7	30·2	9·5	7·3	2·9	10·3	12·1	73·3
1957–8	26·5	8·0	8·6	3·6	9·6	10·1	66·5
1958–9	32·0	10·0	9·0	3·9	10·6	13·2	78·7
1959–60	31·7	10·3	8·6	3·5	10·8	11·8	76·6
1960–1	34·6	11·0	9·8	3·3	10·7	12·7	82·0
1961–2	35·7	12·1	8·0	3·6	11·6	11·6	82·7
1962–3[d]	31·9	10·8	9·6	3·9	10·8	11·4	78·5
1963–4[d]	36·9	9·9	9·1	3·7	10·6	10·1	80·2
1964–5[d]	39·0	12·3	9·8	4·5	11·0	12·4	89·0
1965–6[e]	30·6	10·7	7·5	3·6	9·8	10·0	72·3

[a] Reliability of foodgrain production data has been much discussed, and some review of this is given in the World Bank 1964–5 Report No. I, Appendix I.

[b] Figures for 1949–50 to 1959–60 are adjusted, while those for 1960–1 are fully revised estimates.

[c] Figures may not total because of rounding.

[d] Partially revised estimates.

[e] Final estimates.

Source: Ministry of Finance 1967b: T. 1.4.

In passing, it may be noted that similar considerations affected non-foodgrain output, although not as markedly (Table 3.9). Some areas were diverted to non-foodgrain crops and this possibly occurred for other inputs as well, although this was not evident in published statistics. Nevertheless, the anticipated levels of output for 1965-6 were still below target for cotton and oilseeds. The total restraint represented by lagging inputs no doubt affected these crops too.

TABLE 3.9

TARGETS AND PERFORMANCES: SOME NON-FOODGRAIN CROPS

| Commodity | Unit (million) | Second Plan | | Third Plan | | |
		Target 1960–1	Actual production 1960–1	Target 1965–6	Anticipated production 1965–6	Actual production 1965–6
Cotton	Bales	6·5	5·3	7·0	6·3	4·7
Jute	Bales	5·5	4·1	6·2	6·2	4·5
Oilseeds	Tons	7·6	7·0	9·8	7·5	6·1
Sugar cane	Tons	7·8	11·2	10·0	11·0	11·8

Source: Planning Commission 1961, 1964; Ministry of Finance 1967b. Targets for Second Plan are revised targets quoted in the Third Plan; 1965-6 anticipated figures are those quoted in the Fourth Plan Memorandum.

The Third Plan is widely regarded as a failure in that targets were not achieved. The charge is literally correct, but ought not to be left without discussion. In the first place, the targets can be criticised as unrealistic, especially the original foodgrain objective of 102 million tons for 1965-6. Second, the Plan ought not to carry the responsibility for the droughts, although it can be said that storage and buffer stock targets were below requirements for the period. Finally, the input and general policy performances were below the production objectives, and here the Indian government, if not the Planning Commission itself, is open to criticism.

A production target may be unrealistic because, although related to 'need', it is not capable of achievement: the marshalling of resources and the farm changes required may be impracticable. This, to the author, was the situation, at least in relation to the original target of 102 million tons for 1965-6, and when writing in 1964-5, to the proposed Fourth Plan target of 120 million tons for 1970-1. However, high targets are understandable in the circumstances of India. A high population growth rate, widespread and dire poverty with which must be associated a high income elasticity of demand for food, inadequate food alternatives available, and a highly restrictive import situation, must point to a high target for foodgrain production.[10] Statements by the Indian government often referred to an objective of 510 grams (18 oz) of foodgrains per day; but 454 grams (16 oz) seemed nearer to their practical hopes, though undue weight cannot be attached to the data of per capita consumption (Table 3.4); however, 'need' clearly pointed to high targets, especially when self-sufficiency (i.e. independence of imports) was, for economic reasons, also an aim. Nevertheless, a desirable objective may take longer than one Plan period to achieve, and, no matter how it was rationalised, a target of 102 million tons for 1965-6 was not really consistent with likely performance under the Plan.

It is true, too, that a policy of importing grain in lieu of additional production has not been a practical proposition for any government in India: the continuing annual increase in foodgrain requirements was much greater than the practical possibility of increased exports (which in any case were needed to finance general economic development) or aid. Available foreign exchange is more economically invested in inputs or materials and components required for industrial production of farm inputs in India. If, for example, the annual shortfall in grains is assumed to amount to 10 million tons, of which 5·5 million tons were available free under aid, the cost of the balance could amount to about US$360 million (c.i.f.) at present prices, or about 21 per cent of 1964-5 exports. If the requirement rose to 15 million tons, of which 6 million tons were free, the balance could cost the equivalent of 43 per cent of 1964-5 export values (Crawford 1966, 1968). Those who talk easily about reliance on food imports overlook the high rate of population increase and the effect of rising money incomes in a situation in which expansion of export industry in the short term (say 5 years) cannot match the need. Moreover, they overlook the vital part agricultural expansion must play in India's general economic growth (Crawford 1966: 161-2). The Indian government of the Third Plan period cannot be criticised for emphasising home production in its agricultural policies; it can be criticised for not acting vigorously enough when the situation was discerned, as it was when the Third Plan was drawn up. The inadequate implementation of

[10] See Crawford (1966); Panse, Amble, and Abraham (1964); and Sukhatme (1965). In the author's own work and simply to give some guide as to need, it was assumed that the objective should be not less than 16 oz per head per day (and not more than 18 oz), or, if starting from a year in which apparent consumption was, say, 15 oz, a formula of increase in supplies required was obtained from the equation $s = p + i.d$, where s = growth rate in supplies required, p = population growth rate, i = % increase in per capita income, and d = income elasticity factor. Thus if $p = 2\cdot8$, $i = 2\%$, and $d = 0\cdot6$, $s = 2\cdot8 + 2\cdot0 \times 0\cdot6 = 4\cdot0\%$ p.a. For some time to come, the rate of increase required for agricultural production is not likely to fall significantly below $4\cdot0\%$ p.a.

frequent declarations of high priority for agriculture, made before 1964 by federal and state ministers alike, emerges all too clearly when the record is examined.

MAJOR DETERMINANTS OF THE GROWTH PERFORMANCE

We have seen that trends in foodgrains production, even in the 1960s (if we discount for the exceptional droughts) were positive although decelerating. The decline in additional area brought into production each year offers a major part of the explanation; but we need to look beyond this. It was clear to those concerned that yields per hectare had to be raised even beyond the level already achieved (Table 3.5). It was also evident that, within the technology existing in the early 1960s, it was profitable for many farmers to invest more labour and capital for increased output: there is no other explanation possible for the result shown in Table 3.5. But, since these results were inadequate, we may well ask whether more could have been expected or whether additional incentives were required to ensure a more satisfactory national result. Both questions can be answered in the affirmative, with the second constituting the 'new approach' first seriously accepted by the Indian government in 1964-5 under the general direction of the then Minister for Food and Agriculture, Mr C. Subramaniam.

It is neither practicable nor sensible to look to single factors to explain a complex situation or to produce a desired change in it. In our own analyses, given in our share of the World Bank Mission Reports, my colleagues and I preferred to try to recognise the many facets of the problem. We also recognised that in the short run some factors—fertilisers, better seeds, more water, easier credit conditions— could produce significant results within the general land-use constraints operating; but that for sustained results at the high level of annual increase dictated by population and income growth factors, there must be concurrent attention paid to research and extension and agrarian reform, factors which, because of their difficulty, might not necessarily yield quick results. As it has turned out, a combination of Indian and international research has given promise of quick returns; but there is little doubt that this research effort must be continued if higher rates of growth are to be sustained.

It is also clear that in the 1970s extension services and agrarian reform will have major roles to play in continuing the momentum recently established on the basis of the 'new technology'. We can best discuss these views through a brief review of agricultural policy as it has been developed, with particular reference to the Third Plan period. Generally speaking, this policy section has to be read as applying to the pre-drought, pre-High Yielding Varieties Programme period, which will be reviewed in a later section. It is designed to stress that even within the old technological conditions growth in productivity could be stimulated by appropriate policies. The same policies, however, are shown to be both imperative and more reward-ing when a more advanced technology (represented by high-yielding seeds which call for more water and fertilisers) becomes available. The critical factors, including market incentives, production inputs, farm credit, research and extension, institutional programs, and government administration, will be briefly surveyed in turn.

Market Incentives

There are indications over the decade to 1963-4 that incentives for expanded production were weaker than perhaps intended by the government of India, and were certainly not at a level likely to induce a dramatic response from farmers. Broadly, terms of trade were against the farmer over the period. There was first an unfavourable trend in the early 1950s (Table 3.10), away from a relatively favourable initial situation around 1950. A period of fluctuating terms of trade then followed which, however, at no point repeated the initially favourable situation. Movements in terms of trade between prices of individual items of manufactures and prices of cereals showed some variation from the overall pattern, but with the exception of kerosene

TABLE 3.10

RATIO OF WHOLESALE PRICES OF SELECTED ITEMS TO WHOLESALE PRICES OF CEREALS

(1952–3 = 100)

Year	Iron and steel	Fertiliser	Cement	Cotton manufactures	Kerosene	All manufactures
1950	91	96	98	97	91	—
1952	96	102	104	102	100	117
1954	131	104	114	127	114	101
1956	140	93	115	124	97	132
1958	138	91	123	108	90	107
1960	140	91	128	121	89	108
1961	144	94	136	125	97	119
1962	149	87	140	122	94	125
1963	145	83	137	120	117	122
1964	126	69	117	102	101	101
1965	125	63	116	96	99	101

Source: Ministry of Finance 1967b: Pt 1, 102-3.

and, most important, fertilisers, they were against the interests of the farmer over the period.

Whether fully intentional or not, the failure of farm prices to move upward with other prices and costs was consistent with the emphasis in the early plans on the welfare of the urban consumer, an emphasis also apparent in the PL480 import programs of the late 1950s and early 1960s. This welfare is adversely affected by high food prices, but on the other hand, low prices are no incentive to production. This stress on consumer needs was seen clearly enough in the First Plan (Planning Commission 1953: 173) in which it was stated:

> . . . a well-defined food policy for the period of the Plan is an essential condition for the successful implementation of the Plan.

It went on to point out the vulnerability of the economy on account of the inadequate production of foodgrains:

> These [food prices] hold pivotal place in price structure and must be held stable at a level within the reach of poorer community.

It can be assumed that in the early years of the First Plan, prices (for reasons like the Korean War boom) were in farmers' favour. By the time of the Third Plan they had moved the other way, so that grower needs received more attention. In this Plan it was stressed (Planning Commission 1961: 323) that:

> For achieving the high targets of agricultural production set for the Third Plan, it is important that growers should have full confidence that the additional effort and investment which are called for will yield adequate return.

The statement then went on to support 'minimum remunerative prices for important cereals and cash crops. . . .' It was not until late 1964 that any real effort was made to resolve this perennial dilemma of balancing producers' and consumers' interests through a policy which was to depend partly on minimum guaranteed crop prices (fixed on the advice of the Agricultural Prices Commission) and partly on buying and selling operations of the National Food Corporation.[11] In fact the action came after and behind a rise in prices as the pressure of demand began to tell on the lagging rate of growth. It has remained behind, although, with the recovery of production in the 1967-8 season (especially in wheat), the policy begins again to have relevance.

This new policy of price support has not changed in respect of the Fourth Plan, though it is understandable that recent and current

[11] This was an important part of the 1964-5 World Bank Mission's study. Regrettably it is impossible to repeat it here.

D

Table 3.11

DOMESTIC PRODUCTION, IMPORTS, AND DISTRIBUTION OF FERTILISERS, 1960–1 TO 1970–1 (PROJECTED)

('000 tons)

Year	Nitrogen (N)			Phosphatic Acid (P$_2$O$_5$)			Potash (K$_2$O)	
	Domestic Production	Imports	Distribution	Domestic Production	Imports	Distribution	Imports	Distribution
1960–1	120	172	212	54	0·1	53	25	29
1961–2	154	143	292	65	0·6	64	30	28
1962–3	194	229	360	88	8	81	44	37
1963–4	219	198	426	108	12	121	64	52
1964–5	243	257	492	130	12	147	57	72
1965–6	233	323	583	117	21	134	94	90
1966–7[a]	308	617	847	132	141	242	120	114
1967–8[b]	500	888	1,350	250	359	500	300	300
1968–9[b]	760	940	1,700	n.a.	n.a.	650	n.a.	n.a.
1969–70[b]	1,000	1,000	2,000	n.a.	n.a.	800	n.a.	n.a.
1970–1[b]	1,700	700	2,400	n.a.	n.a.	1,000	700	700

[a] Estimated.
[b] Projected.

Source: Fertiliser Association of India 1965, 1967; material made available by Ministry of Finance.

high prices for foodgrains (greatly aggravated by drought) have again brought urban interests to the fore. Policy remains one of assuring minimum or floor prices to the farmer for his output which, while not unreasonable to consumers, will give him the incentive to invest for expansion of production. It is evident, however, that the greatly improved wheat supply position is already testing the government's ability to support prices. For rice, the test is less immediate if we are to judge by the *Economic Times* of Bombay indices which show rice prices still at an inflated level. The whole price situation is further complicated by zonal restrictions and by the need to build up buffer stocks: it is difficult to interpret official indices and the *Economic Times* price data without full data on the marketing structure as affected by government regulation.

Inputs

Fertilisers.[12] In noting the rapidly increasing use of chemical fertilisers in Indian agriculture in recent years (Table 3.11), it is well to keep in mind the extraordinarily low usage per hectare when expressed in absolute terms. Between 1952-3 and 1963-4 inclusive, the growth rate in distribution of nitrogen (N) for consumption was approximately 18 per cent per annum; for phosphate fertilisers (P_2O_5) 32 per cent per annum; and for potassic fertilisers (K_2O) 22 per cent per annum, all rates compound. These dramatically high rates unfortunately reflect a very low absolute base in 1952-3. Even in 1963-4 India used only 2·51 kg of N per hectare (2·24 lb per acre) for all land used in agriculture, only 0·72 kg of P_2O_5 and 0·30 kg of K_2O per hectare over the same area. Quite clearly, of course, these averages mean little. A small percentage of farmers used, for example, 10 kg of N or more per hectare, while a great majority used practically none at all.

More particularly it can be seen (Table 3.12) that during the Third Plan, available supplies of all fertilisers fell considerably short of targets. Failure was particularly notable in the shortfall in domestic production of nitrogenous fertilisers. Potentially, this would have meant increased reliance upon imports. These, however, also fell short of the mark—a reflection both of foreign exchange scarcity and of inadequate conviction at this point of their urgent necessity. Droughts and the 'new technology' were dramatically to change this attitude after 1965.

[12] Green (organic) manures are not reviewed in these comments. They have a definite and growing place in Indian agriculture despite some technical and economic problems. It should not be thought, however, that the amount of green manuring practised in any way destroys the validity of the strong case developed here for a sustained and large-scale fertiliser production and usage program.

Table 3.12

AGRICULTURAL INPUT TARGETS AND ACHIEVEMENTS IN THE THIRD PLAN[a]

	Unit	Plan targets	Achievement 1965–6	Per cent achievement†
Irrigation				
Major and medium additional utilisation (gross)	m. hectares	5·2 (12·8)	2·2 (5·5)	43
Minor additional (gross)	m. hectares	5·2 (12·8)	5·3 (13·1)	102
Land reclamation				
additional	m. hectares	1·5 (3·6)	1·7 (4·2)	117
Soil conservation				
additional	m. hectares	4·4 (11·0)	4·0 (9·8)	89
Double cropping area	m. hectares	n.a.	26·3 (65·0)	—
Area under improved seeds	m. hectares	82·2 (203·0)	48·6 (120·0)	59
Plant protection	m. hectares	20·0 (50·0)	16·6 (41·0)	82
Fertilisers:				
Nitrogenous—consumption of N	m. tons	1·0	0·6	60
—domestic production of N	m. tons	0·8	0·2	25
Phosphatic —consumption of P_2O_5	m. tons	0·4	0·15	38
Potassic —consumption of K_2O	m. tons	0·2	0·09	45
Organic and green manures				
Urban compost	m. tons	5·0	3·4	68
Green manure	m. tons	41·0	21·5	52
Other inputs				
Tractors	thousand	10·0	6·2	62
Power driven pumps	thousand	150·0	150·0	100
Diesel engines	thousand	66·0	85·0	129

Note: Bracketed figures are acres, as in source.

a Targets and achievements are expressed as total additions during a five-year period for the first three items, as total cumulative achievements measured at five-year intervals for the next two items, and as yearly figures for the remaining four items.

Source: Planning Commission 1961, 1966a.

The low fertiliser usage which has characterised traditional agriculture is not explained even partially by unresponsiveness of Indian soils. Even under the long-established constraints of standard farm practices, availability and reliability of water supplies, types of seed available, and credit systems, etc., considerable potential existed for the profitable use of additional fertilisers. It is true that in some districts farmers were approaching optimum levels of application for the seeds available, but despite such cases, under existing conditions, the difference between actual and profitable usage of fertilisers was sufficient to ensure a steady investment response to increased fertiliser and water availability on the part of farmers with reasonable access to capital. It was this fact that explained the productivity increases achieved in many states before the so-called 'new technology' arrived.

Nor do factors of social inertia offer more than a partial explanation of the failure to use adequate quantities of fertilisers. While investment was profitable for many, for others it was not. There were, of course, a great many who could have responded but did not: the extension service was not immediately equal to the task. Nevertheless, it must not be overlooked that inputs fell short of the planned amounts. Given the yardsticks in general usage (Ministry of Food and Agriculture 1965b: 12)—for example, 1 ton of N yields an additional 10 tons of foodgrain—the shortfall in fertiliser supplies (N and P_2O_5 especially) in 1965-6 (Table 3.12) and the lag in water development clearly suggest that any norm established on the basis of planned inputs for 1965-6 would, in an average season, have been several million tons in excess of that possible with the actual supplies of inputs available. This factor alone made the mid-term downward revision of the target for 1965-6 from 102 million tons to 92 million tons inevitable. In the event drought obscured the position, but it is doubtful if the output in an average season, given the inputs available, could have been expected to exceed 90 million tons.

Under the conditions prevailing in the early years of the Third Plan, there was a basis for expansion. It was thwarted by inadequate incentives in the early part of the period, by poor credit arrangements and by shortfalls in input supplies. It was clear that the targets were too high; but, given proper attention to the level of inputs, it was apparent that the growth trend in productivity could have been raised. What has been said of fertilisers is also true of other inputs. Of these, water and quality seed supplies are among the most important. Some further observations are therefore in order, if only to provide the basis for later comment on the critical constraints operating under the 'new technology' now being encouraged in India.

Water.[13] In 1951 there were about 20·8 million hectares (51·5 million acres) under irrigation and 1·8 million hectares were being double-cropped, giving a gross utilised area of 22·6 million hectares (56 million acres). This total was divided between 9·5 million hectares (23·5 million acres) in major and medium-sized irrigation schemes and 13·2 million hectares (32·5 million acres) in minor schemes, though the distinction between major and minor is not clear-cut.

Over the period of the three Plans, gross irrigated area expanded 60 per cent, in other words the irrigable area was enlarged by about 14 million hectares within the fifteen years, or at an average of nearly one million hectares per annum. Even so, the achievement fell short of targets (Table 3.12). Moreover, official figures, although indicative as orders of magnitude, exaggerate actual achievements. For instance, in some estimates supplied by certain states, the figures may correspond not to areas actually watered artificially but to the total areas commanded by major canals or supplied by tubewells. The typical discharge of state tubewells was considered more likely to supply, at best, only one-quarter or one-fifth of its command area (or area allegedly served by the system). This underlines the extensive nature of the major systems which were intended primarily to supply supplementary irrigation during the monsoon season and so to act also as a degree of insurance against drought. They were not designed to provide water for intensive cropping and hence maximum utilisation of each hectare. This offers one explanation of the small percentage of land that is double-cropped.

In 1951-2 the area under minor irrigation comprised 55 per cent with surface (canal) water systems and 45 per cent with groundwater schemes. These proportions are now roughly inverted, with more and more emphasis being put on deepening and boring fresh wells during the last fifteen years.

Investment in new large-scale works is virtually at a standstill. Stimulated by drought, by the realisation that there are large resources of groundwater to be tapped, and the pressure to use water intensively, the major thrust is now on small-scale tubewells, both state-owned and directed and privately owned, and on efforts to supplement (by similar means) the water available from regular canal schemes.

The story of irrigation cannot fairly be presented in a few paragraphs. It is a mixture of some fine engineering achievements and a failure to link engineering skill with careful study of the best use of water available, resulting in inefficiency in many schemes. The

[13] Acknowledgment is due to L. Garnier for his detailed analysis of the water situation in the 1964-5 Report of the World Bank Mission.

marriage of the engineering and agricultural professions is now under
way. Still lacking until very recently has been adequate survey work
to assess groundwater reserves. But the real change towards the end
of the Third Plan was the recognition of these deficiencies and an
understanding that India's agricultural problems called for more
intensive use of water in areas where water in relation to land supply
could be made relatively abundant. This new approach is behind the
more marked emphasis on minor schemes, including the supplementing
of old canal systems (Planning Commission 1966a, 1966b). As it has
turned out, the new policy is essential to any program for spreading the benefits of the new high-yielding dwarf varieties of rice and
wheat and of some of the new hybrids.

It has now been strikingly demonstrated that a more assured water
supply will add to output even if other inputs remain constant. Combined with new varieties, more fertilisers, and better farming, it offers
the prospect of higher yields for each crop, and in addition, prospects
of double- and triple-cropping. It is in this direction that increased
water supply offers itself as a key factor, offsetting the lack of additional land as a means of adding to production.

Other inputs and farm credit. It can be said of the Third Plan that
much was understood and said about new and improved seeds (i.e.
good quality traditional varieties), about the need for more efficient
power for farm work, better implements, and improved farm practices generally. But a well-organised seed industry hardly existed and
there was little manifestation of an organised attempt at pest control.
The development of electric and diesel power was more evident but
was limited by constraints of credit availability and farm structure,
not to mention traditional conservatism. Table 3.12 suggests inadequate performance (except for diesel pumps) against target. Once
again the story seems to have been a failure of drive or co-ordination
of effort to secure the result planned, reflecting both scarcity of
resources and the absence of an 'organised will' to achieve targets.

Much the same has to be said about credit.[14] Only towards the
end of the Third Plan did there emerge a more thorough understanding of the steps needed to establish adequate credit facilities.

The 1951 Rural Credit Survey (Reserve Bank of India 1955)
revealed that 63 per cent of all rural families were in debt, and it is
doubtful if the situation has greatly improved since. Of the borrowings by the cultivators, 47 per cent were claimed to be for current

[14] Very fully reviewed by my colleague, W. Ladejinsky, in the 1964-5 Report of
the World Bank Mission. The few paragraphs here are dependent on his writing,
but there is no pretence that the complicated credit structure can be outlined.
However, it is important to stress useful developments since 1965, noted in the
1967 Report of the Mission and confirmed by later advices.

and non-current agricultural purposes and 53 per cent for consumption purposes.

There are many reasons why an Indian peasant must borrow, but the major reason is the prevailing poverty of Indian cultivators. It arises from the small holdings, their subdivision and fragmentation, from frequent crop failures, from loss of stock, from land purchases, from expenditures on marriage and other social ceremonies, and from lack of employment opportunities in the off-seasons.

Historically the money-lender has been at the base of the agricultural credit structure. Though in bad repute, until a decade or so ago he provided nearly 70 per cent of total rural borrowings. With the rise of institutional or co-operative credit his share declined, but it is still probably close to 60 per cent of the total. The money-lender has remained an indispensable source of credit, and from the cultivator's point of view he has many advantages over the credit co-operative, despite interest rates often in excess of 15 per cent per annum. He is easily accessible; he has simple, elastic methods of business; he maintains personal contact with the borrower; he has local knowledge and experience. He seldom inquires about the purpose of a loan, provided he can collect the high interest rate (often 25 per cent or higher).

It is understandable that the early Nehru administration should have looked to co-operatives which had in fact come into existence in 1904. Despite limited achievements in the first fifty years of operation (they provided only 3 per cent of rural borrowings in 1950), it was felt that they should be developed as the principal source of credit for the growing needs of agriculture.

The co-operatives have been uneven in performance, many being too small to be viable. They have been most useful in crop loans, and by 1961-2 accounted for about 24 per cent of total farm borrowings (Reserve Bank of India 1964). They have served as distributors of some supplies, especially fertilisers. Their inefficiency, especially in small villages, has been rightly attacked; but the real controversy has been about their apparent monopoly position and their inadequacy as sources of credit for farm development. A good deal of the ideological fervour associated with the co-operatives in the early fifties had gone by the end of the Third Plan, and new credit instruments were being built up, with openings being made cautiously for private institutional credit. Generally speaking, there was a recognition of the critical importance of farm credit, and a more pragmatic approach was evident. The process of change remains far from complete; the money-lender is still of major importance in production credit. However, forms of banking (co-operative and otherwise) are now being expanded to provide credit for farm improvement and development.

A major stimulus to credit expansion has been given by Reserve Bank support of the Land Bank structure as well as by central government budgetary support directed to ensure that a new technology can be developed. As we will have occasion to observe, however, it is fortunate that much of the impetus for new developments which has come so markedly since the Third Plan period ended (especially in private irrigation), has apparently been supported by the private resources available to the farmers directly concerned. This reliance on 'credit-worthy' farmers with significant resources of their own will, however, not suffice. It is when new credit needs arise, as the new technology spreads to less affluent farmers, that the newly developing credit structure will be fully tested.

Research. The story of research in India has yet to be told fully.[15] The research effort has been considerable, although uneven in direction and emphasis. There were considerable gaps in effort, especially in the areas of soil and water management and pest control. Often the performance of experimental stations lagged behind that of good farmers in their area (Hopper 1965). Relationship between research and extension services was weak and often ineffective. On the whole it was not a greatly inspiring picture, and even in 1964 it was difficult to see investment in research as the quick way to a breakthrough in Indian agriculture. And yet it was the optimistic view of the then Minister of Food and Agriculture in 1964 (based on information submitted to him by an advisory panel of agricultural scientists) that on existing knowledge and research tried out in India the yields of wheat, rice, sorghum, millet, and barley could be raised fourfold by the appropriate use of available improved strains, fertilisers, water, and seeding practices. He did admit, though, that such an increase could not be expected to occur overnight, owing to the time needed for communication of the information required (Subramaniam 1964). As we shall note, considerable progress has in fact since been made in the breeding of high crop varieties by a process of Indian-based research and in the adaptation of overseas materials and ideas. The emphasis had been on the breeding of drought- and disease-resistant varieties, rather than on those that would yield better, given assured water, fertiliser, and artificial plant protection. In the context of a stable agriculture the research work was successful, but in the circumstances of the developing Malthusian problem a change in emphasis

[15] In the work of the 1964-5 Bank Mission, members had the advantage of full and helpful discussion with Indian government research organisations, and with the Rockefeller group with headquarters in New Delhi. An unpublished report of a joint Indian-Rockefeller group was freely drawn upon. It is difficult to get a full appreciation of the problems concerning research from any single government document.

was necessary and was eventually provided. Today the emphasis is on adaptive research, combining the high-yielding qualities of exotic and some Indian varieties of seed with the still desirable qualities of drought and disease resistance. The problem is not now any lack of sense of need and purpose, but whether adequate funds will be provided to support research.

Similarly, extension work has been rather lack-lustre.[16] This has been partly because research workers had relatively little to offer. More significantly, however, the story of the extension worker (with a few shining exceptions) has been marked by poor training, heavy tasks of paper work, lack of personal status and incentives, and confused lines of administration. These defects of the system are all recognised and, once again, change is under way, becoming marked by the end of the Third Plan period. Unsatisfactory extension work was a less critical handicap in the early stages of the new move forward in agriculture than it will be in the future. The first impetus given by the better-off, more experienced farmers needs to be supplemented by services to those farmers more in need of guidance and assistance. This problem may prove far more critical in the early 1970s.

General Comment on the Third Plan

Little mention has been made of community development or of agrarian reform in the Third Plan. Some progress was made in the solution of these problems,[17] but inadequately in terms of the hopes of many. Official community development and agrarian reform programs made little direct contribution to agricultural advancement in the Third Plan, but, as with extension activities, their comparative lack of contribution was not crippling at this stage. It is clear, however, that if progress does not gather faster pace, frustration will develop in farm communities. as roads and other amenities fail to materialise, and great social tensions will develop between the farmer 'haves' (especially those able to exploit the 'new technology') and the 'have nots' (especially the small and insecure tenants and debt-ridden small owners).

Altogether, the Third Plan period is marked by downward revision of foodgrain targets and failure to reach the revised targets. Drought explains the low crop in 1965-6, but the need to revise output targets downward resulted from failure to provide the inputs planned for

[16] Again discussed at length by Ladejinsky and Hopper in their contributions to the Bank Reports and by Hopper in published work. Ladejinsky's writing on the Community Development Program is highly pertinent, but regrettably unpublished.

[17] Again the author acknowledges Ladejinsky's writing.

the Third Plan period. By the end of the period, nevertheless, despite the extreme setback represented by drought, there was strong evidence that considerable change was at hand. There was a decided lift in morale of administrative personnel and scientific workers alike (Crawford 1966, 1967, 1968), and there was more evidence of real priority being accorded by governments for agriculture's needs. The evidence was both concealed and stimulated by drought experience: it is to the story of the 'new technology', which is the one unchanging plank of policy in the 'on and off again' Fourth Plan, that we now turn.

The *Draft Outline* of the Fourth Plan (Planning Commission 1966a) gave effect to the new policies which were in fact already in the process of being implemented. One of the most important changes was a shift of emphasis to agriculture, to accelerate its rate of growth, and the initiation of a well-conceived program to accomplish this. The program envisaged a new strategy for agriculture which had earlier been presented in a small pamphlet. To quote:

The new policies could be described as follows:

(a) to apply scientific techniques and knowledge of agricultural production at all stages, particularly in the fields;

(b) to select a few areas with assured rainfall and irrigation for concentrated application of a package of practices based on improved varieties of seeds responsive to heavy doses of fertilisers and availability of inputs and to fix special targets of production of foodgrains for such areas, the area proposed being 32·5 million acres [13·2 million hectares] and the additional yield expected in 1970/1 being 25·5 million [metric] tons;

(c) to achieve high production of subsidiary foods both through intensive production programmes and overall development; and

(d) to base implementation of imported projects under the Plan on 'Schedules of Operations', specifying the responsibilities and roles of the Central and State Governments and other agencies so that programmes may be operated in the light of a clearly defined understanding between the Centre and the States. [Ministry of Food and Agriculture 1965a: para. 2.2]

As already observed, 'high priority' had been officially accorded agriculture in the past as, for example, in the Third Plan, with somewhat hollow results—an experience which could well provoke scepticism about the new declaration. However, there are several factors now operating which point to a more realistic attitude towards agriculture on the part of the planning and executive authorities in India.

First, the droughts of 1965-6 and 1966-7, with their grim threats of widespread malnutrition and even starvation, have had a salutary effect on attitudes to planning. Second, there is a realisation, reinforced by the drought, that an expanding agriculture is vital to general economic growth or, in negative terms, a stagnant agriculture is a serious drag on the rest of the economy. For too long there has been evident in India a feeling that industrialisation and agricultural expansion were antithetical forces, or worse, a feeling that agricultural expansion could take care of itself. It is now clear that if agriculture stagnates, food price inflation occurs to the damage of sound economic growth policies in the industrial and commercial sectors (as witnessed in 1965-7). On the other hand, if agriculture is encouraged to expand and succeeds, higher agricultural income will provide a rising market for industrial products as well as a much needed additional source of investment funds. If agricultural inputs are produced in India an expanding agriculture will also absorb these as well as consumer goods. A mutually beneficial relation is possible and important to all sectors. This is now realised, and emphasis is rightly being given to agriculturally-oriented industrial development. For to ignore agricultural needs in domestic industrialisation would be to throw an insupportable strain on the foreign exchange sector if priority for agriculture is to be a reality. As Sir Arthur Lewis has put it, 'agriculture will have to be seen to be important before any considerable progress will be made' (cited in Crawford 1966: 160).

Third, there is a conviction shared by political leaders, civil servants, scientists, and farmers alike that a new technology is available which can give quick results even within traditional constraints which limit the growth of the agricultural sector. All these factors have produced a determination, evident at last in the allocation of scarce foreign exchange resources, to try to give agriculture the resources it needs, even if this squeezes other strong claimants.

Two major aspects about the new strategy—the High Yielding Varieties Programme (HYV Programme)—need to be noted. The first is that the Programme naturally highlights the production of foodgrains, but it is not confined to them. Cash crops important for export earnings or for import savings are to be subject to a somewhat similar package (intensive area) program. The second is that gains from the new technology are not limited to the increased output directly associated with higher-yielding varieties of rice, wheat, maize, and so on. In important cases, crops which enable the same land to produce larger yields also have short-duration growing periods, so facilitating double- and triple-cropping. There is here a clear realisation that policy must be directed towards maximising yields per hectare over a whole year of crop cycles. The new technology (including improved water management) holds considerable promise in this respect—a promise likely to yield higher dividends as research and extension officers and farmers gain experience.

As noted above, the new strategy calls for intensive production in areas selected for their favourable productive qualities, but more particularly for their location in areas of assured rainfall or irrigation, or both. The plan calls for the selection of some 13 million hectares (32·5 million acres) or 11 per cent of normal area under crops concerned (rice, wheat, maize, sorghum, and millet). The additional production originally projected from this area was 25·5 million tons in 1970-1 (Ministry of Food and Agriculture 1965a). In other words this presupposed that yields in selected areas would be three to five times those in unselected areas. Expectations are properly now more modest, with the realisation that farmers will not move immediately to maximum yields per hectare, even if, as is unlikely, input resources were fully available. Furthermore, undue importance should not be attached to target areas. Given the availability of seed, they could easily be exceeded if the accompanying demand for fertiliser, water, and other inputs can be met. There are certainly more than 13 million hectares suitable for the new technology. Any attempt to police the selection of the prescribed area too closely is likely to fail. What may well happen is a bigger area coverage with offsetting lower additional yields as fertiliser is spread more thinly than the official plan calls for.

The overall production target for foodgrains at the end of the Fourth Five Year Plan (1970-1) has been set at 120 million tons.[18] In official terms this was to be achieved partly from increased output under the intensive area schemes of the new technology, and partly from increased output arising under the old technology applied to the rest of the crop area (i.e. from old varieties plus a share of the increased supply of inputs, especially water and fertilisers).

The target has two facets: one is its relation to 'needs', and the other its reality in terms of inputs presumed to be available. In relation to need, the target could perhaps be challenged on the grounds that it will yield a per capita availability of 500 grams (18 oz) of foodgrains per day, which, if other elements of a balanced diet were adequately available, could be too high. If productive capacity does expand to the level which could yield 120 million tons, some diversion of resources to other agricultural commodities might occur

[18] This tense is adopted only because the status of the Fourth Plan is in doubt. There is no need, however, to assume that less importance will be given to agriculture when a formal plan is finally launched.

(through a decline in foodgrain prices relative to prices of, say, milk, eggs, fruit, and meat, or to export commodities). This would not be an undesirable outcome.

The question whether the target is feasible in relation to planned additional inputs of water, fertiliser, plant protection materials, etc., is a more difficult one to judge. The base production figure used is 90 million tons, an adjusted, anticipated output for 1965-6. Clearly the realised output for the 1965-6 season of 72 million tons, which was so heavily influenced by the drought, would be an inappropriate base. As already observed in commenting on the 1964-5 results, the base of 90 million may be a little high, but is appropriate to a 'norm' of, say, 87 million tons for 1964-5. What we are judging is in fact whether a combination of effects of the new and old technologies together would produce an increased foodgrain output of 30 million tons in five years. As indicated above, the anticipated additional output from the intensive HYV Programme was 25·5 million tons. Additional output from the old technology would therefore be only 4·5 million tons. In these terms, the target of 30 million tons is conservative in relation to the original Third Plan targets and the hopes aroused by the new technology. However, there are two reasons for thinking it too high. One stems from doubts as to the availability of the anticipated supply of nitrogenous fertiliser. The second arises from doubts as to whether farmers will apply the high doses of fertilisers recommended for either maximum or optimum yields under the HYV Programme. Some of the new technology may well spread beyond the areas planned for the official HYV Programme; but in this case the 'spill-over' would substitute for some of the efforts anticipated from farmers under the old technology. The net effect is that some reduction in expectations in respect of final output would seem wise at this early stage of introduction of the new technology. This author would rest well content with an achievement of some 15 million tons additional production from the planned areas under the new varieties. This caution is based not merely on the improbability that farmers will quickly move to maximum or even optimum output, but also because the earlier projections assumed a too smooth operation of the Programme. Thus, while in the first two years (1966-7 and 1967-8) the winter (*rabi*) crops of new wheat varieties have done very well, the performance with new rice varieties has been far more uneven.

Nevertheless, there does seem a good case for expecting total additional output from both 'new' and 'old' agriculture of 20 to 25 million tons. The major limiting factors appear to be fertiliser and water availability, although other elements of policy, such as pest control and credit supply, will prove almost as important.

As far as the new technology itself is concerned, there need be no doubt that a great potential exists. Absolute yields are higher, compared with old varieties, for the new varieties of wheat, rice, maize, sorghum, and millet at given levels of fertiliser applications, and the maximum yields possible with increasing applications are considerably greater in every case. The improved wheats are based on imported Mexican seeds (associated with Mexican/Rockefeller programs) and the development of Indian varieties; rice exotics come from Taiwan and the Philippines and there are important Indian strains now in use, while the hybrids have mostly been developed in a joint Indian/ Rockefeller Foundation program. It is difficult to quote average or typical results but experiments, demonstrations, and actual farm experience with less than maximum yield applications of fertilisers have given data suggesting that it would be reasonable to expect the following:

Wheat: dwarf varieties: 40-60 per cent increases in yields over standard local non-dwarf varieties.

Rice: probably 80-100 per cent in favour of new varieties under similar conditions of water control.

Maize: probably 40-50 per cent better.

(In all three cases, assumed applications of fertiliser would be greater than would in practice be used for old varieties.)

Sorghum: possibly at least 100 per cent higher at any likely level of fertiliser application, and some 40 per cent better without fertiliser.

Millet: possibly 60-80 per cent better at any likely level of fertiliser application, and again significantly better (perhaps 25-30 per cent) at nil application.

It is important to note again that much of the favourable result is due not only to the high uptake of fertilisers by the new plants, but particularly to the genetic ability of the varieties to yield under conditions of extreme crowding; that is, the yield increases are to a large extent due to the high populations per hectare.

It should also be noted that, while assured water conditions are vital to the most effective results for rice, wheat, and maize, the new millet and sorghum varieties hold promise without fertiliser and under non-irrigated conditions. On the other hand, hybrid crops call for more skilled farm management.

Data on costs and returns are not available in any systematic way, but in the major wheat-growing areas of Punjab, Northern Rajasthan, and Uttar Pradesh, dwarf wheats in 1965-6 gave an average increase in yield of 54 per cent over old varieties.

At present prices and allowing for a substantial risk discount, the optimal dosage of nitrogen for the high-yielding dwarf wheat is around 120 kg per hectare, for wet-season rice it is probably about 70 kg per hectare, and for rice grown during the dry season (when there is a higher sunlight intensity) it is about 100 kg per hectare. For the more vegetative maize, sorghum, and millet varieties, dosages of 100 to 130 kg of nitrogen per hectare are probably close to optimum. Research on phosphorus alone and in combination with nitrogen is still tentative, but available evidence points to a substantial response to this nutrient as well. There are few data on potash response, but these appear to indicate a low return to this plant food.

At an input of 120 kg of nitrogen per hectare to the dwarf wheats, yield would be increased by about 1,376 kg. At present farm prices of around Rs75 per quintal (100 kg), this extra yield would sell for about Rs1,032, a gross return gained at a cost of Rs258 for the nitrogen. Even on local wheats a 'recommended' dose of 40 kg of nitrogen per hectare would add 506 kg of grain to yield and give a monetary return of Rs380 for an outlay of Rs86. (It is of interest to note that at prices prevailing until mid-1963, the local varieties would provide a gross of Rs200 at a cost of Rs70, not a highly attractive rate of return when discounted for risk.) It seems clear that under present price relations even the limited response to nitrogen of the older plant types provides an attractive opportunity to investment in nitrogen use. It should be anticipated that today's demand for fertiliser for wheat will come from all cultivators having access to irrigation, not merely from those who can procure seed of the high-yielding varieties. While less easily documented, a similar statement would hold for farmers growing any of the other crops under adequate moisture conditions.[19]

The new agricultural strategy was launched in 1966-7, so that at the time of writing it has been in operation for two years. However, the first year was naturally affected by exceptional drought which, by severely reducing the total crop output, rather concealed the favourable results on the very small areas under the new varieties. The results are in any case not well reported in ordinary statistics, but it seems that the 1967 winter wheat crop did show a significant response.

Advance estimates for 1967-8, a good season climatically, put the foodgrain output at close to 100 million tons. Compared with the level of output in 1964-5, another favourable year climatically, the annual increase implied amounts to 3·96 per cent per annum, which is well above the level which could have been expected from the declining growth trend under the old conditions discussed earlier in this chapter. Some of the increase may be seasonal (1964-5 was also 'good'),

[19] This and the preceding paragraph are drawn directly from Hopper's work in the 1967 Report of the World Bank Mission to India.

but much must be attributed to both the new varieties and the expanded usage of fertilisers and underground water.

The most promising aspect of the new Programme[20] so far has been the large-scale introduction of dwarf wheat varieties in Punjab, Uttar Pradesh, Haryana, and Rajasthan. Under favourable conditions of well-irrigated and drained fields, adequate fertiliser, and improved practices, it is possible for the new varieties to yield twice or three times as much as the traditional varieties, but this level of performance cannot be projected as a reasonable expectation of average result. However, as already noted, other gains can be looked for. Some of the new varieties can be sown later than the old varieties they replace, thus facilitating multiple-cropping. In addition, the recent release of amber-seeded types is helping to break consumer resistance to dwarf wheats, while rust disease problems have not yet proved serious. Dwarf wheat varieties coverage seems to have been $1 \cdot 6$-$2 \cdot 0$ million hectares (4-5 million acres) during the 1968 winter season, and, according to advice received by the author, could well spread to 4-$4 \cdot 5$ million hectares (10-11 million acres) in 1969—the bulk of the irrigated wheat area. Improvements in input use (e.g. seed drills, land levelling, and timely irrigation) will determine how widespread uniformly high yields will become.

The introduction of new fertiliser-responsive dwarf varieties of paddy has given an important boost to the rice economy. These varieties, under favourable conditions, can also yield two to three times as much as the traditional Indian varieties. Yet, even with the potential from the new varieties and the considerable scope for progress with the traditional varieties through increased fertiliser, double-cropping, and better practices (Subramaniam 1964), there has not yet been a breakthrough of the same order for rice as for wheat. The 1967-8 dwarf variety coverage may have been $1 \cdot 6$-2 million hectares (15-18 per cent of irrigated rice area), but further increases may prove difficult. It has been established that the imported exotic paddy varieties are less adapted to the Indian soil and climate than the wheat varieties. Furthermore, summer (*kharif*) cultivation of two varieties, TNI and IR8,[21] has been discouraged in some central and southern

[20] A good deal of the material in parts of this section is drawn fairly directly from observations made by Bank officials in 1968 and communicated to me. The views are supported in similar advice from other friends in the field. The official 'watch' on and statistical recording of the new Programme are inadequate, although some helpful and realistic assessments for particular crops have been issued in particular areas, e.g. Ministry of Food and Agriculture, Community Development, and Co-operation (1967).

[21] TNI, or Taichung (native) I, is the main imported dwarf rice in the HYV Programme, while IR8 is a strain developed at the International Rice Research Institute in the Philippines.

areas as a result of blight disease and gall midge attack. A further complication is the preference of Indian consumers for particular grain shape (long) and cooking quality—characteristics so far absent from the high-yielding varieties. Fortunately, the problems encountered seem well within the capacity of adaptive research to handle. Provided research is supported there is no need to do other than recognise the necessity to avoid pitching short-term targets and expectations too high (see earlier comments).

High-yielding hybrid varieties of maize, sorghum, and millet may have covered 1·2-1·6 million hectares in 1967-8 (4-5 per cent of aggregate area under these crops). But performance has not been up to the reasonable expectations of those who developed these largely new varieties. Area expansion under hybrid sorghum has been held back by the high susceptibility of the released hybrids to shoot fly attack and, in addition, these varieties have suffered from bird depredation. More important, an inadequate policy framework for the private hybrid seed industry has resulted in poor quality seed, which has set back the hybrid program. Moreover, it is probable that insufficient recognition has been given to the level of farming skill required by the new hybrids, and finally, as will be stressed later, the amount of research support has been inadequate.

The Importance of a Comprehensive Policy

It is all too easy to review recent progress, and especially raised hopes, in terms of a single factor—the availability of new high-yielding seed varieties. These will certainly prove adequate in the case of wheat; there is much yet to be done in the development and production of equally satisfactory rice varieties; and, in the case of the other crops, best results are likely to come when open-pollinated varieties are available. But even confidently assuming that the scientists, given necessary support, will adequately respond to these challenges, and less confidently predicting that a satisfactory seed industry emerges, there is a great deal more required. The emergence of a new technology calls for the assured continuance of a comprehensive policy, composed of many elements to be concurrently applied to the agricultural sector. At the risk of some repetition from the earlier part of this chapter, the point requires elaboration. As before, brief comment will be offered on a number of particular points.

Price incentives. Current price policy with respect to foodgrains is to assure minimum or floor prices to producers which will provide an incentive for them to invest for expansion, but which will also be reasonable for consumers. An important adjunct is the operation of a buffer stock system which will equalise food supplies year by year and thereby stabilise prices.

The extreme food shortages during the drought years have not permitted these goals to be met. Food supplies, even with large quantities of imported grain, have not been sufficient to build up a buffer stock, and until late 1967 prices were at an inflationary level which required a degree of price subsidy to consumers. With the return of good weather the cereals price index is again declining. It dropped 11 per cent from October 1967 to February 1968, compared with a 17 per cent rise for the corresponding period of the previous year.

It is important that price policy should succeed. Continued high prices at levels which characterised the drought periods would clearly be detrimental to consumer interests and to the operation of the economy as a whole. In the long term it is probably true to say, however, that, even without droughts, a threat of upward pressure on prices will continue to prevail so long as the growth trend for food-grains is below 4-4·5 per cent per annum. Rising money incomes add heavily to the pressure exerted by the population growth rate, which is currently at approximately 2·5 per cent per annum.

In these circumstances it is imperative that the Indian government manage its obvious dilemma. It must, if need be with the aid of imports, build buffer stocks to deal with the threat of high prices whenever production, especially of wheat or rice, falls below the necessary trend. On the other hand, it must have the administrative capacity to step in and offer to buy whenever good crops bring prices down towards the 'assured' floors. Prices must not be allowed to pierce the floor if farmers are not to be discouraged.[22] It is fair to admit on the basis of past experience that governments may succumb to indifference to this problem under the influence of one or two good crops. Hopefully the impact of drought will remain vivid long enough to offset this danger.

Production of high-yielding seed. Reference has already been made to the critical importance of a good seed industry. The author has made no secret of the high importance he attaches to the development not only of properly organised and administered standards, but also of a major private (and/or co-operative) seed production industry. The story of the industry is rather a sorry one. Fortunately, however, recent signs point to significant improvement. The charter of the National Seeds Corporation puts it in a strong position to influence standards if it is properly equipped to exercise that influence.

[22] This is discussed in some detail in the 1964-5 Report of the World Bank Mission. The problem of making price incentives effective must be rated as one of the elements critical to the success of the new Programme.

Again, as a result of developments within some agricultural universities (as at Puntnagar, U.P.), and of the clear indication that trained and experienced farmers can make highly efficient seed producers, there is ample evidence of the possibility of building an effective seed industry. Naturally the problems are rather greater in respect of the hybrids than for wheat and rice. What is said later about research is also relevant to this point.

Fertilisers. The availability of adequate supplies of fertilisers (Table 3.11) is almost certainly the most critical aspect of the Fourth Plan for agricultural growth. The new strategy for agriculture rests heavily on increased applications of fertiliser required by the high-yielding varieties. The strategy will collapse if starved of fertilisers or, as is more likely to prove the case, it may not be fully effective since it will be impossible to give absolute priority in supplies to farmers belonging to the favoured group in the new strategy, if this means a widespread denial of supplies to the great bulk of farmers who must rely on a less favourable technology.

The dwarf and hybrid varieties require two to three times the fertiliser dosage recommended for best results from traditional varieties. India's traditional improved varieties yield almost as well as the dwarfs and hybrids when both are subjected to low levels of fertiliser application. In addition they require less attention and less water, produce relatively more fodder, and fetch a better market price. Thus the incentive to use high-yielding seed is directly related to the ease of obtaining fertiliser and to the security given to the higher-level investment by assured water supply.

Nitrogenous fertiliser availability has risen by more than 40 per cent per annum since the introduction of the new HYV Programme, but despite this the pressure of demand has remained high because of the increased fertiliser uptake of the new varieties. A public distribution system predominates in the allocation of nitrogenous fertiliser, but there has been pressure on this, with the HYV Programme accounting for the greater part of increased supplies of nitrogen, and the continuing expansion of area and crop coverage planned for the Programme.

However, the fertiliser supply situation, as seen in June 1968, should ease relatively in 1968-9. The 1967 kharif season suffered from late arrivals and slow unloading of imports, but availability of nitrogen improved during the rabi season in 1968 and there could be a carry-over into 1968-9 of some 300,000 tons. The Indian government has already sanctioned imports of about 800,000 tons, while home production is likely to exceed 600,000 tons (a figure below Fourth Plan target but much in advance of recent performance). Supplies of phosphate and potash are also likely to ease following improved import arrangements. In view of the importance of fertiliser it is fortunate that 'priority for agriculture' has been officially interpreted to mean a willingness to obtain imported supplies to make up for shortfalls in domestic production.

Water. Because of the nature of priorities established in the Fourth Plan, irrigation development promises some worthwhile returns during the currency of the Plan, as well as a useful trial of better-integrated policies for irrigation development. The main points are completion of major schemes carried forward from the Third Plan, 'right up to the cultivators' fields (i.e. including field channels)'; commencement of a few new major and medium schemes if economic circumstances permit; a relative upgrading of minor irrigation development; and special attention to the integration of agronomic and administrative services in all irrigation development (Planning Commission 1966a: Ch. 12).

Areas which theoretically enjoy an assured water supply are chosen for the implementation of the new strategy. However, these chosen districts often do not have it in practice (most depend on a satisfactory rainfall) and this almost certainly has been a limiting factor in the penetration and spread of the new HYV Programme.

An assured water supply is necessary (with the exception of sorghum and millet) if the new Programme is to spread. It is unfortunate that more than half the area under irrigation (some 36 million hectares) is fed by minor tanks, minor diversion works, and shallow wells, which provide insufficient water for long dry spells when water is most needed. An extensive canal system serves much of the remaining area, but this is designed to give protection during drought rather than to provide for intensive year-round cultivation. Yet, despite some clearly discouraging evidence, the results of assured water and the new varieties have become so obvious that it is not surprising that a rapidly increasing number of farmers is investing in minor irrigation. Moreover, in areas of reasonably assured groundwater supplies considerable gains in productivity have been evident, even under the old technology.

It is difficult to obtain reliable data on the progress of minor irrigation, but statistics, as well as field observations, point to a boom, especially in private on-farm investment. About 400,000 electrically-powered wells were in operation before the commencement of the new strategy, but by 1965-6 the annual rate of connections increased to 105,000, while for 1966-7 and 1967-8 it has levelled off at about 140,000. However, supply is still lagging far behind demand, with the waiting list totalling about 250,000, and only lack of factory finance is said to prevent the annual installation rate rising to about 180,000. It need not be assumed that all the new wells serve the HYV Programme.

In the Indo-Gangetic and coastal plains the private demand for tubewells is also greater than private and public capacity to provide the necessary materials, supplies, and supporting services. A major task of public policy will be to step up the level of these services considerably.

In particular it should be noted again that enlargement of utilised irrigation areas will undoubtedly yield results in increased output (through higher yields and double-cropping), even if no additional supplies of fertiliser are available. But it must be expected that the readily evident benefit of assured water will further stimulate the demand for new high-yielding seeds, plant protection materials, and particularly fertilisers.

Plant protection materials. Plant protection can be regarded as one of the four principal elements in the HYV Programme, ranking with water, fertilisers, and the seeds themselves. It is indeed recognised as such officially, and an active program is being stimulated and co-ordinated by the central government. Arrangements are being made with domestic and overseas firms for manufacture of chemicals and equipment, and it can be assumed that constant touch will be maintained with overseas experimental work and local adaptive research will be widely extended. The principal bottleneck in organisation seems to be the inadequacy as yet of mobile teams equipped with power sprays for acting quickly over large areas.

Credit. The Planning Commission has recognised, and recent governments have accepted the fact, that the enlarged program for agricultural development outlined in the Fourth Plan requires a considerable expansion of agricultural credit. It is clear from the *Draft Outline* (Planning Commission 1966a: 139, 176-7) and from subsequent action by central and state governments, that serious efforts are being made to meet the need.

In the field of short-term credit there is a continuing effort to make co-operatives more viable and to provide special facilities where co-operatives are, and will remain, too weak to cover demands of producers. There also seems to be a strong move to reshape co-operative lending in ways which will give the very small (virtually subsistence) farmer more scope. Nevertheless, it appears unlikely that co-operatives will be able to meet much more than half of total short- and medium-term credit requirements.

The problem of ensuring a large enough supply of long-term credit (five years or more) is even more severe. A major part of the agricultural program depends on private initiative in dugwell and tubewell development, soil conservation, land levelling, and grading. All these activities need capital equipment, and therefore credit. Fortunately, substantially increased financial services to the progressive farming sector are becoming available through more flexible procedures and additional branch facilities established by Agro-Industries Corporations, Land Mortgage Banks, and commercial banks. Loan advances made by the Land Mortgage Banks for medium- and long-term credit needs of owner-cultivators have risen from about Rs120 million in the early 1960s to Rs560 million in 1965-6 and Rs830 million in 1967-8. The credit target for 1968-9 is Rs1,040 million. There has also been a gradual re-orientation of lending towards productive purposes.

Credit supply remains a bottleneck; but in contrast to performance under the Third Plan, government action in the years since that Plan, through institutions and via the Budget itself, has been far more consistent with recognition of a major problem.

Research and extension. The very drama of the rapid emergence of a new technology may be reason for some concern, if energies and attention are not devoted to the great logistic problem of securing the inputs, especially fertiliser. By the same token there remains the critical need to mount a concurrent and vigorous research program, without which confidence already expressed in the work of the scientists could be misplaced. Research has been the means of innovation, but continuous innovation is necessary. New varieties must be evolved to replace those now being used to launch the new program, either because they are better yielders or because they have other desirable characteristics such as shorter growing periods, better resistance to insects and disease, or more tolerance to less than highly-skilled cultural management. The HYV Programme has run into and will continue to meet problems which will require early attention and solution if the momentum now launched is to be sustained. The new technology tends to aggravate old gaps in research, and produces new ones as well. It gives rise to a more pressing need for effective integration of research and extension.

Given this need for continuous research, it is a matter for concern that the *Draft Outline* of the Fourth Plan gave so little attention to it and apparently none at all to the financial resources needed. This is not to say that no provision is made for research, but there appeared to be a signal lack of recognition of the high priority deserved. The apparent lack of adequate resources would suggest an excessive preoccupation of administrative attention with the immediate logistics of the new agricultural strategy and perhaps, too, an unwitting reliance on external research results for solving India's problem. The latter would be a costly error. Since the publication of the *Draft Outline*, nationally co-ordinated plans of research related to the foodgrain crops have been launched—but hardly with evidence of organising drive and public understanding of the importance that is essential to their success.

Three particular further comments are in order. Not only must the problems of the new seeds be attended to, but it is imperative that research be greatly expanded in soil and water management, water and soil engineering, entomology, plant physiology, crop rotational practices, and the economies of farm practices, fertilisers, etc. It is also necessary to heed the problems of cultivators outside the present scope of the new technology. The real cause for concern about research in India is no longer the ability (never much in doubt) or willingness of the research scientist to apply himself to the practical problems and needs of the farmer, but rather the level of support he will get. With respect to expenditure on research it is clear, even yet, that research programs in agriculture fare poorly alongside those in atomic energy. Moreover, it appears that research related to tobacco received more from the central government in 1965-6 than all the foodgrains put together. However, there is no doubt that the present approach to the organisation of Indian research is right (Indian Council of Agricultural Research 1966, Rockefeller Foundation 1967).

The position with regard to extension is far from satisfactory. The much greater sophistication of the new technology requires reconsideration of the workings of the extension service and in this connection a few comments are offered in rather categorical terms. First, the old system with the village-level worker at its centre is inadequate. A much higher level of training is required, and those spreading the new techniques should have links with research centres. Second, the system of national demonstrations, now in its infancy, is important for the morale and efficiency of the research scientists associated with the work. The limited experience of these demonstrations certainly supports the case for associating extension more closely with the best research centres than in the past.[23] Third, at the state level, the agricultural universities should be strongly encouraged with a view to their becoming dominant in agricultural research and extension. At least this principle of linking research and extension can be usefully adapted to India from American Land Grant College experience.

CONCLUSION

In the circumstances of the last fifteen years, it would have been difficult to conclude three years ago that the objective of a 5 per cent average annual increase in agricultural output (Fourth Plan target)

[23] Those concerned with these matters would profit considerably from reading the work (much of it in Bank Mission Reports) of Hopper and Ladejinsky earlier cited.

was realistic. There were, of course, considerable opportunities for increased output at rates above the trends evident in the early 1960s, but below this high target, if full attention and high priority had been given to a number of critical factors. Enough Indian farmers were clearly willing and able to expand output, given incentives and inputs. While there was evident in 1965 the beginning of a more advanced technology, a major advance did not seem immediately possible. There was heavy reliance on the undoubted scope for getting more from old varieties by pushing them nearer to the limits of their capacity to use water and fertiliser. A considerable obstacle seemed to be that the 'organised will' was not present in central and state administrations to ensure priorities in use of scarce resources for agriculture. In short, it was clear that the growth rate in agriculture could be accelerated, but it was not clear that the concerted efforts of all parties, necessary to realise as great an acceleration as possible, could be expected.

Despite the great difficulties ahead and the heart-breaking proportions of the two droughts, the morale of administrators and scientists is now much higher and, of course, the proportion of farmers eager to avail themselves of the new technology is embarrassingly high in the short term. Yet it would be false to look at the scene other than with clear glasses; the critical factors restraining growth are numerous and, indeed, are rendered the more difficult by the very needs of the new technology. This chapter has reviewed, albeit sketchily, the need for a sounder organisation of the seed industry; noted the highly critical shortage of, and problems of lifting, fertiliser supplies; noted the importance of plant protection materials; noted that added supplies of water will lift output but will also aggravate the demand for fertilisers; briefly surveyed a potentially critical problem of credit supply; stressed the importance of a stable price policy in the interests of consumers and costs in the general economy while protecting incentives to the farmers; and, not least, argued for a more ample program of research and extension, being no longer worried about its relevance or profitability.

Many of the points are not new, but several have a different order of urgency in the new technology from in the old. Fertiliser supply remains the main problem. This especially, however, turns into a great and rising pressure on the critical balance of payments situation. It would not be difficult to argue that the greatest single bottleneck in Indian agriculture is foreign exchange. The new program calls for a large investment in fertilisers and other inputs for which home production in most cases continues to be short of need. As observed earlier in the chapter, the alternative of food imports to match the rapid growth in total demand is not really open to India. The cost of agriculture's input requirements could rise to 20 per cent, even 25 per cent of export earnings over the five-year period. But even at this, the cost is very much less than the cost of commercial imports of foodstuffs otherwise necessary. Aid directly related to the foreign exchange needs of Indian agriculture will be far more fruitful if applied to support a policy of self-help in India than it could possibly be if confined to PL480 and equivalent non-commercial supply programs.[24]

There are no serious alternatives open to India; the costs of not succeeding in agriculture are far greater than the present and continuing strain of finding the resources for an effective program of production of fertiliser and other inputs. Fortunately, the prospects of success are now greater and it is the author's belief that India can, given even only the modest availability of resources it has planned, go much nearer to achieving its targets by 1970-1 than would have been possible under the old technology and the circumstances of three years ago. The difficulties ahead are so great that one cannot feel justified in believing that all these can be so satisfactorily resolved within the next few years as to assure full achievement. If, however, as seems likely, India reaches 110 million tons of foodgrain output or better early in the 1970s, there will be established a momentum not likely thereafter to falter significantly.

One difficulty arising from the approach used in this paper—a review of performance and prospect in terms of aggregates—was earlier stressed. The approach too readily conceals the wide variation in performance between individuals and between areas. Moreover, the HYV Programme relates only to a small, although highly significant, proportion of the whole. Its very success will produce, and is producing, social tensions, particularly between those favoured and those not able to participate in the new Programme. Technologically, the distinction between large- and small-scale cultivators need not occur, although it must, for the present, occur between those with assured water and those without. In practice, however, the ability to participate in the new technology tends to favour those already in the stronger position under the old technology, because of the readier availability to them of all the elements in the new policies. Thus existing disparities between rich and poor farmers may widen and tensions increase in the short run.

There are two obvious approaches to this problem, both feasible, and both likely in the longer run. The first is to ensure that all the elements in the new policies—price assurance, inputs, credit, high-yielding seeds—are made as available to the small-scale cultivator, both

[24] The whole question of the foreign exchange bottleneck is reviewed by L. Sonley and the author in the 1967 Report of the World Bank.

owner and tenant, as to the larger-scale operators. Likewise, the research programs must be directed to the problems of farmers not likely to have access to assured water (irrigation) supplies.

The second approach is to tackle more vigorously the problems of tenancy, for these problems will be exacerbated if owners see profit to be obtained under the new technology by evicting tenants or by refusing to share the results equitably. Fortunately, while success in the HYV Programme will stimulate dissatisfaction on the part of non-participants, the newly-found political strength of the great mass of farmers may well provide a greater likelihood of action to pursue agrarian reform more vigorously than it has been in most states.

There remains only one major observation to make. On the test of performance the policies of the Indian government should not be judged by the results of production in drought years, but by the evidence of advance where conditions have been normal, and, not least, by its recent determined efforts to give priority to agriculture in the use of scarce resources (of which foreign exchange is the most critical), to orient its industrial program towards agriculture's needs, and to develop policies likely to give more adequate scope to private initiative both in agriculture itself and in the industrial sectors of direct importance to agriculture.

The recent evidence is that the government has performed in accordance with its declared intentions to give highest priority to agriculture. The constraints on performance will, however, continue to be considerable, even after good seasons return. While not wishing to minimise the gravity of the constraints the author is satisfied that both policies and performance are now better related than they were in the period of the Third Plan.

4

Thailand

T. H. Silcock

THAILAND has been praised for devoting a relatively large amount of attention to the rural sector, and its rural economy is indeed growing fairly rapidly.

SIZE AND CHARACTERISTICS OF THE RURAL SECTOR

We shall define the rural sector in Thailand to include agriculture proper (i.e. rice, tree, garden and field crops, livestock farming) and also forestry and fisheries, though no detailed consideration will be given to fishery problems. We must ask first how important the rural sector is in the whole Thai economy during the 1960s.

Table 4.1 shows the relative importance of the rural sector in the whole economy during the early sixties in terms of three important indicators: its contribution to the export trade, its employment of the labour force, and its proportion of the total gross domestic product. The most casual glance at this table shows that rural incomes are much lower than other incomes. About four-fifths of the work force, producing nearly nine-tenths of the export earnings, earn little more than one-third of the national income. We shall need to investigate the significance of this further. First, however, we may consider in more detail the different columns of the table.

The rural sector produces 88 per cent of Thailand's total exports. Industry in Thailand (other than the milling of primary products) has developed almost solely for the home market, and almost the only other significant export is mining products. It is, however, a little misleading to compare exports of these rural products with the gross domestic product derived from agriculture and forestry as shown in the statistics. Rice is milled, and is exported in sacks most of which are produced in Thailand. Timber is at least sawn and roughly shaped. For comparability between the trade figures and the GDP figures (Table 4.1), the workers in rice mills and saw mills should be counted in the rural sector, but their income is in fact listed under manufacturing. The proportion of total income involved is small for rice, probably much larger for timber (Usher 1967).

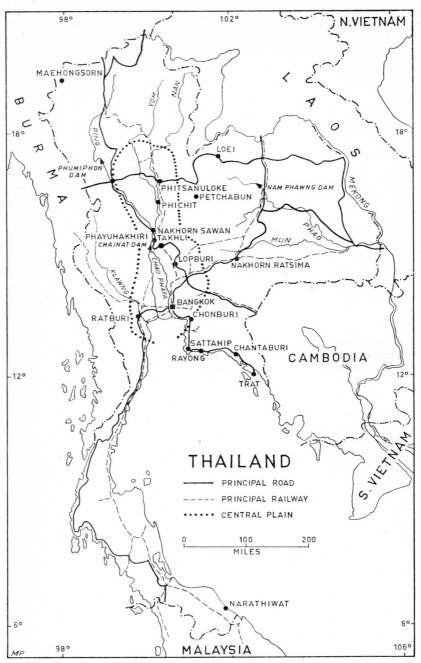

THAILAND

— PRINCIPAL ROAD
--- PRINCIPAL RAILWAY
•••• CENTRAL PLAIN

0   100   200
MILES

DEPARTMENT OF HUMAN GEOGRAPHY, A.N.U.

Map 4 Thailand

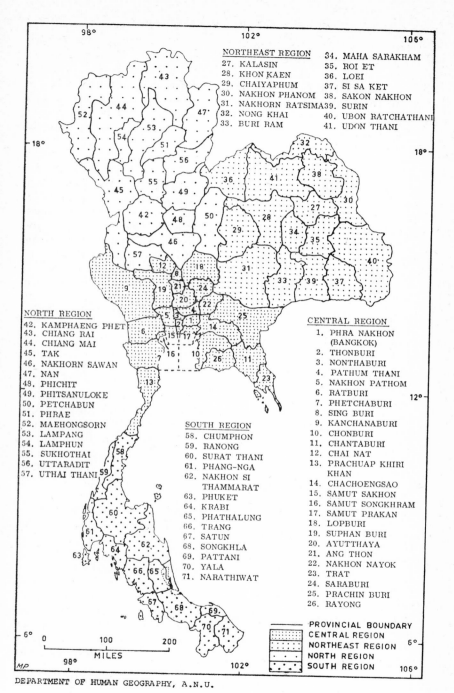

NORTHEAST REGION
27. KALASIN
28. KHON KAEN
29. CHAIYAPHUM
30. NAKHON PHANOM
31. NAKHORN RATSIMA
32. NONG KHAI
33. BURI RAM
34. MAHA SARAKHAM
35. ROI ET
36. LOEI
37. SI SA KET
38. SAKON NAKHON
39. SURIN
40. UBON RATCHATHANI
41. UDON THANI

NORTH REGION
42. KAMPHAENG PHET
43. CHIANG RAI
44. CHIANG MAI
45. TAK
46. NAKHORN SAWAN
47. NAN
48. PHICHIT
49. PHITSANULOKE
50. PETCHABUN
51. PHRAE
52. MAEHONGSORN
53. LAMPANG
54. LAMPHUN
55. SUKHOTHAI
56. UTTARADIT
57. UTHAI THANI

SOUTH REGION
58. CHUMPHON
59. RANONG
60. SURAT THANI
61. PHANG-NGA
62. NAKHON SI
    THAMMARAT
63. PHUKET
64. KRABI
65. PHATHALUNG
66. TRANG
67. SATUN
68. SONGKHLA
69. PATTANI
70. YALA
71. NARATHIWAT

CENTRAL REGION
1. PHRA NAKHON
   (BANGKOK)
2. THONBURI
3. NONTHABURI
4. PATHUM THANI
5. NAKHON PATHOM
6. RATBURI
7. PHETCHABURI
8. SING BURI
9. KANCHANABURI
10. CHONBURI
11. CHANTABURI
12. CHAI NAT
13. PRACHUAP KHIRI
    KHAN
14. CHACHOENGSAO
15. SAMUT SAKHON
16. SAMUT SONGKHRAM
17. SAMUT PRAKAN
18. LOPBURI
19. SUPHAN BURI
20. AYUTTHAYA
21. ANG THON
22. NAKHON NAYOK
23. TRAT
24. SARABURI
25. PRACHIN BURI
26. RAYONG

PROVINCIAL BOUNDARY
CENTRAL REGION
NORTHEAST REGION
NORTH REGION
SOUTH REGION

0   100   200
MILES

DEPARTMENT OF HUMAN GEOGRAPHY, A.N.U.

Map 5   Regions and provinces of Thailand

It is a matter of some interest that only two-fifths of the exports of the rural sector consist of rice. This is a result of a very great increase in the export of other crops in the post-war period, which is discussed later. The export value includes a rice premium which is equal to some 40 per cent of the world price, so that rice as a proportion of farmers' total income from exports would be appreciably less; probably no more than one-quarter of all their export income comes from the sale of rice.

In terms of employment the predominance of the rural sector is rather less than in terms of trade (Table 4.1). In 1960 almost 88 per cent of the total population of $26 \cdot 3$ million lived in rural areas and the proportion of the total work force classified as agricultural was 82 per cent. This latter estimate is now likely to be an overstatement for two reasons. First, though the drift to the towns is less in Thailand than in most of the less-developed countries, the proportion employed in agriculture has probably declined since 1960 (Caldwell 1967). The labour force surveys undertaken in 1963 and 1964 unfortunately included only municipal areas (National Statistical Office 1963a, 1963b). The only real evidence is the (comparatively slight) trend toward urbanisation shown in previous censuses: between 1947 and 1960, for example, the urban population almost doubled to $3 \cdot 3$ million, while the total population increased by about 50 per cent. With the great improvement in transport since World War II this trend is more likely to have accelerated than diminished.

The second reason for the overestimate is that a much higher proportion of women, mainly working as unpaid family workers, are included in the productive labour force in agricultural work than in other work. Usher (1966: 433-4) has argued that this is a statistical bias, due to the difficulty of distinguishing 'productive' from 'unproductive' work, particularly of women, in a subsistence economy. This is true also of men's labour, since there is much less specialisation between the sexes in Thai rural life than in most traditional rural economies. The family is a team doing the work both of the house and of the farm (Sharp *et al.* 1953: 123-4). Moreover, it is not known what share of farm families' productive work is agricultural and what should be attributed to other activities, such as house construction, cottage industry, and petty trading. The proportion of the male labour force working in the rural sector in 1960 was $78 \cdot 5$ per cent, and this is probably a truer indication of that sector's relative importance in productive employment.

Turning to the relative importance of agriculture in the gross domestic product, we find that the figures understate its true importance, since rice is valued, not at its international price, but at a price almost halved by the operation of the rice premium and other taxes.

TABLE 4.1

AGRICULTURAL EXPORTS, WORKFORCE, AND PRODUCTION

	Average annual domestic exports (1961–5) (m. baht)	(%)	Economically active population[a] ('000)	(%)	Average annual gross domestic product (1961–5) (m. baht)	(%)
Rural sector						
Rice	3,830·5	36·1	11,187	81·2	7,924[b]	11·3
Other crops and livestock	5,103·1	48·1			13,236[b]	18·9
Forestry	340·4	3·2	36	0·3	2,014	2·9
Hunting and fishing	84·9	0·8	111	0·8	1,675	2·4
Total	9,358·9	88·2	11,334	82·3	24,849	35·6
Rest of economy	1,245·9	11·8	2,438	17·7	44,999	64·4
Total	10,604·8	100·0	13,772	100·0	69,848	100·0

[a] 1960 Census: includes 11-year-olds and over.

[b] The division between rice and other crops is not reliable; for the 1964–5 detailed figures of GDP were not available at the time of writing. Figures for agricultural crops were calculated for 1961–5 and divided between rice and other crops according to percentages calculated for 1961–3. Probably relatively higher prices for rice approximately offset relatively lower increases in output. For detailed comparison the new national income figures to 1966 should be used.

Source: National Statistical Office 1965b; NEDB 1964b; Ministry of Agriculture 1955– (1958–65).

This lowers the price not only of rice but of other crops which compete with rice for resources or for the farmer's purchasing power. Probably most of the crops which are not exported sell at a lower price than they would do without the premium; rubber and other export crops would not be affected, though one or two of them pay minor duties themselves. A revaluation of rice at the international price less milling and marketing expenses, and a revaluation of other crops at the figures at which they would sell if rice reached this level, would probably raise the share of the rural sector in the gross domestic product to about 40 per cent (Silcock 1967b).[1]

The figures in Table 4.1 show that Thailand is no longer a monoculture economy, depending almost wholly on rice; but they may go too far in belittling its importance. Rice is still the basic subsistence crop of Thailand, and a very high proportion of the farmers grow at least some rice. The area under rice is still more than twice as great as the area under all other crops taken together, and we can probably safely say that rice uses twice as much labour as the whole of the rest of Thailand's rural sector.

Another measure of the relative importance of the rural sector is its role in the public finance of Thailand (Table 4.2).[2] First, rice is of great importance as a source of revenue. The rice premium yields more than the whole income tax, and more than twice as much as the income tax on individuals. Tobacco is, as in most countries, a source of revenue; in Thailand most of the raw tobacco is locally grown. This revenue is collected through a government monopoly, though a few private individuals are allowed to prepare tobacco for sale on payment of duty, and probably many more prepare it for sale or use without permission or payment.

The forests are also an important source of revenue. By no means all the revenue accrues to official government funds. It will be seen from the table that even if the Forestry Industry Organisation sold only teak, the total yield from teak in 1965 would be under 32 million baht. However, Usher has analysed the cost structure of teak production and shown that the total yield per cubic metre to the government should be 2,500 baht (Usher 1967: 225-7). The total export of teak in 1965 was 45,000 tons. Even assuming a substantial underestimate of costs by Usher, the revenue ought to be some 90 million baht; between half and two-thirds of the revenue that should be collected is going astray. It is impossible to estimate whether there are similar losses on the rest of the forest revenue.

[1] This will be discussed more fully in T. H. Silcock (in press).

[2] Generally speaking, Thai financial figures, though quite detailed, are insufficiently annotated to enable reconciliation of figures from different sources, but figures in Table 4.2 do suffice to show important facts about the economy.

TABLE 4.2

DIRECT REVENUE FROM, AND EXPENDITURE ON, THE RURAL SECTOR IN THE THAI GOVERNMENT BUDGET, 1965 (million baht)

Revenue (actual)			Expenditure (estimates)	
Total taxes		10,283	Economic expenditure	3,370
Sale of goods and services		313	Defence, internal and external	2,850
Revenue from government enterprise		224	Public health and welfare	1,865
Other revenue		377	Education: primary	1,211
Total revenue of which		11,197	other	917
			General administration	934
Revenue from rural sector			Debt service	688
Rice premium	1,257		Other	584
export duty	197		Total expenditure of which	12,420
milling duty^a (approx.)	80			
Total		1,534	Expenditure on rural sector	
Rubber export duty	164		Roads	1,028
licences	19		Railways	99
Total		183	Other rural transport^c	335
Tobacco: total revenue from local tobacco^b (approx.)		700	Irrigation	696
			Other agriculture	264
Livestock			Forestry	87
export duty on hides	5		Fisheries	37
epizootic control levy	7		Total	2,546
Total livestock		12		
Forestry				
teak export duty and royalty	18			
other duties and royalty	107			
sawmill duty^a	40			
profits and sale of forest produce	28			
Total forestry	193			
Fisheries: licences	6			
Total Direct Revenue		2,627		

a Total business tax on milling divided 2 : 1 between rice-milling and saw-milling.

b Total revenue from government tobacco monopoly and private producers, reduced by 10 per cent to allow for use of some imported leaf (approx. 10 per cent by weight).

c All air transport excluded, but all other transport expenditure included.

Source: Office of Prime Minister 1962– and 1965–. Actual expenditures are not broken down appropriately in the Budget Documents, and in 1965 actual expenditure appears to have differed relatively slightly from the original estimate.

In comparison with rice, tobacco, and forest products, rubber—the untaxed export value of which was about two-thirds that of rice in 1965—is relatively lightly taxed. In that year it contributed to the government official budget less than 10 per cent of export value. Like forestry products, however, rubber pays a number of unofficial levies. Chinese committees levy one baht for every five sheets for the support of Chinese schools—the rubber merchants are all Chinese. A further 300 baht is paid to customs officers for every ship carrying rubber cargo. Further unofficial payments, in addition to the official ones, are paid in respect of the many licences required in the rubber business.[3]

It could be argued that the unofficial levies on teak and rubber, which lower the real incomes of producers, are not part of the taxation system. This argument, however, fails to do justice to the comparatively effective workings of the Thai government (Evers and Silcock 1967). These payments in fact influence the expenditure pattern in reducing indirectly the general level of salaries paid to Thai civil servants. It is impossible to estimate the total of unofficial levies in all sectors. The available evidence suggests, for instance, that there are considerable opportunities for irregular income in communications and in the industrial sector (IBRD 1959: 198-200), and these probably have an influence on development policy.

Some 25 per cent of official revenue is derived more or less directly from the rural sector as defined above, and some other large items which cannot be clearly allocated—e.g. 900 million baht from petroleum products, 400 million from imports of rough cotton cloth, and 150 million from the state lottery—can be expected to fall predominantly on the rural sector.

How far is the expenditure in fact directed to rural development? Not enough explanation of the budget figures is published to enable one to go much beyond the breakdown given in the *Budget in Brief* (Office of the Prime Minister 1962-(1962): 18-19). The expenditure section of Table 4.2 does not attempt to show all the government funds spent in the countryside, any more than the revenue section can show all the funds derived from it. Of the 3,000 million baht spent on primary education and on public health and welfare, almost certainly more than half would be spent for the benefit of farming communities, even though the standard of service in the country is lower than in the towns. Within the economic sector an approximate allocation is easier.

The approximate equality between the amount taken directly from the rural sector and the amount spent on it is coincidental and of no real significance. More important, about three-quarters of the economic

[3] Moreover, in the remote southern areas most rubber producers and traders pay taxes to the communist armies there.

expenditure is devoted to the rural sector, and the economic sector as a whole absorbs more than one-quarter of all the revenue.

Perhaps the most striking feature is that over half of the economic expenditure is devoted to roads and irrigation. This can be explained in terms of the past history of Thai economic development. Road development had been deliberately held back under the absolute monarchy to keep the railways solvent and ensure the servicing of the foreign loans borrowed to pay for them (Ingram 1955: 194-5). Irrigation was held back, although suitable plans had existed for forty years, because the monarchy was extremely sensitive to incurring loan obligations for anything that did not produce an automatic revenue, and was unwilling to charge for irrigation water.

Yet under the absolute monarchy the railways had effectively unified the country and some division of labour had developed. There was a keen demand for transport. Moreover, the Royal Irrigation Department had acquired a high level of competence during its forty years' existence and this could be put to use when foreign loans became available for large dams.

LINKAGES WITH THE REST OF THE ECONOMY

The importance of the rural sector may also be assessed in terms of linkages. To what extent do agriculture and forestry provide raw materials for Thai industry? What relations exist between prices in the agricultural sector and the rest of the Thai economy?

Rice, timber, and rubber all need substantial mechanical operations before they normally enter international trade. Changes have been made in these operations during the twentieth century, as a result of the introduction of industrial processes. About one-fifth of all those engaged in manufacturing are employed in rice mills, saw mills, the manufacture of gunny-sacks, and in rubber milling, the products of which enter international trade as raw materials (Table 4.3). These industries have grown with the expansion of the crops on which they depend, but there have been other developments also.

Rice milling for export was fully mechanised by World War II.[4] In the main, steam power was used, with rice chaff as free fuel (Usher 1967). The chief post-war development has been the decentralisation of much of the milling, even for export. Decentralisation was due to the disturbance of transport, first by bombing during the war, and later by government restrictions on the movement of rice. There are medium-scale mills with a capacity of 20 tons or more per day in the larger rural centres, and small mills with a capacity of only about 3

[4] Information in this and succeeding paragraphs relies partly on personal inquiries by the author in Bangkok.

TABLE 4.3

EMPLOYMENT IN MANUFACTURING WITH AGRICULTURAL LINKAGE,
1960[a]

Wood and cork products (excluding furniture)	59,664
Food products (n.e.i.)[b]	55,812
Rice mills	32,983
Cordage, rope, and textiles (n.e.i.)[c]	27,220
Saw mills and planing mills	26,769
Cotton textiles—spinning, weaving, and finishing	19,815
Tobacco and snuff products	16,442
Sugar	10,560
Rubber products	4,948
Leather and leather goods (excluding footwear)	1,242
Oils and fats (vegetable and animal)	498
Total with linkage	255,953
Total all other manufacturing	214,195

[a] Includes employed population 11 years of age and over, and those looking for work.
[b] Includes meat and dairy products, canned foods, and grain mill and bakery products.
[c] Probably most in this category are employed making gunny-sacks.
Source: National Statistical Office 1960.

tons in many rice-growing villages. Farmers normally have their rice
for home consumption milled by machinery now, and this saves a
good deal of family labour. Probably the time saved by mechanical
milling and mechanical transport has contributed to the increased
interest in supplementary crops, by freeing more family labour.

Thailand's production of gunny-sacks from locally grown kenaf is
a post-war phenomenon. There are several different government
factories affiliated to different ministries (Silcock 1967a: 308-16), and
they enjoy a protected market. Quality of production has so far been
poor but is improving, and most of Thailand's own needs are now
locally met.

Saw-milling has shown little technical development recently; it was
mechanised before World War II but stagnated, partly because of
restricted supply, during the post-war period. The comparatively
large number of employees in other wood manufactures (Table 4.3)
reflects some diversion of private saw-milling interests to hardboards,
parquet flooring, etc., and includes the Thai Plywood Company, a
subsidiary of the government-owned Forest Industry Organisation.
The small profits accruing to the government from this last operation suggest that there may be some inflation of the numbers employed
in the government sector, for purposes of patronage.

Mechanical methods have been increasingly adopted in the extraction of timber (Samapuddhi 1957). The figures of employment are not, however, sufficiently detailed to enable us to assess whether this has increased productivity.

The great expansion in rubber output since World War II has led to the establishment of a number of rubber mills in south Thailand, and this in turn has made possible some manufacture of toys, shoes, and other rubber goods in south Thailand and in Bangkok. As both the rubber mills and the manufacturers are included in the total of under 5,000 employees, it is clear that the process has not gone very far.

No discussion of linkages between the rural sector and the rest of the economy would be complete without an attempt to assess the principal effects of the low internal price of rice created by the rice premium and export duty. The cost of the main item of diet always has a relatively large effect on the money price of labour, but this effect is accentuated in Thailand partly because of the particular dietary importance of rice and partly because the practice of paying a proportion of the wage in kind survives, despite the comparative sophistication of the Thai economy. Thus labour is relatively cheap in the Thai economy in comparison with internationally traded goods.

This cheapness is accentuated by the fact that, in rural areas, the cheapness of rice encourages the production and discourages the consumption of vegetables and other components of the Thai cost of living (Usher 1967; Silcock 1967b). On the supply side, labour is pushed out of rice-growing to seek other sources of income, including cultivation of fruits, vegetables, etc.; on the demand side, reduced income and substitution of rice for other dietary items tend to reduce the consumption of these other items.

In terms of comparative static analysis, of course, since trade must balance, this distortion would produce a situation in which Thailand (which enjoys a comparative advantage in rice) was deprived of some of the advantages of the division of labour, since the income of rice producers would be reduced in terms of international trade goods. Thailand would produce instead other crops and also industrial products, which it could have obtained more cheaply in exchange for rice. This is not, however, the relevant analysis for Thailand's current situation. In dynamic terms, the encouragement of new exports and the discouragement of imports provides a favourable balance for the import of capital goods. The stimulus to change to new crops and new industrial products gives the inducement to import this equipment. Much of the advantage generated, directly and indirectly, by the depressed prices of rice and of labour in terms of imported goods, has undoubtedly been wasted in consumption of luxury goods for a small and politically powerful middle class. More is probably wasted in imports of relatively inappropriate capital goods (Silcock 1967c). Yet a good rate of growth is nevertheless being achieved, as will be shown in a later section.

LAND RESOURCES AND UTILISATION

Of Thailand's total area of 51·2 million hectares, some 9·6 million hectares are under watershed forests. Of the remaining 41·6 million hectares of land area, just under one-third, or about 12 million hectares, are held in individual holdings (Samapuddhi 1957: 3). The rest is either exploitable forest (16 million hectares) or waste. The area in individual holdings has nearly doubled since World War II but, though there is still land to spare, the government's policy at present is to limit the expansion of cultivation—for climatic and other reasons—to rather less than half the total area (NEDB 1964a: 70).

The area of land occupied per head of population cannot be estimated year by year with any precision, but in Table 4.4 the area under all crops, as given in the agricultural statistics, is used for comparison with the total population. Coverage of the crops actually grown has improved during the period in view—cassava figures on a national basis, for example, are available only from 1956, and fruit and vegetables from 1957. The true area per head in 1950 may have been 0·34 instead of 0·32 hectares. It appears, however, that over the post-war period the area per head has remained approximately constant, with a slight net fall in 1957, a rise during the period of Sarit's rule, and a subsequent slight fall again. From 1947 to 1963 the population outside municipal areas rose from 15·76 to 24·98 million, the municipal population having doubled (Silcock 1967b). The use of these growth rate and area data shows that between 1950 and 1965 the area in agricultural holdings per non-municipal inhabitant has remained at about 0·36 hectares.

MAIN CROPS IN THE ECONOMY

It will already be apparent that rice is much the most important crop in Thailand, though its importance is diminishing. It is grown in every region—probably in every rural district—in the country. About half is still consumed on the farm (Silcock 1967b: 235-6). Though the proportion of rice marketed and consumed in the rural area itself has more than trebled in the period since World War II, and despite a sales expansion of local rice in the towns with growing urbanisation, the heavy taxation of rice has probably delayed market specialisation. The economy is not a subsistence economy in the sense that money

Table 4.4

SOWN AREAS OF PRINCIPAL AGRICULTURAL CROPS, 1950–65

('000 hectares)

Year	Rice[a]	Rubber	Maize	Kenaf	Tobacco[b]	Oil-seeds[c]	Cassava[d]	Garden[e] crops	Fruit[f]	All non-rice crops	All crops	Area per head in hectares
1950	5,541	338	37	5	30	194	13	n.a.	n.a.	752	6,293	0·32
1951	5,960	352	42	6	42	216	13	n.a.	n.a.	828	6,789	0·34
1952	5,368	368	45	11	43	221	14	n.a.	n.a.	851	6,219	0·30
1953	6,173	384	48	10	54	222	14	n.a.	n.a.	890	7,062	0·33
1954	5,557	400	53	6	54	251	14	n.a.	n.a.	947	6,504	0·29
1955	5,770	416	56	8	56	256	14	n.a.	n.a.	979	6,749	0·30
1956	6,024	430	82	18	58	274	40	n.a.	n.a.	1,070	7,093	0·30
1957	5,075	444	98	13	62	307	38	38	70	1,334	6,416	0·27
1958	5,758	456	126	21	61	302	45	43	69	1,390	7,147	0·29
1959	6,066	469	200	45	62	307	62	48	77	1,568	7,632	0·30
1960	5,920	481	285	141	59	358	72	61	109	1,894	7,814	0·30
1961	6,179	493	307	275	42	342	99	90	152	2,067	8,246	0·30
1962	6,659	502	328	114	42	390	123	118	147	2,021	8,680	0·31
1963	6,602	524	418	154	40	405	141	123	174	2,378	8,978	0·31
1964	6,539	528	552	218	83	403	106	117	173	2,538	9,078	0·31
1965	6,478	531	576	384	72	445	102	117	242	2,872	9,350	0·31

[a] Crop years 1950–1 to 1965–6.
[b] Calculated by adding area planted with local varieties of tobacco to area planted with Virginia tobacco.
[c] Includes castor beans, groundnuts, sesame, and coconuts.
[d] No figures available for cassava for the whole kingdom until 1956. Figures before 1956 are for Chonburi province.
[e] No figures available for garden crops until 1957. Garden crops include Chinese kale, cabbage, cauliflower, Chinese cabbage, string beans, egg-plant (long type), egg-plant (crisp variety), tomatoes, pumpkins, sweet potatoes, potatoes, yam beans, and other potatoes.
[f] No figures available for fruits until 1957. Fruits include pineapples, water melon, sweet, Numwha, lady finger, and other bananas.

Source: Ministry of Agriculture 1955 (1965: 29, 48, 54, 58, 84, 176), (1955: 68).

exchange is unimportant; at least since the revolution of 1932, Thailand has been a reasonably unified economy with some local specialisation. Yet farmers try to avoid having to buy rice. They trade in other things, and will grow a surplus of rice partly as an insurance against poor weather and only partly for cash income. It is more accurate to say that Thai farmers keep the growing of their basic rice crop separate from their cash transactions than to describe them as subsistence farmers.

Not every province grows a surplus of rice, though most do. The main deficit areas are Bangkok and some provinces to its southeast and southwest, and also the extreme south of Thailand. The flow of rice is thus mainly southward (Usher 1967: 208). The far north and most of the northeast, however, are predominantly areas which grow glutinous rice, which the growers themselves consume, exporting only a relatively small surplus (Ministry of Agriculture 1962: 62). Bangkok's main rice supply is from the Central Plain.

Rubber, the next most important commodity, is mainly grown in the far south, where it is almost wholly a smallholders' crop.[5] It is planted mainly by people of Malay race. Largely as a result of the pre-war international restriction schemes—which led to much corruption—there is a tradition of illegal planting in the south, which delayed the introduction of a replanting scheme. Without government aid it is difficult for smallholders to replant with modern high-yielding materials, for technical considerations prevent the replanting of a very small area, and replanting of half the holding or more means excessive loss of income. A scheme of assisted replanting has been operating since 1960, but has not yet had a very marked effect.

Some rubber is also produced in parts of the southeast with favourable rainfall conditions (in the provinces of Chantaburi, Rayong, and Trat). Here some of the planting is carried out by fairly large-scale plantation methods.

Maize is now very nearly as important as rubber. The area planted is already officially greater, and with the recent fall in the price of rubber, the value of output for 1967 may also prove to be greater. The soft variety is grown in many parts of Thailand for local consumption. The hard Guatemala variety is grown for export, and the expansion of this variety (not separately distinguished in the agricultural statistics) has occurred since World War II. Of the total crop, some 70 per cent is now grown in a small group of provinces at the northern end of the Central Plain, the most important being Nakhorn Sawan and Lopburi.

The fourth main crop, kenaf, was deliberately developed for the government gunny-sack factories and as a crop for the northeast of

5 For a fuller discussion of Thai rubber, see Silcock (1967; in press).

Thailand (IBRD 1959: 75). It has now spread from its original province of Petchabun (McFarland 1944: 514) over the whole Northeast Region. By 1965 half a million tons were produced and about two-thirds of the crop is now exported, the balance being used in local manufacture. It grows well in the relatively dry Northeast Region, but quality is adversely affected by lack of water for retting. Attempts are being made to overcome this problem by stimulating improved water storage on farms and by establishing centralised retting factories.

There are no other export crops yet comparable with these four main ones, but there are others which show signs of rapid growth (e.g. mung beans, castor beans, and cassava).

Sugar cane, groundnuts, soya beans, and sesame are grown for local consumption and for export. Sugar cane, which was a nineteenth-century export, was produced almost wholly for home consumption during the twentieth century, mainly in Chonburi. Since 1958 increasing quantities have been exported, and until 1966 exports were stimulated by a subsidy similar to that in Australia, derived from a tax on local sales; this has helped to raise exports to 84,000 tons in 1965. In 1966 the increasing cost of the subsidy caused the government to abandon the system, and exports declined to 55,000 tons.

Production of groundnuts is not localised; they are grown in the northern valleys, where the typical holding is only about two hectares, but where irrigation enables the planting of secondary crops in the rice fields. They are also grown in most other areas, wherever cultivation of secondary crops has developed. Under 20,000 tons—less than one-fifth of the crop—are exported (Department of Customs 1962: 234; 1963: 244). Soya beans are widely produced, but the chief concentration is in Sukhothai at the extreme north of the Central Plain, and in the northern valleys. The main concentration of sesame is in Phichit and Phitsanuloke, in the same general area as the maize and mung beans.

There are a few important specialised centres growing crops for local consumption. Cotton from Loei in the Northeast Region is manufactured within Thailand. Ratburi is an important centre for chillies and onions. Garlic—protected by an import duty—is grown both in Ratburi and in the northern valleys. Fruit-growing is also widely distributed in Thailand.

SPECIAL CHARACTERISTICS OF THE RURAL ECONOMY
RELEVANT TO DEVELOPMENT

A notable feature of the Thai rural economy is its flexibility. The anthropologists describe Thai rural society as 'loosely structured' (Embree 1950: 181-93). Villages, for example, are not closely-knit political entities exercising strong control over their members' activities (Wijeyewardene 1967). Although the sense of status is strongly developed, there is very free movement up or down the social ladder. Moreover, the attitude to land is much more commercial than in most peasant societies. The market is not well organised, with any clear institutional pattern of rates of return, professional agents, or formal valuation, but land is readily bought and sold.

Related to this is a considerable volume of local migration (Chapman and Allen 1965). In Thailand there is not a very marked flow into urban centres (Caldwell 1967: 44-9), and indeed, apart from Bangkok, there are no large towns. Bangkok's share in the total urban population has actually increased since World War II, mainly owing to a flow from the neighbouring Central Plain. Yet within the rural areas there is considerable movement: in the northeast it is from the more populous river valleys into less densely settled, drier areas; in the north it is from the more crowded to the less crowded valleys; it is also from the south to the north of the Central Plain, and from the northeast, both to the far north and to the north Central Plain.

There is also an obvious willingness to try new crops. The recently-developed crop reports for individual provinces show much greater variation in area planted for particular provinces than for the whole country, which reflects a great deal of experimentation. This is related to the fairly rapid development of the transport system and to the accompanying growth of new market opportunities.

Because the floodplain of the Chao Phaya River was easily able to produce a rice surplus, Bangkok rapidly became an international centre of the rice trade, shipping the surpluses partly to the expanding industrial market of Japan, and partly to the raw-material-producing colonial areas of Southeast Asia. The Chinese entrepreneurs of Bangkok developed a distribution network adapted to encouraging subsistence farmers to sell their surplus. The entrepreneurs who deal with subsistence farmers will devote their energies to promoting maximum development only where this promises greater returns than consolidating a position of monopoly based on their superior marketing skill. On the whole, conditions in Thailand favoured a system of supplying the limited consumption needs of subsistence farmers on credit, buying rice cheaply before the harvest, and building up a secure position (Narkswasdi 1963: 97-105). Consequently, the opening up and unification of Thailand created some of the conditions for a market economy, but kept in being a large reserve of rice-producing farmers selling their surplus mainly to meet obligations. It is probable that the adaptation of the distribution system to subsistence farming led to some expansion of rice-growing in areas such as the northeast, where encouragement of specialisation in other crops would have been more beneficial; the structure of government and enterprise was not adapted at this time to more radical transformation of the economy.

As part of the program of unifying and modernising the country, education had been spread from the palace outwards until, in the last decade before the 1932 revolution, a nation-wide primary education system was achieved. By the end of World War II at least a majority of Thai farmers could read and write. Moreover, a Thai nationalism, fostered by the last two absolute kings and by the politicians who succeeded them, generated some opposition to the network of Chinese enterprise. This did not, in the end, lead to the displacement of the Chinese entrepreneurs, but helped, after World War II, to create some of the conditions for more rapid growth (Evers and Silcock 1967).

GROWTH BEFORE WORLD WAR II

Hitherto, partly because of unreliability of the statistics and partly because rice was of such overwhelming importance, the growth of the Thai economy before World War II has been interpreted mainly in terms of the output of rice, supplemented by some consideration of teak, rubber, and tin. Available figures for rice, rubber, and teak are included in Table 4.5.

It may be seen from these figures that in the thirty years from 1915-19 to 1945-49 the area under rice doubled. This slightly exceeds the rate of growth of population. Since the accelerating population growth had increased the number of dependants, and since some other occupations, such as commerce and transport, had expanded relative to population, the amount of land cultivated per farm worker must have increased considerably.

Marketed rice production expanded in the outlying areas, with the growth in population and the development of transport (Ingram 1955: 45); for example, there was a considerable expansion in the north and northeast. However, in the north, rice is grown in irrigated mountain-valley holdings where the area per head is smaller than in the Central Plain. There must therefore have been a great deal of land taken up in hitherto unused parts of the Central Plain. What political, technical, or social forces drove Thai farmers to spread out into new and less productive areas without significantly increasing the intensity of cultivation of their best land?

One political factor was a marked tax inducement. New land was not merely made available free for the opening up, but farmers were taxed at a lower rate during the pioneer phase.[6] A favourable technical

[6] This tax incentive was introduced by King Mongkut in 1857.

TABLE 4.5

GROWTH IN THE RURAL SECTOR, 1910–14 TO 1945–9,
AVERAGE ANNUAL PLANTED AREA

('000 hectares)

Period	Rice	Tobacco	Maize	Cotton	Sesame	Pepper
1910–14	1,840·0	11·4[a]	6·4[a]	3·5[a]	n.a.	n.a.
1915–19	2,220·3	9·4	5·2	4·7	1·1	3·7
1920–24	2,606·7	9·6	6·3	4·9	1·5	2·4
1925–29	2,888·8	9·4	7·7	3·6	1·6	1·6
1930–34	3,213·3	11·1	6·8	3·3	1·1	1·6
1935–39	3,395·4	10·0	7·8	6·2	1·1	0·9
1940–44	4,146·0	14·6	11·7	27·6	4·6	0·4
1945–49	4,645·9	20·7	20·6	35·4	8·6	2·2

Period	Rubber Exports[b] (metric tons)		Production ('000 metric tons)	Teak Exports[b] ('000 metric tons)	
	Year Book	McFadyean	Mahapol	Year Book	Ingram
1910–14		66			90·0
1915–19		51			68·0
1920–24		1,193			83·0
1925–29	7,299	5,022		91·8	92·0
1930–34	9,216	7,620		67·3	67·0
1935–39	38,557	36,942	173·1	84·8	85·0
1940–44	19,521		84·6	27·0	27·0
1945–49	45,208		134·3	34·0	55·3[c]

[a] 1911–12 only.
[b] Annual averages.
[c] 1947–49 only.

Source: National Statistical Office 1965b; McFadyean 1944: T. 3; Mahapol 1954:
App. IV; Ingram 1955: T. VI, T. X.

factor may have been the steel ploughshare, which was probably introduced and adopted during the late nineteenth and early twentieth
centuries. This would have tended to increase the area that one man
could cultivate.

A social factor that almost certainly operated was the growth of
indebtedness to middlemen, and an increase in the share of the proceeds of the rice trade going to the Chinese entrepreneurs who controlled the collection of rice for export (Skinner 1957: 104-5). The
Thai social and political system in the latter years of the absolute
monarchy was not equipped to resist monopoly, and one of the
advantages of the post-war movement out of rice has been the substitution of entrepreneurs with more financial interest in development

for those who aimed to secure control over a section of a mainly subsistence economy.

On the whole there was insufficient growth in the rice sector of the Thai economy to contribute significantly to development up to World War II. The one period of rapid growth was immediately after the development of the main railway system, in the second decade of the twentieth century. It may be significant that at this time the new Chinese entrepreneurs were just building up their position, and were keenly competitive.

Rubber, by contrast, grew at a relatively rapid rate (McFadyean 1944: 226-9). There is great uncertainty about the figures even of exports, because most rubber went to Malaya for processing before World War II, some overland, and some by ship from Kentang to Penang. Inducements to smuggle varied from year to year, and customs control was very inadequate.

Although the detailed figures in Table 4.5 cannot be completely trusted, it is clear that exports expanded discontinuously, in sudden spurts. The year-by-year figures given by McFadyean (1944) show a rapid rise to a peak of 102 tons in 1912, followed by several years at a lower level. Between 1920 and 1925 exports rose from 84 tons to 5,400 tons, at which approximate level they remained for seven years. Then between 1932 and 1936 they rose to nearly 35,000 tons and were over 40,000 at the beginning of World War II.

These discontinuous expansions must have been partly the result of price stimuli occurring seven or more years earlier, but current conditions appear also to have had some effect. The expansion in the early twenties was also certainly a response to new planting during the period of very high prices in 1910, rather than during the war when tapping was delayed by shipping shortages. Similarly there was probably a good deal more tappable rubber than was utilised during the heavy slump of the early thirties.

Rubber brought discontinuous bursts of prosperity in the south of the country and contributed something to the expansion of government revenue before World War II, but it was not until after the war that it had any real effect on development.

Teak output and export showed relatively little development in the period up to World War II. The period of rapid expansion was from 1875 to 1910, when exports rose from an annual average of about 5,000 to over 120,000 cubic metres. At the end of this period, however, the forests were being heavily overcut and the recently-established Royal Forest Department renegotiated concessions with the foreign firms involved, and instituted a more conservative policy (Ingram 1955: 110-11; Mahapol 1954: 13). This provided for regeneration, but not for any significant development. During the period from 1910 to World War II exports remained fairly stable, in the neighbourhood of 85,000 cubic metres, except during World War I and the depression of the early thirties.

Output of the main crops thus did not expand rapidly enough to stimulate any significant growth before World War II. Such information as we have about the minor crops suggests that they, too, showed no significant increase until the period of World War II. Some growth was beginning in the years just before the war; for example in five years the area under maize and sesame doubled and the area under cotton increased threefold.

The comparative failure of these other crops to expand before World War II is in marked contrast to the changes that occurred during and after the war.

TRENDS IN CROP YIELDS

Ingram (1955: 48-9) has shown that the yield of rice up to World War II declined almost continuously. There were periods when the decline was temporarily reversed, but the trend was unmistakable. After World War II this trend continued for rice almost up to 1960. It is too early to decide how much of the reversal of this trend in the last few years is due to chance factors, but undoubtedly it has been reversed.

Several factors have contributed. Improved strains have been developed and are being more readily accepted. Mechanised ploughing has spread over much of the Central Plain and some other areas, mainly using hired tractors. Irrigation has probably cut down losses due to flooding or drought in the main irrigated areas since the construction of the Chainat dam and the more recent improvement in distribution channels. The adverse factors, however, still continue; population growth is still causing an extension of the margin of cultivation, and the artificially low price of rice discourages modern investments.

The other crops which have expanded show a clear contrast. Here, in spite of increases in the area under cultivation proportionately far greater than that in the area under rice, yields per hectare have in almost every case increased strikingly since World War II. Maize yields have risen from an average of 800 kg per hectare in 1947-9 to 1,918 kg in 1963-5, kenaf from 1,231 in 1950-2 to 1,406 in 1963-5, mung beans from 663 in 1947-9 to 1,113 in 1963-5, and castor beans from 781 in 1947-9 to 1,056 in 1963-5 (Ministry of Agriculture 1955-(1965: 54, 56, 65, 84)).

People producing a new crop for sale seem, on the whole, more inclined to experiment with new methods than those who continue to grow the traditional crops. This may not, of course, signify causation. Those who would have adopted new methods as they became profitable may well be those who are most ready to move from rice to another crop when one is available. Yet at least the prevailing atmosphere of optimism which this fact generates among the growers of new crops tends to be self-reinforcing.

RELATIVE MOVEMENT OUT OF RICE SINCE 1940

There has been a relative transfer of land, and probably a slightly smaller transfer of labour, from rice to other crops since the beginning of World War II, a trend which was initiated by the wartime closure of the rice trade. It was continued by the post-war rice monopoly and rice premium. A comparison of the averages for the years 1951-3 and 1962-4 shows this shift from rice to other crops. The population during this period increased by 38·6 per cent but rice land only by 13 per cent (National Statistical Office 1965b: 41, 164). Rice comprised a declining proportion of the total planted area. In the first period it was 87 per cent; by 1962-4 it had fallen to 74 per cent.

CURRENT GROWTH RATES

It is interesting to compare the growth in real agricultural product for different parts of the rural sector with that of the economy as a whole (Table 4.6). All non-rice crops taken together grew twice as fast as the economy as a whole. Crop production as a whole, however, grew only about two-thirds as fast as the whole economy, because of the low rate of growth of rice. Indeed, if any earlier terminal three-year period had been chosen the effect would have been even more marked, for rice would then have shown no increase in yield.

It is clear that a small group of export crops has been responsible for a high proportion of the growth. These have not merely greatly increased in area under cultivation but also have shown much the highest rate of increase in yield. The most natural explanation is that they have responded to the combined stimulus of high international prices, a low internal cost of living, and the simultaneous opening up of new land by transport and irrigation. This explanation will, however, be further analysed later.

One notable feature is the large increase in the area under crops for local use and the comparatively small improvement in yield. In fact, any increase in yield of these crops represents a remarkable achievement. Highly capital-intensive methods have been developed for chillies and onions in the Ratburi area, and the spread of these crops to the north and elsewhere could be expected to lower overall

TABLE 4.6

GROWTH RATES FOR PRINCIPAL AGRICULTURAL CROPS,
1951–3 TO 1962–4

(compound, in percentages)

| | Planted area ('000 hectares) | | Average annual rates of growth | | |
	1951–3	1962–4	Planted area	Yield	Production
Rice	5,833	6,600	1·13	1·33	2·41
Other export crops[a]	151	857	17·10	4·13	21·97
Rubber[b]	271	425	4·17	1·90	6·11
Semi-export crops[c]	186	274	3·58	3·11	6·98
Crops for local use[d]	90	446	15·62	2·12	18·04
Kapok and coconut	120	268	7·61	n.a.	—
Tobacco	47	44	−0·46	1·67	1·20
All non-rice crops	865	2,315	9·36	2·97	12·63
All crops	6,698	8,916	2·63	1·59	4·31
Gross Domestic Product (1956 constant prices, billion baht)	33·3	64·0			6·12

[a] Includes maize, mung beans, cassava, castor beans, kenaf, jute, and ramie.

[b] Tappable area.

[c] Includes sugar cane, groundnuts, soybeans, and sesame.

[d] Includes cotton, chillies, onions, garlic, other vegetables, pineapple, water melons, bananas.

Source: Ministry of Agriculture 1955–(1964). For kapok the change from 1951–3 to 1962–4 was estimated on the basis of an extrapolation of the trend from 1957 to 1964. Chillies, onions, and garlic figures are available only from 1952, and the average of 1952–3 was substituted for the average of 1951–3. Cassava figures before 1956 are available for Chonburi province only; an estimate for the whole kingdom was made on the basis of subsequent figures for Chonburi and for the whole kingdom.

yields. However, the improved local market, generated partly by post-war industrialisation around Bangkok and partly by the development of new cash crops for export, has stimulated a great deal of local development, with both the government and private enterprise publishing handbooks on the cultivation of these crops, and with widespread adoption of pesticides, weed-killers, fertilisers, and seed selection.

The method of aggregating growth which is used here weights the different crops according to the area cultivated, not to the value of crop produced. This is unsatisfactory except as a rough measure, for the true significance of a crop's contribution depends on its value. However, it avoids one difficulty which affects all Thai agricultural statistics. The value of rice within Thailand is kept down by the rice taxes to a little over half its international value. This means that in measuring changes in Thai national income the relatively slow rate of growth in rice is given a low weight and the relatively fast growth of other crops and of industry a high weight, and this would probably give an unduly favourable picture of the growth of Thai national income.

MAJOR DETERMINANTS OF ECONOMIC DEVELOPMENT

Infrastructure Investment

It has already been shown that there were special policy reasons inhibiting expenditure on transport and irrigation before World War II, and that some two-thirds of the economic expenditure in the Thai budget is now devoted to these two objectives. We have seen that, within the general pattern of post-war growth, agriculture has made a significant contribution, but that it is the non-rice sector, and particularly the new crops (and sporadically rubber), which have dramatically increased yields, much more so than rice cultivation. It has been argued that the effect of the rice taxes in keeping rice prices low has partly influenced the direction of development; but we must now ask how far such development has been due to the investment in irrigation and transport.

The main irrigation area of Thailand is the Central Plain, where, as the analysis in the IBRD Report (1959: 38-44) shows, the long-run objective was a reliable supply of water for two crops of rice per year over a large part of this area. This was not, however, an objective that the Chainat dam alone could achieve. The dam can impound some surplus floodwater and release some water to relieve drought; it can also help to spread the natural inundation through widespread channels, reducing the annual rise and fall of the level. But it is not a vast storage dam that can eliminate either the annual inundation or even its unreliability. Van der Heide's vision sixty years ago was a higher Chainat dam (Ingram 1955: 82-5), but even this could have provided only limited double-cropping, and could not have completely eliminated the unreliability of the system.

To eliminate this unreliability, two conditions are essential: greatly improved maintenance of the irrigation channels, and deep storage dams in the northern mountains impounding water from the Chao Phaya's main tributaries, the Ping, the Yom, and the Nan. One of the mountain dams—the Phumiphon dam on the Ping River—has now been completed, but the project for improving the distribution channels is still under way. Double-cropping (except in the northern valleys) has actually been of very small importance in Thailand, even though it has received considerable publicity. In the northern valleys it has probably not yet reached $2\frac{1}{2}$ per cent of the crop and nowhere else has it reached 0·3 per cent (Ministry of Agriculture 1962: 48).

Irrigation improves the reliability of the water supply within the irrigated area, and is probably therefore partly responsible for the recent increase in yields. As yet, however, there is no evidence that it has diminished the proportion of the total area damaged by flood and drought, either in the kingdom as a whole or in the Central Plain. The years 1957-9 were bad years for flood damage and the average percentage of fields damaged has fallen since then, but not below the percentages recorded in the early post-war years (National Statistical Office 1965b: 164). The proportion of the total rice area irrigated has risen very slightly in recent years, but is still barely above one-quarter.

It seems very probable that transport investments have had a greater effect than irrigation. The growth of the road network has been remarkable in the post-war period. Up to World War II there were no real trunk roads, and long-distance transport was by rail, sea, or canal. Now every provincial capital except Maehongsorn in the far north and Narathiwat in the far south is served by either sealed or good gravel roads. Moreover, the road system is far less centralised on Bangkok than the rail system. New links have been or are being built linking the northeast with the southeast and the north. The last decade has also seen a great expansion of feeder roads, so that few villages are now unconnected with any road, and most roads carry bus services and regular goods transport.

It has proved very difficult to establish statistical relations between rural capital expenditures and any direct gain in productivity. The Royal Irrigation Department has recently attempted, but not published, some cost-benefit studies, but these have not been more than forecasts unchecked against subsequent performance. American interest in highway construction has led to some cost-benefit studies of highways, mainly in the north and the northeast (USOM/Thailand 1960). The figures produced by these studies do not demonstrate that the roads studied were worth while; but it is pointed out that other economic gains, which cannot be adequately quantified, almost certainly create a sufficient justification for constructing them. Indeed the impression of greatly stimulated activity as a result of road building can hardly be avoided by anyone who visits Thai rural areas.

Centres of Growth

One reason for attributing the agricultural expansion in Thailand since World War II mainly to transport, is the comparatively great concentration of the rapidly expanding export crops in areas where good modern transport developed early after the war. The two main centres of growth of the new crops have been the provinces around Nakhorn Sawan in the northern part of the Central Plain and the province of Chonburi in the southeast, with an important subsidiary growth centre in Nakhorn Ratsima in the northeast. Nakhorn Sawan has been the main growth centre for maize and mung beans. Chonburi has been the main growth centre for cassava and sugar. Nakhorn Ratsima has been a subsidiary growth centre for cassava and several minor crops.

The extent to which the growth of these crops has been localised is not usually appreciated in Thailand. From 1959 to 1964 the area under maize for the whole kingdom increased from 200,000 to 552,000 hectares. During the same period the maize area in the Central Plain rose from 104,000 hectares to 498,000 hectares, so that in the rest of the kingdom the area actually fell by some 44 per cent. Even within the Central Plain, the increase was heavily concentrated in the two provinces of Nakhorn Sawan and Lopburi, which together registered an increase of 289,000 hectares or 83 per cent of the increase for the whole kingdom (Ministry of Agriculture 1959-: Product Series 1 and 2; Ministry of Agriculture 1955-(1964: 54-5)).

For mung beans the position is similar. The total area under this crop rose from $7 \cdot 3$ to $17 \cdot 6$ thousand hectares, and in the Central Plain from $5 \cdot 9$ to $16 \cdot 1$ thousand hectares, between 1959 and 1964; but for Nakhorn Sawan province alone the increase was from $0 \cdot 7$ to $9 \cdot 2$ thousand, some 82 per cent of the total increase for the kingdom.

Nakhorn Sawan and Lopburi are not areas in which there was an abnormal degree of switching from rice to the new export crops. Both are areas in which the rice crop itself has expanded exceptionally rapidly over these five years. Nakhorn Sawan's output expanded 122 per cent, while Lopburi's rose by 89 per cent, compared with an increase of 42 per cent for the whole kingdom. Thus, although the rate of growth of maize and mung beans is much greater than that of rice, it is not a growth of new crops at the expense of rice, but rather a rapid new development of agriculture in previously unused or under-used land.

Fortunately, unusually detailed crop statistics are available in the Nakhorn Sawan Provincial Government *Annual Report* (1964: 133-208), which enable us to pinpoint even more closely where the growth occurred. It is clear from Table 4.7 that the two districts of Takhli and Phayuha Khiri, with just under one-quarter of the population and just over one-quarter of the total area of the province, planted two-thirds of the maize and three-quarters of the mung beans

TABLE 4.7

AREA UNDER MAIZE AND MUNG BEANS BY DISTRICT,
NAKHORN SAWAN PROVINCE

District	1960 population ('000)	Agricultural area, 1962–3 ('000 hectares)	Area under maize, 1964 ('000 hectares)	Area under mung beans, 1964 ('000 hectares)
Muang	119·8	38·8	11·6	0·7
Krok Phra	26·9	18·5	5·0	1·9
Chumsaeng	62·6	39·2	9·3	1·6
Nong Bua	39·3	29·6	1·3	0·3
Banphot Phisai	84·8	57·5	7·7	3·0
Takhli	104·8	65·6[a]	71·5[a]	27·7[a]
Tha Tako	57·3	48·0	2·8	0·4
King Phaisali	30·7	26·5	5·9	0·6
Phayuha Khiri	53·6	34·1	31·0[a]	12·5[a]
Lat Yao	69·9	42·8	11·3	2·4
Total	647·6	400·7	156·2	51·0

[a] The total area under agriculture in 1962–3 is based on *Census of Agriculture* figures. The area under maize and mung beans in 1964 is based on Ministry of Agriculture returns. It is probable that this fact in itself would lead to some inconsistency. However, it is by no means impossible that the area under maize and mung beans in these two districts could have at least doubled in a year and a half. This discrepancy is not greater than could be explained by mere lapse of time.

Source: Changwat-Amphoe Statistical Directory: 5; Nakhorn Sawan Provincial Government 1964: 133–208; National Statistical Office 1965a.

in the province. National and provincial output figures both show that output per hectare in Nakhorn Sawan is above average for the kingdom. Since total output of Nakhorn Sawan province was small in 1959, it appears probable that these two districts alone accounted for about one-third of the increase in maize output and half the increase in mung beans output of the whole kingdom.

These two districts are described in the provincial *Annual Report* as having very convenient transport both by land and water. They are traversed by both the main railway and the main road, which intersect in this area; the two districts are also located near the confluence of the main rivers, the levels of which have been stabilised by the Chainat dam, enabling regular water transport through the year. This is an area in which census figures, analysed by Chapman and Allen (1965), show considerable inward migration from the lower Central Plain area and from the northeast; and the *Annual Report* refers to much casual labour which is under-reported in the census.

It is noteworthy that there was a lag of a few years after the improvement of transport facilities before the main growth occurred. The improvement in the availability of freight cars on the railways occurred in the early fifties; slightly later came the improvement of the main road, associated with the building of the Chainat dam which was completed in 1956 (IBRD 1959: 38). Improvements in water transport followed the completion of the dam. The main expansion in output occurred after 1959.

The province of Chonburi has been the main centre for the production of cassava since World War II. Up to 1956 no figures are available for production in other provinces. The crop was originally introduced into south Thailand from Malacca, but there is no record by provinces before World War II. It seems likely that the factories processing cassava into tapioca flour were set up during the short-lived industrialisation period in the early part of the war. Most of the expansion between 1959 and 1964 occurred in Chonburi province, where the area under cassava increased by about 50 per cent from 41,000 to 60,000 hectares. The area in the rest of the kingdom, however, doubled from 22,000 to 45,000 hectares, though part of this may be due to under-recording in 1959, when records of cassava appear to have been incomplete. About half the increase in the rest of the kingdom occurred in Rayong (adjacent to Chonburi) and in Nakhorn Ratsima. Nakhorn Ratsima is the terminus of the American Friendship Highway, completed in 1958.

The area under sugar cane in the whole kingdom expanded between 1959 and 1964 (147,000 to 161,000 hectares). The area outside Chonburi province contracted by 7 per cent (116,000 to 108,000 hectares), while Chonburi's area expanded by over 70 per cent (31,000 to 53,000 hectares).

The reason for the concentration of growth of these crops is probably the processing facilities in Chonburi town, though this in turn may be attributed to the excellent transport. The fine highway to the southeast appears to have been built originally to open up the seaside resorts—patronised first by the royal family and later by the American advisers and the expanding Thai middle class.

It may be significant that all three of the main growth points are in neighbourhoods where there are substantial concentrations of Americans. There are large bases at Takhli and at Sattahip in Rayong province. Nakhorn Ratsima is not merely the terminus of the first American highway, but is itself another base.

This may, of course, be mere coincidence, but one possible connection may be in terms of capital and enterprise. If capital is the bottleneck, the expenditures of American servicemen on goods (and services) in the countryside may make funds available, or the circulation of money there may attract enterprising—mainly Chinese—merchants, who, once there, will tend to initiate further development. These explanations are not mutually exclusive: the lack of funds for development in rural areas may be largely due to unwillingness of capitalists to invest in remote areas, and American bases may both attract and help to create capitalists.

A quite different explanation may be that the presence of foreign officers in a neighbourhood may, by making it more difficult for graft and irregularities to go undetected, improve the quality of economic administration. Highway construction and transport services both offer attractive sources of corrupt income, if visiting Thai officials can be persuaded to turn a blind eye. This is more difficult, even in country areas, if the scandals become notorious to foreigners.

The Role of Government

The main contribution of government, as already indicated, is the provision of transport routes and irrigation. Railway improvement received international assistance in the immediate post-war years because of the world shortage of rice. Much international aid has also been attracted for road building. The original American program was probably mainly strategic in intention, but the economic impact of the Friendship Highway has stimulated other overseas aid, from Australia, Japan, West Germany, and elsewhere. The construction of roads is advantageous almost everywhere, because of the previous policy of holding back road construction. Similarly, irrigation is in most cases largely implementation of schemes long known to be beneficial but held back for political reasons.

The effort in irrigation has been considerable. The main Chainat dam has been built to control the flooding of the Central Plain. The Phumiphon dam, which is mainly a power project, has also been completed and helps to impound water for the Central Plain system. The Nam Phawng project in the northeast has been completed, as one of the tributary projects under the international Mekong scheme (United Nations/ECAFE 1964: 87-90; NEDB 1966: 37). The Klawng River scheme in the west is expected to provide additional irrigated land in the Central Plain, and also to improve the already very productive fruit and vegetable area in the southwest. On the whole, however, irrigation seems so far to have had relatively less impact than transport. There have been serious delays in completing the distribution systems, especially in the Central Plain area. It is rightly felt that because of the heavy taxation of rice, the rice-growers are already paying more than the cost of the irrigation, and for this reason no charge is made for irrigation water. However, if no charge is made, the problem remains of unequal benefits to farmers from irrigation water. This hampers co-operation where farmers are expected to maintain the facility. It is particularly evident in the small tank projects in the northeast (Phanupongse 1964).

In time the benefits, forecast in terms of engineering criteria, will probably be translated into economic realities. This will, however, require more detailed local co-operation between the Royal Irrigation Department and local administration than is at present evident. Yet although not all of the promised benefits have been achieved, irrigation has undoubtedly contributed to the increased yields of recent years.

Foreign aid has also enabled the government to improve very greatly its research and extension work. Improved strains of rice and maize have been developed, and there is widespread interest in using them. Perhaps the most important change has been the adoption of the hard Guatemala variety of maize. In the important province of Nakhorn Sawan, for example, an *ad hoc* inquiry in 1958 showed that all farmers interviewed grew the traditional soft variety for human consumption (Chuchart *et al.* 1962: 18). In 1964 the provincial *Annual Report* (Nakhorn Sawan Provincial Government 1964: 138-208) showed that all farmers in the province had switched to the hard variety.

The impact of efforts to improve the breeds of cattle and pigs is evident in the rural areas, though marketing difficulties, owing to political factors, prevent much of the benefit reaching the farmer.

There is considerable local evidence that, in the last five years, Thai farmers have become noticeably more willing to experiment with scientific methods as well as with new crops. This is due not only to the cumulative effect of government efforts but also to changes introduced by private entrepreneurs.

Since the war it has no longer been profitable to consolidate a position as patron and financier to a group of subsistence farmers, buying their surplus rice, selling them imports, and maintaining control. Farmers are now mostly literate; rice is relatively unprofitable. Profits can, however, be made by encouraging improved methods, fostering new crops, and taking a share of the increased productivity. Merchants in the country centres have become active in distributing new varieties, particularly of vegetables, selling irrigation pumps and insecticides, and hiring out tractors.

The marketing studies initiated by Kasetsart University, and now carried out also by the Ministry of Agriculture itself, probably have made it more difficult for new monopoly positions to emerge. The diversification of agriculture away from rice has also probably helped to maintain a fluid and competitive situation.

THE NEED FOR AGRICULTURAL GROWTH

Demographic Pressure

The relative concentration of Thai development effort on the agricultural sector is partly a cause and partly a consequence of the relatively slow urbanisation of the Thai population. On the whole, however, the explanation of the tendency of Thais to remain in the rural areas is probably social rather than economic, arising from such factors as the barrier of the Chinese lower middle class, the centralisation of the Thai economy on Bangkok under the absolute monarchy, and the fact that education was indigenous and not colonial in character. These factors create population pressure in rural areas themselves, and the result has been large-scale migration to new land.

Though good land is still available, it is not unlimited; and the decline of the teak industry, due to overcutting in the post-war period, combined with fears about the effects of deforestation on Thailand's climate and water table, led the government in 1956 to begin extensive forest reservation and to lay down a policy of limiting agriculture to half the land area. This has focused increasing attention on raising yields.

The area cultivated per head seems to have remained fairly constant. If the rate of expansion of the total area cultivated is to fall, the area per head in the rural areas will then depend on trends in urbanisation. This in turn partly depends on industrialisation policy. At present Thai industrialisation—at least in the sectors promoted by the government—is very capital-intensive. Since the rate of saving of the Thai middle class is not high, and the government has set limits on the proportion of foreign borrowing, it will be difficult, even in the future, to provide employment for more than a small influx from the country to the towns. If existing methods of industrialisation are used capital will just not be available to absorb many into industry.

Population Policy and Growth

Thailand is thus faced with the prospect of a rapid growth of the rural population at a time when it is concerned with restricting the opening up of new land. It may be expected that in time this will lead to a population policy of encouraging the growth of family planning (Caldwell 1967). However, this is not the basis of present policy, as the army group which controls the government is (for mistaken military reasons) lukewarm about family planning. But even a rapid change of population policy would not affect the immediate pressures.

It appears that at present the situation is partly being met by migration and increased specialisation. A good rate of growth of rural incomes is not merely a consequence of this increased specialisation. It helps to maintain flexibility in the economy and so encourage more specialisation; for there is less resistance to change when new markets are developing.

POSSIBLE DIRECTION OF GROWTH

External Diversification

The post-war period has seen the development of several 'miracle crops', which have greatly diversified Thailand's rural economy. The expansion of rubber may have come to an end. The long delay in introducing a replanting scheme, and the diversion of some of the replanting fund's resources to repairing hurricane damage before the scheme was properly under way, prevented effective replanting during the period of relatively high prices for natural rubber, up to 1966. Maize, however, has continued to expand, and kenaf appears to have overcome setbacks which followed the rapid expansion in 1960-1. A further spread of maize into other parts of the north Central Plain can be expected as Japan's demand for animal feed expands. Later, Thailand may expand its own livestock industry, if the present political obstacles can be overcome. Kenaf takes more out of the soil, and there are dangers in planting it further and further afield on new land in the northeast. Though proper application of fertiliser can prevent soil exhaustion, this is a risky adventure for farmers near the margin of subsistence, in a region where rainfall is very uncertain. Kenaf is likely to expand further, but new crops will need to be sought in the long run, for kenaf can hardly be expected to compete indefinitely with synthetic packing materials. Sorghum is at present being tried in some northeastern areas.

Other probable growing points in the export trade are mung beans, cassava, castor beans, and perhaps groundnuts. Thailand's exporters are keen to develop new markets, but quality control is likely to be a problem. Machinery for control exists, but it easily leads to monopoly when effectively enforced. The close relations existing between politicians and the leading exporters persuade Thai economists generally to favour as little interference with the export trades as possible.

Internal Specialisation

While the recent period of rapidly developing transport has been a period of increasing concentration of the main export crops, a different trend is evident in the crops produced mainly for local consumption. In many areas improved transport facilities have led almost immediately to experiments with new crops, either for a market in a nearby town, or for national distribution. Instances already mentioned include the spread of chillies and onions from the southwest to the northern valleys; water melons are spreading from the neighbourhood of Bangkok further out into the Central Plain and the northeast; ground-nuts are also being planted in more and more areas (Ministry of Agriculture 1959-).

This development is partly due to improved transport; it may also, however, partly result from the increase of cash income from exports as diversification has proceeded. Rice cultivators, in the main, either grow kitchen garden crops for their own needs or exchange rice for a few supplementary foods. Their rice surplus is commonly sold in advance to meet credit needs for household consumption, so that little is available to sell for cash. The switch to export crops has probably meant that more cash is available for buying supplementary foods.

The pattern of planting of these crops for local use changes every year, but generally the proportion of the whole area planted in the three leading provinces has declined over the last few years for which figures are available.[7] Clearly there is a good deal of experimentation, but as yet the influence of specialisation is not sufficient to offset the two factors making for dispersion—increasing rural incomes and the increasing area which is accessible to the large Bangkok market.

In time, specialisation is likely to develop again for these home-consumed crops on a nation-wide basis. This can be expected to raise rural incomes further. At present a large number of new areas are being brought within reach of the national market, and this is the dominant influence.

Changes in Techniques

The most striking recent technical development is the spread of the use of insecticides, mainly for cultivation of vegetables.[8] Both private traders and government agencies play a part there, though the government's main concern is organising resistance to pests, such as locusts, grasshoppers, or rats, that are attacking a whole area. Usually those who take up vegetable growing use fertiliser, and there is some evidence that these are the farmers who are more willing than others to experiment with fertiliser on the rice fields.

There is a good deal of publicity for double-cropping of rice in areas where irrigation is available. The Ministry of the Interior is actively fostering such double-cropping through local government agencies. It is not as yet making a significant impact, and the Rice

[7] This is discussed more fully in Silcock (in press).

[8] Field investigation by the author in eight different provinces of Thailand in 1967.

Department argues that it diverts scarce resources of expertise from the main rice crop. At the same time rice uses relatively large amounts of irrigation water. An expanding number of farmers use private irrigation pumps to grow cash crops in the off-season, especially in the Central Plain, but these private irrigators generally grow another second crop in preference to rice.

The last decade has seen a great expansion in the import of fertiliser, from 20·6 thousand tons in 1955 to 88·9 thousand tons in 1965. It is not possible to trace how much of this represents government distribution, mainly to rice farmers through farmers' groups, irrigation co-operatives, etc., and how much is used on vegetables and the non-rice export crops.

Rural Mechanisation

Mainly during the last five years the use of tractors for ploughing has spread over a fairly large part of Thailand. The figures in the *Census of Agriculture* of 1963 probably give a seriously inadequate picture of the degree of mechanisation, even at that time, because so many tractors are hired on a piece-work basis to undertake a particular piece of ploughing. In some instances relatively large tractors are used for transport in the wet season in areas without roads. The Agricultural Department is also, however, now trying to promote the sale of small tractors to the wealthier farmers for use in orchards or for the cultivation of upland crops.

Irrigation pumps are widely used. A common variety is the wooden endless chain pump, adapted from the type developed in the nineteenth century by Chinese tin-miners in Malaya but driven by an engine instead of by water power. These pumps are manufactured in Thailand.

The most highly mechanised sector of Thai agriculture is undoubtedly the vegetable-growing area in Ratburi province. The vegetables are grown on raised banks between canals, dug out by earth-moving equipment and flooded by mechanical irrigation pumps. Mechanical equipment for fertilising, watering, and spraying the crops is mounted on small power-driven boats, so that virtually only the harvesting is done by hand.

The spread of mechanisation is likely to increase the area that can be profitably operated as a unit. It will also, however, strain the capital resources of the small farmer. In these conditions the increase in tenancy that has been noted among farmers in the Central Plain, and increasingly elsewhere, may be a natural and even economically useful response to new conditions, and not merely evidence of the submergence of the Thai farmer, as is often alleged on political grounds.

MEASURES TO PROMOTE GROWTH

Follow-up of Government Investment

It has already been explained that Thailand achieves its fairly
satisfactory rate of growth partly by the stimulus of diversification and
industrialisation provided by the rice taxes, and partly by expenditure
both of its own funds and of foreign aid funds on roads, irrigation,
and other public investment for the benefit of agriculture. The rice
taxes resulted not from any deliberate long-range policy but from a
sequence of intelligent adjustments to external pressures (Silcock
1968); rural investment was relatively straight-forward because a
number of obviously productive schemes had previously been held up
for political reasons.

For two reasons it is now becoming more important to scrutinise
carefully the machinery for assessing the effect of government investments (Silcock 1967b). One reason is that the relatively obvious
preliminary works have been undertaken. There is no lack of possible
schemes to achieve further improvement, but it is becoming much
more difficult to arrange them in order of priorities. The second reason
is that the rice taxes make it unwise to rely either on the working
of the price system or on highly sophisticated planning techniques,
without a good deal of prior analysis.

The Chainat and Phumiphon dams have been built, and once the
ditches and dykes scheme is completed, choices must be made among
a number of other projects, none of which is clearly essential to the
others. The main road system will soon be completed, making a
unified network covering the whole kingdom. Even this may then
involve choice between feeder roads and maintenance expenditures.

The basic techniques of cost-benefit analysis have been applied both
to roads and to irrigation, though so far without much success. It
seems important that, in any future schemes, close attention should be
given to the effects of the rice taxes. One method which has some
attractions is to adopt a uniform price for rice based on (though
not necessarily identical with) the international price, which could
be used in policy calculations related to output, subsidies, taxation, etc. Differences between the actual price and this national
price could be allowed for in all policies affecting the cost of production of rice or rice substitutes. This would enable a number of
schemes to be assessed not in terms of the gains to the participants
but of the gain to Thailand from the rice produced.

This suggests that research should be undertaken into the type of
change which is likely to be economically most effective. An incidental
advantage of such research would be the guidance it would give in
allocating the scarce resource of trained extension workers. At present

there appears to be a general attempt to allocate staff fairly uniformly, except for areas of acute political disturbance. Correct assessment of the time at which adequate staff should be provided, in relation to such facts as the arrival of a new road in a district, would enable change to be achieved more effectively.

Training and Research

The Thai government has, during the last five years, become more acutely aware of shortages of different grades of skilled manpower. This has led both to attempts to increase supplies by special training and inducements and to a policy of screening planned developments according to their manpower requirements.

It would seem worth while to consider whether the supply of research and extension workers might be improved by a system of training in which apprenticeship in field operations would play a much larger part in the initial selection and training, while the necessary academic knowledge would be achieved after the initial commitment to research had already been introduced (Silcock 1967a: 317-22).

Promotion of Particular Crops

Some of the special problems of particular crops that are receiving consideration are the following:

Rubber is unlikely to become competitive at probable future prices unless the greatly improved clones now available can be substituted for existing unselected rubber. It seems unlikely that Thai civil servants will at this stage be able to overcome the racial barriers and the suspicions based on past malpractice sufficiently to institute smallholder development schemes like those in Malaysia. Consideration is being given to initiating more development of the plantation type, and this may well be the only way in which low enough costs can be achieved without unacceptable reductions in earnings.

Maize exporters have from time to time suffered from the monopsonistic position achieved by Japanese buyers of Thai maize. Suggestions of formal cartel organisation among Thai exporters seem, on the whole, dangerous in Thai conditions, for they might well destroy the comparative flexibility of the Thai market. More research directed to overcoming the moisture problem, which prevents Thailand sending maize to more distant markets, seems a more promising type of defence.

Kenaf has been a most successful crop in providing supplementary incomes in the northeast, though long-term uncertainty of demand and possible exhaustion of the soil make it a doubtful prospect in the long run. One of the chief achievements of kenaf in the northeast

is its help in focusing attention on the importance of specialisation
on crops more suitable to the climatic conditions, rather than on trying
to devise irrigation systems to maintain it as a subsistence rice area.
Historical influences have inhibited trade and specialisation in this
area, but the kenaf boom, combined with studies by Kasetsart University (Chuchart *et al.* 1959a, 1959b, 1961, 1962), following a lead given
by the IBRD Report in 1959, have now concentrated more attention on
marketing, transport, and extension work with new crops in this area.

Land Policy

Ideologically Thailand is strongly committed to the pattern of the
independent landowning Thai farmer, with ample land for all who
need it and who are prepared to open it up.[9] In some areas, such as
the north and northeast, this may still be appropriate policy. It is
doubtful, however, whether it is an economically sound policy in
either the Central Plain or the south where there has been a growth
of tenancy. This is not because tenancy is at present harmless in these
areas. Wherever the farmer is a tenant there is always a danger that
the owner will have less knowledge of, and less interest in, practices that
improve the land than would the farmer himself. The landlord may
have acquired the land as security for a loan, at a bargain price, and
may use it simply to dominate his former tenant, ensuring permanent
control of his surplus but undermining any incentive to produce a
larger surplus. There is little doubt that this does sometimes happen
in the Central Plain, though this is less clear in the south.

There are, however, disadvantages in ownership also. The present
farmer is likely to invest savings in extra land, and to be short of
working capital, and it takes a long time for an efficient farmer to
get as much land as he can profitably work, or for an inefficient one to
lose the land that he is wasting. In the context of development, when
land is scarce, there are advantages in tenant-farming if the owners
are interested in development and in obtaining good farmers as
tenants. Which method works best in any given context depends on
several different factors. There are arguments based on incentive,
and on the difficulty of distinguishing current costs from investments
in good farming practice, which favour ownership; but arguments
arising from capital shortage, from the need to develop fairly large
areas at once, and from the need for rapid adaptation to change,
tend to favour tenancy. It is by no means obvious that the small
landowning farmer is the best agent of development in all parts of
Thailand, even though he probably is in some parts.

9 In 1963, 85 per cent of the arable land area was farmed by owners (National
Statistical Office 1965b: 172).

In the south, as we have seen, the technical needs of the rubber industry probably require either group settlement schemes or plantation development. In the Central Plain it is doubtful whether it is economically wise merely to resist the spread of tenancy. Resistance on principle makes it more difficult to devise suitable institutions for a real estate market which could ensure that farmers obtained a reasonable price for their land. Such institutions could, in addition, open up the possibility of large-scale land development, perhaps by subsidiaries of banks. Rising land values, combined with mechanical aids which increase the area that can be profitably farmed, suggest that it may be more efficient to provide a framework in which the farmer of proved skill can obtain control of more land, and also of farm machinery, on a rental basis, while the less efficient ones can sell at a fair price and perhaps move to areas where skill commands a lower premium.

CONCLUSION

Most of the above suggestions affect relatively minor features of the economy. In spite of a situation in which there is much corruption in the government and an industrialisation policy which has obvious faults, Thailand has, during the last decade, succeeded in achieving a rapid rate of growth in its agriculture.

Part of this is certainly due to foreign aid. Aid has been important chiefly in limiting maladministration and introducing technical skills. Thailand's own foreign exchange reserves are approximately equal in value to the total aid that it has received, and it has therefore benefited more from techniques and skills than from the financial effects of aid.

It is also worthy of note that much of the initiative for improvements, not only in technique, but in administration, has come from Thai civil servants who have used foreign aid to limit the harmful effects of corruption and inefficiency in their own government.

F

BURMA

BURMA PROPER:

UPPER: Magwe, Sagaing, Mandalay

LOWER: Arakan, Tenasserim, Irrawaddy, Pegu

STATES: KACHIN
KAYAH
KAWTHOOLEI (KAREN)
SHAN

CHIN SPECIAL DIVISION

——————— PRINCIPAL ROAD
– – – – – PRINCIPAL RAILWAY
·············· ADMINISTRATIVE BOUNDARY

0   50   100   150   200
MILES

DEPARTMENT OF HUMAN GEOGRAPHY, A.N.U.

Map 6   Burma

The Union of Burma

H. V. Richter

DEVELOPMENT and modernisation of the economy have not thus far been such urgent problems for Burma as for most other poor countries of the world. Geographic, social, and economic factors have combined to afford it time to adjust to changing world conditions and to fit changes in its economic structure into historic patterns. If independence and self-sufficiency were preferred to development, then near stagnation could be tolerated—and can indeed still be tolerated for many years to come in the outlying frontier regions where many of the minority peoples of the Union live.

Three main factors seem to account for this pause before pressure for development has become strong or widespread. First, the country possesses an exceptionally favourable land-people ratio, which, with its ample water resources, has allowed it to accommodate a hitherto slowly-growing population by expanding farm production in traditional ways. This still provides Burma's 26 million people with export surpluses of foodstuffs and farm sizes larger than elsewhere in Southeast Asia. Second, there is the historic isolation of the country behind the great mountain ranges which form its frontiers, and also the traditionalism of its village life, which have rendered Burma less subject than more open societies to the restlessness engendered by rising expectations. Third, there is the nature of development over the past century, which has demanded from the indigenous peoples an expansion of traditional activities, rather than the undertaking of new ones.

The period of British rule introduced radical changes to Burma, in stimulating a breakaway from a near-subsistence to a mainly monetary economy. But foreign capitalism played a narrower role here than in some other colonial territories. It largely by-passed estate-type production of industrial crops to concentrate on the extractive industries, trade, and banking. The expansion in agricultural surpluses which it stimulated came from smallholder production of rice and other foodstuffs, so that the cultivators' break with subsistence agriculture was not complete.

Not only has the need and the desire for economic growth been less strong in Burma than in many other countries of the region, but such development as has taken place since independence has been subject to two types of political constraints: those created by the hostility felt towards foreign elements within the community and to foreign influence generally, and those imposed by the discord between the indigenous races of the Union. Strong nationalist attitudes, mainly the inheritance of colonial rule but also in part a reversion to the isolationism of former times, have limited the contribution to economic growth which could be accepted from non-indigenous Burmese residents or from foreign countries. Nationalist policies have intensified over the years. Only very limited foreign aid has been sought; foreign debts do not compromise the country's independence; and its currency is relatively strong. Minimum use is made of the scarce skills or foreign trade links of expatriate citizens.

The recurrence of the historic discord between the races of Burma has been a major hindrance to development. Many of the minority races living in the states and hill regions prefer to opt for independence from the central government even if this entails lower living standards. Insurrection and dacoity have diverted scarce resources to defence, destroyed capital equipment, inhibited investment in mining in the hill areas and to a lesser extent in forestry, and have reduced economic activity generally outside the settled districts.

Subject to these constraints there is, however, a growing pressure for development, especially in Burma Proper, where 83 per cent of the population of the country live. This arises principally, as elsewhere in Asia, from the recent decline in mortality rates which has upset the population/resource balance and which is already causing population pressure on the limited new land which can be brought into production in Burma Proper by present techniques and expenditures. But the demand for development and diversification of the economy is also social, created by expansion of the educated middle classes. While it is clear that change must come if living standards are not to fall, there is as yet no consensus about the direction change should take.

When Burma achieved independence in 1948 it endorsed enthusiastically the principle of developing the economy on socialist lines. The ruling group, springing from the Dohbama Asiayone (Thakin Organisation) of the 1930s, mostly envisaged agricultural development in a Robert Owen type of setting, carried through by prosperous individual smallholders working together on co-operative principles. A minority saw such co-operation as a stage on the way to the collective organisation of farming. But the policy-makers assigned no clear role to the agricultural sector. Experience under colonial rule had left behind an aversion not only to capitalism as such, but also to the notion that agriculture should continue to play the dominant part assigned to it by the British in indigenous economic development. While the right of the peasant to justice and higher living standards was stressed, so was the need to draw on the surplus of the farm sector to finance industrialisation and diversification of the economy. State investment which would assist the farmer was relatively neglected. This ambivalence of government policy towards rural development has characterised all post-war governments, parliamentary and revolutionary. It has affected particularly the rice sector, which has been the prime object both of state welfare policies designed to raise the level of the poorest of the cultivators, and of restrictive official price policies detrimental to profit margins for the paddy farmers and farm funds for investment.

Economic development in Burma, and attitudes in independent Burma towards the direction of change, have been profoundly affected by the colonial period. It is therefore proposed to outline here the main trends in the agricultural sector during this time. The colonial period has been well documented by Furnivall (1931, 1948), Harvey (1925, 1946), Andrus (1947), Christian (1942), Cady (1958), Tun Wai (1961), and others. The summary which follows is drawn largely from these sources.

AGRICULTURAL DEVELOPMENT IN COLONIAL BURMA

The British annexed Burma in three portions: Tenasserim and Arakan in 1824-6, the remainder of Lower Burma in 1852, and Upper Burma in 1885. Burma had been a largely self-sufficient economy both by virtue of its varied natural resources and because of the mercantilist policies of its rulers. Land was plentiful, subsistence farming was the primary occupation of the people, and commerce consisted chiefly of barter of the products of cottage industry and crafts. The population remained low because of recurrent warfare between its peoples and very high child mortality, and is thought to have been between four and five million at the beginning of the nineteenth century. At this time about three-quarters of the population lived in Upper Burma, only a sparse population and huge areas of waste cultivable land having been left in Lower Burma, following the fall of the kingdom of Pegu to that of Ava in 1757. Foreign trade was a monopoly of the King of Ava and exports of rice were normally forbidden, to maintain domestic stocks as insurance against famine.

Under the colonial impact the traditional and largely self-sufficient Burmese economy was transformed into a dualistic export-orientated economy. This was true not only in that enclaves of foreign controlled capital-intensive export sectors were superimposed on the traditional structure, but also in that what Furnivall terms an 'industrial agriculture' grew up in Lower Burma, where a highly commercialised indigenous rice sector developed to produce paddy mainly for export, which was handled by expatriate enterprise. In Upper Burma the direct colonial impact on the economic organisation of the rural sector was less strong, but here, too, there was increasing specialisation. The region, which had previously been self-sufficient in rice save in exceptionally bad seasons, gradually shifted to growing more cash crops, making up its rice deficit by imports from Lower Burma. This expansion and concentration of agricultural production was accompanied by a progressive decline in domestic cottage industries, many of which had been ancillary activities of farm families, which succumbed to competition from imported goods.

The British opened up Burma to large-scale foreign trade and established the conditions of law and order in which an expansion of production, both indigenous and alien, could take place. British, other European, and Indian companies concentrated on primary sectors not previously exploited on a significant scale, providing resources lacking in Burma for the establishment of capital-intensive sectors in forestry, petroleum, and mining, and for the marketing of their products overseas. Efforts were at first concentrated on the extraction, milling, and export of teak from the coastal areas controlled after the 1824 annexation. Following the take-over of the rest of Lower Burma in 1852 the procurement, milling, and export of rice assumed the major role. After 1930 it came to be rivalled in value by the production and export of petroleum from Upper Burma and of mineral ores and concentrates, especially of tungsten, tin, lead, zinc, and silver, from the Shan and Kayah states and other frontier regions.

The expansion of agricultural production was left largely to Burmese enterprise, with the exception of small rubber estates in Tenasserim. The aim of the British administrators was to help create a class of prosperous peasant proprietors. Land policy from the mid-1850s was designed to encourage the bringing into cultivation under rice of the huge area of swamp and jungle in the delta of the Irrawaddy-Sittang Rivers. This delta region, as well as the low-lying coastal areas of Arakan and Tenasserim, was composed mainly of alluvial soils ideally suited to rice cultivation owing to heavy and dependable rainfall and impermeable subsoils.

Initially only slow progress was made in efforts to encourage Burmese settlement in the delta. Policies designed to encourage migrants from India to take up smallholdings and to promote the development of estates by making land grants to European interests failed. No large-scale farms were established by Burmese or foreign

TABLE 5.1

GROWTH OF POPULATION AND RICE ACREAGE, 1856–1966

	Population (million)					Area sown to paddy (million hectares)					
Date	Lower Burma[a]	Annual increase (%)	Upper Burma[b]	Total	Annual increase (%)	Date	Lower Burma	Annual increase (%)	Upper Burma	Total	Annual increase (%)
1856[c]	1·46		3·60	5·06		1855	0·40				
1872[d]	2·59	3·65	n.a.			1875	0·96	4·47			
1891	4·41	2·84	3·31	7·72		1890	1·78	4·18	0·55	2·33	
1901	5·41	2·06	5·09	10·49		1900	2·66	4·11	0·80	3·46	4·04
1911	6·21	1·41	5·90	12·12	1·46	1910	3·16	1·73	0·87	4·03	1·53
1921	6·86	1·00	6·35	13·21	0·87	1920	3·48	0·96	0·71	4·19	0·38
1931	7·77	1·24	6·90	14·67	1·05	1930	4·01	1·45	1·00	5·01	1·81
1941	8·88	1·36	7·94	16·82	1·38	1940	4·02	nil	1·01	5·03	0·05
1954[e]	n.a.	n.a.	n.a.	20·04[e]	1·36	1954	n.a.	n.a.	n.a.	4·21	−1·08
1966[e]	13·32[e]	n.a.	11·93[e]	25·25[e]	1·94	1966	n.a.	n.a.	n.a.	5·01	1·47

[a] Includes Kawthoolei.

[b] Including states, except Kawthoolei.

[c] Rough estimate; source: Tun Wai 1961: 32.

[d] 1872–1941 Census data. Area of census progressively extended, mainly up to 1901.

[e] Official estimate, based on partial censuses.

Source: (To 1931) Andrus 1947: 23, 43; (1941 to 1966) CSED 1952–(1967); UBRC 1964–(1966–7).

enterprise. Burmese smallholder settlement, however, proceeded gradually. Settlers usually worked for a time for others, then used their savings to clear virgin land, for which the British type of alienable land titles were commonly granted if land revenue was regularly paid. The pioneers, accustomed to fairly extensive farming practices based on draft animal power rather than on manual labour alone, normally reclaimed individual plots as large as could be worked with one or two pairs of bullocks or water buffaloes (i.e. between four and eight hectares). They often borrowed money to purchase cattle and to pay for labour hired to help with the reclamation of land and for seasonal work. Traditional Upper Burma wet-rice-growing techniques of sowing, transplanting, reaping, and threshing were used, although these techniques were often less carefully applied, since farm sizes were larger than in Upper Burma and water control necessarily less sophisticated because of lack of storage other than bunds.

A continuous extension of smallholder paddy holdings, mainly but not wholly in Lower Burma, was induced by the higher rice prices which followed the opening of the Suez Canal in 1869, improvements in steam shipping, and the entry of Burmese rice into European markets. The area sown to paddy throughout Burma (Table 5.1), which had been about 400,000 hectares in 1855, rose to 1·6 million by 1886 and to nearly four million by 1910 (Fisher 1964: 436). At the height of the land boom between 1890 and 1900 over 100,000 hectares of new land were brought into cultivation annually, mostly by Burmans[1] and Karens moving down into the delta of the Irrawaddy and the valleys of the Sittang and Salween Rivers from Upper Burma and the hill districts of Lower Burma. The production and export of rice rose roughly in proportion.

By the beginning of the twentieth century foreign trade was five to six times larger than in the late 1860s. Rice alone brought in over two-thirds of export earnings; other products of indigenous enterprise, such as raw cotton, dyestuffs, precious stones, hides and skins, contributed a further 7 per cent (Furnivall 1948: 551). From 1900 to the mid-1920s the value of exports of rice and other rural products rose steadily and of the products of Western enterprise more sharply, so that Burma became one of the most export-orientated of the world's economies.

But a decline in rice prices (Table 5.2) began in 1927 and accelerated in the world depression of the 1930s. This depression affected rural exports more than the products of Western enterprise.

By the 1930s the composition of foreign trade had been modified, both by the growth of Western enterprise and by a decline in the

[1] Throughout this chapter the common practice is followed of using the term Burman to designate race and Burmese to denote nationality.

unit price of rural exports. Rice remained the principal export item, contributing 47 per cent of the value of exports between 1937 and 1941 and other exports of indigenous enterprise 5 per cent, but 45 per cent of export earnings in this period were brought in by products of Western enterprises (petroleum and products 26 per cent, minerals and ores 11 per cent, timber and rubber 8 per cent).

TABLE 5.2

RICE PRICES, RANGOON

(Rs per metric ton)[a]

	Paddy		Boat paddy
1855	21·6	1927	87·2
1865	24·0	1928	81·0
1875	31·2	1929	76·2
1885	45·5	1930	66·1
1905	50·3	1931	37·4
1915	59·9	1932	47·9
1920	91·1	1933	31·6
1925	89·6	1934	26·4
1930	63·7	1935	39·8
1935	45·1		
1939	47·9		

[a] Data converted from baskets of 46 lb.

Source: University of Rangoon n.d. (c. 1959): 22; Tun Wai 1961: 102.

The striking expansion of commercial activity in the colonial period was accompanied by three serious problems for the Burmese. Ownership and control of the commercial sector of the economy passed increasingly into non-indigenous hands; the newly established primary sectors created little secondary or tertiary activity and few opportunities for employment or for learning new skills; and the cash crop sector rapidly became highly competitive, leading to small-holder indebtedness, especially in the monocultural rice districts of the delta.

Because of their capital-intensive nature, control of forestry, oil, and mining enterprises was from the beginning in expatriate hands. But the British not only introduced foreign capital into Burma; they also introduced foreign labour, which, because of its greater experi-ence (or greater assiduity in regular employment), soon dominated the administration, banking, mining, milling, wholesale, and foreign trade. Moreover, the colonial authorities, faced with a shortage of unskilled local labour, encouraged the immigration of Indian and to a lesser extent of Chinese seasonal workers. By the mid-1920s some 400,000 Indian seasonal workers came into Burma annually. Some of these stayed permanently to find occupations as casual labourers on the docks, railways, public works, mills, and farms. Thus between the two world wars the big enterprises, petroleum, teak extraction and milling, mining, railways, Irrawaddy shipping, the port rice mills, and the big import-export houses were controlled principally by British and other Europeans, while Chinese and Indian traders took a large share of medium- and small-scale enterprises in foreign and retail trade and provided many services. Finance and credit for agriculture was provided mainly by Chettyars, a caste of Indian money-lenders. Although the proportion of aliens in the total population never exceeded 10 per cent, their share in commercial activity may be gauged by the fact that in 1931 they formed two-thirds of the population of large towns and one-quarter of that of small towns (1931 Census, quoted in Furnivall 1948: 118). During the 1930s some 6 per cent of gross domestic product was expended annually in remittances abroad of company profits and migrants' savings.

The Burmans, and especially other indigenous races, remained overwhelmingly in rural areas and largely in agricultural occupations. In 1931 only 6·6 per cent of the indigenous population of 13·4 million lived in towns; 66 per cent of the total labour force worked on farms, and a further 3 per cent was engaged in forestry, fishing, hunting, or animal husbandry. Two-thirds of women workers were classified as agricultural, the remainder being engaged mainly in cottage industry and petty trading. Only about one-fifth of Burmese males were engaged in non-agricultural occupations, principally unskilled labour, trading, crafts, and clerical work. Burmese male employment in mining or in manufacturing industry was not significant outside rice-milling.

At the same time the official British policy of fostering Burmese peasant proprietorship was failing lamentably. The rapid extension of cultivation which had taken place had been founded on credit, which was made freely available to smallholders by Chettyars and other money-lenders after 1880. Such credit was granted, either to meet the cost of reclamation and purchase of land or to meet seasonal cultivation expenses, mainly against the security of the British-introduced land titles, which were negotiable instruments, and to a lesser extent against promissory notes or the advance sale of crops. The amount of agricultural credit granted rose steadily, so that by the late 1920s it roughly equalled the harvest value of all crops (Aye Hlaing 1958: 16). The Chettyars had at this time some Rs300 million out in longer-term agricultural loans and Rs200 million in seasonal loans, at interest rates ranging from 15 to 36 per cent per annum according to the security offered. Other money-lenders, Burmese, Chinese, and Indian, lent some Rs50-100 million annually in short-term loans, often made and repaid in kind, generally on poor security and at very much higher interest rates.[2] Such interest burdens were impossible for the smallholder to carry if a large proportion of assets was mortgaged. Any failure of production, loss of cattle, or fall in produce prices put him further into debt. This usually led to foreclosure of mortgage for the owner-cultivator and to the loss of his holding for the tenant-cultivator.

The 1931 census data showed that the proportion of the agricultural work force who were owner-cultivators had declined from 51 to 37 per cent during the previous decade while the share who were agricultural labourers had risen from 27 to 40 per cent. An official banking inquiry in 1930 estimated that only about 14 per cent of cultivators throughout the country were free of debt, owned their own cattle, and had sufficient paddy to last until the next harvest, while one-third of all cultivators had debts equal to more than half of the value of their land (Andrus 1947: 306).

Conditions deteriorated in the great depression of the 1930s, when rice export prices dropped to less than half of the record levels reached in the mid-1920s, land prices fell, and land alienation accelerated. Although export prices of other rural products fell about equally during this period, the rice sector was naturally most strongly affected, since a greater proportion of its output was marketed. By 1939 nearly half of the farmed area of Lower Burma—some 2.2 million hectares—had passed into the hands of non-agriculturalist owners and most of the rest was mortgaged.[3] The tenancy problem was most acute in Lower Burma, where, in 1938-9, 59 per cent of farm land was under tenancy, almost all of which was let on a fixed rent basis (either cash or paddy). In Upper Burma less than one-third of the cultivated area was under tenancy at this date, well over half the rented area was owned by other farmers, and three-quarters of rented land was let on a crop share basis.

The absentee owners, whether Chettyar or Burmese, had little interest in the productivity of the land, since their rents were fixed in cash or kind, not levied as a share of the crop. A bidding up of

[2] The rupee, introduced from British India, retained a constant official relationship to sterling of 13.33 (1s 6d) to November 1967. The currency was renamed the kyat in 1952, and its par value with the International Monetary Fund has since 1953 been K4.76 to the U.S. dollar.

[3] By 1939 the Chettyars had unwillingly accumulated under foreclosed mortgage over one million hectares of the best rice land in Lower Burma which they were unable to resell. These unrealisable assets reduced Chettyar loan capital, so that they were able to lend out only about Rs100 million in agricultural credit by the end of the 1930s.

rents developed, and with it a great instability of tenure, so that by the end of the 1930s nearly half of Lower Burma tenants changed holdings every year (Andrus 1947: 70-1). This, like many other agrarian conditions of this time, was not only socially undesirable, but also inimical to sound cultivation.

The problems of rural indebtedness, land alienation, high rents, and insecurity of tenure had long been recognised by the British administrators, conscious too late that many of these problems had been created by the changes that they had introduced by encouraging a too-rapid shift from subsistence to cash farming and substituting Western for traditional law. From the 1890s a series of ameliorative measures was attempted, but these were either abortive, because they were opposed by the largely expatriate merchant communities, or entirely inadequate in scale.[4]

The legacy of the British period to the agricultural sector (apart from its stimulus to expansion) was not entirely negative. Some useful public works were undertaken, principally the strengthening and enlargement of the embankments of the Irrawaddy and its tributaries at strategic points to provide flood protection for 540,000 hectares in the delta by 1940, the cutting of a canal to link the Pegu and Sittang Rivers, the construction of a 3,300-kilometre railway system to carry some 4 million tons of freight annually,[5] and the extension of the irrigated area in the dry zone of central Burma to 630,000 hectares by 1941 (Andrus 1947: 57, 66). Private foreign enterprise invested heavily in Irrawaddy shipping and in large steam mills and installations at the ports, while considerable indigenous capital was invested in smaller rice mills in the interior. Some new crops were introduced, mainly into Upper Burma, notably groundnuts, sugar cane, and an additional range of beans. State services to the farm sector, such as the inoculation of cattle, model farms, the provision of quality seed, improved tools, etc., were efficient, although (except for the veterinary services) on a small scale. The agricultural and forestry departments and state farms set up by the British formed the basis of the post-war departments and state extension services. At no time, however, was much emphasis put by the colonial authorities on agricultural science or education. No large-scale basic research program for rice was developed comparable to those carried out for rubber in colonial Malaya or for raw cotton in Egypt.

Economic activity was disrupted by World War II. Great physical damage was inflicted by the scorched earth tactics of successive cam-

4 An outline of the agrarian legislation drafted or passed between 1891 and 1941 will be found in Aye Maung (1954: 2-3).

5 A large share of this was mineral ores. The waterways, carrying a volume of freight almost equal to the railways, were the principal carriers of farm products.

paigners, especially to petroleum and mining installations and to river and railway rolling stock. It has been estimated (MNP 1951a-(1951: 2)) that, of the national capital excluding land, valued at Rs3,500 million at pre-war prices, approximately half was lost by 1946. Claims for compensation for the destruction of buildings, railway and mining capital were later put at Rs1,710 million ($US517 million). With conditions insecure and foreign markets cut off during the war, rice, cotton, and groundnut production declined sharply, much land fell into disuse, and cattle numbers were depleted. To the smallholder, however, one favourable aspect of the war was that most Chettyar money-lenders fled back to India and that collection of all interest, rent, and land revenue was so lax as no longer to form a significant burden. By 1946 agricultural production was 59 per cent of the pre-war level, timber output was less than half that of 1938-9, and production of minerals was negligible (MNP 1951a-(1951: 2)). According to official estimates (CSED 1952-(1952: I, 53)) gross national income in 1946-7 was 60 per cent of the 1938-9 level at constant prices, and per capita national income at current prices amounted to only K151.

Post-war reconstruction was heavily set back by the communal and communist revolts which broke out shortly after independence was granted in January 1948. Further widespread damage was done to the country's stock of capital equipment, particularly to that of the mines and transport system. Rice production and exports declined and gross domestic product, which had recovered to 72 per cent of its 1938-9 level in 1947-8, fell to a post-war low of 61 per cent of the pre-war level in 1949-50. By the end of 1950 the worst of the insurrection had been contained in the central areas of the country, but in the frontier and hill regions the various rebel factions, although disunited, remained entrenched in sufficient strength to inhibit economic growth.

THE FARM SECTOR AFTER INDEPENDENCE

Because of the concentration of indigenous economic activity on farming and the decline of cottage industry during the colonial period, the lack of a factory sector and of indigenous industrial and management skills, and because political factors inhibited the speedy restoration of war-damaged mine and forest undertakings by foreign or domestic enterprise after independence, the low-yielding traditional agricultural sector was the base from which development had to grow. Agriculture provided employment for over two-thirds of the work force and contributed, together with fishery, nearly 40 per cent of gross domestic product in the four years following independence. It was an important source of domestic savings and the primary source of foreign exchange earnings to finance development. Rice alone contributed 80 per cent of export earnings between 1948 and 1951, an even larger share of the value of exports than at the beginning of the century. Other farm products brought in 6 per cent.

Trends in Agricultural Production

The rate of growth of agricultural production since independence has been slow. Paddy production regained pre-war levels only in 1962-3, while output of other crops has expanded more slowly than population. Economic progress of the farm sector has not, of course, been uniformly sluggish. There has been progress in some sectors and some districts. There have also been occasional spurts of limited general activity and of private investment. Nevertheless the general trend has been one of declining productivity in comparison with the colonial period. There has been a significant reduction in the volume of agricultural production in relation to population growth; a reduction in the cultivated land/man ratio and the size of farms; in plough cattle per farm; in the productivity of farm labour; and in the marketable surplus of the farm sector, both for sale on the home market and, most strikingly, for export.

Data on agricultural production in Burma are limited, especially for crops other than rice, and statistics are admitted to be subject to significant margins of error. But they are probably sufficiently accurate to indicate general patterns and trends. In the first place they show a great concentration on production of food crops, to which 92 per cent of the sown area was devoted in 1965-6. Of these, apart from rice, exports form a significant proportion of output only of pulses (an average of 37 per cent of output through 1962-5) and maize (21 per cent). Among the industrial crops, rubber is primarily an export crop and a share of cotton and jute production is also exported.

They also show (Table 5.3) that output of the main crops has not expanded significantly since before World War II, except for wheat, groundnuts, sugar, and onions. The index of agricultural production reached a high point of 12 per cent over the pre-war level in 1962-3, from which point it had drifted slightly by 1965-6. In the meantime the country's population increased by 50 per cent between 1941 and 1966, so that farm production per head of total population had dropped by over one-quarter.

Expansion of agricultural production, both of rice and of other crops, has been achieved since the colonial annexations largely through extension of area cultivated, rather than through raising yields per hectare or the relatively small share of land which is double-cropped.

The failure to expand agricultural production significantly since independence has not been due to a shortage of cultivable land. The

TABLE 5.3

PRODUCTION OF MAIN CROPS

('000 metric tons)

	1936–7 to 1940–1 average	1946–7	1952–3	1955–6	1960–1	1961–2	1962–3	1963–4	1964–5	1965–6 provisional	1966–7 target
Paddy	7,545	3,826	5,832	5,869	6,789	6,908	7,665	7,783	8,507	8,055	8,097
Pulses	254	n.a.	195	233	267	275	322	344	279	278	343
Groundnuts	184	105	179	207	357	363	432	337	343	288	409
Sesame	46	42	55	45	65	76	85	54	101	58	92
Cotton, lint	21	5	22	18	12	16	19	17	22	15	29
Sugar cane	1,000	701	1,076	856	1,042	1,158	1,292	1,114	1,084	1,448	1,996
Tobacco	45	n.a.	44	40	36	42	59a	55	58	53	72
Wheat	9	{59	{86	{80	{66	21	33	54	72	96	110
Millet	63					73	51	61	41	46	40
Maize and other cereals	n.a.	n.a.	n.a.	n.a.	n.a.	57	80	83	61	57	71
Chillies, onions, garlic	62	52	n.a.	n.a.	n.a.	114	144	117	122	83	133
Rubber	11	9	11	13	11	13	14	13	13	12	12
Index of production											
paddy	100	51	77	78	90	92	102	103	113	107	
totalb	100	55	84	81	99	101	112	109	111c	106c	

a Includes green tobacco from 1962–3 on.

b To 1963–4 weighted average of gross value of out-turn of principal crops at 1947–8 prices.

c Calculated from percentage change in value of production at current prices.

Source: MNP 1951a–(1951–1964); CSED 1952–(1952, 1955, 1964); UBRC 1964–(1966–7: 62–5).

TABLE 5.4

LAND USE

('000 hectares)

	1936-7 to 1940-1 average	Per cent	1961-2	Per cent	1962-3	Per cent	1963-4	Per cent	1964-5	Per cent	1965-6	Per cent change pre-war to 1965-6
Net sown area[a]	7,069	82·0	7,162	71·0	7,704	79·5	7,968	81·6	7,935	77·6	7,899	11·7
Double-cropped area[a]	493	5·7	532	5·3	671	6·9	767	7·9	824	8·0	875	77·4
Total sown area[a]	7,562[b]	87·7	7,695	76·3	8,375	86·4	8,735	89·5	8,759	85·6	8,774	16·0
Fallow[a]	1,553	18·1	2,922	29·0	1,987	20·5	1,793	18·4	2,295	22·4	n.a.	47·8[c]
Occupied area[a]	8,622	100·0	10,084	100·0	9,691	100·0	9,761	100·0	10,230	100·0	n.a.	18·7[c]
Cultivable waste[d]	8,451		6,989		7,382		7,312		6,843			−19·0
Total cultivable area[d]	17,073		17,073		17,073		17,073		17,073		17,073	
Per head of population sown area,[a] hectares	0·46		0·33		0·35		0·36		0·35		0·35	−24·0
cultivable area,[d] hectares	1·04		0·73		0·72		0·70		0·69		0·68	−35·0

[a] Including states.

[b] Total sown area for this period is given as equivalent to 7,756,000 hectares in MNP 1951a-(1951–1964).

[c] Per cent change to 1964–5.

[d] Excludes cultivable areas within the 27 million hectares which have not been surveyed in the Chin Hills, Naga Hills, Shan States, Kayah State, and Myitkyina.

Source: UBRC 1964-(1966–7: 5); CSED 1967: 51.

fallow area, including abandoned land, was larger in 1964-5 than in the late 1930s, being estimated at 2·3 million hectares (UBRC 1964- (1966-7: 5)) as compared to the pre-war 1·55 million. Moreover, there are still large areas of potentially cultivable well-watered lands in Burma (Table 5.4), quite apart from the 27 million hectares of land in the sparsely populated states and frontier regions which have never been properly surveyed. According to pre-war surveys, Burma Proper had nearly 42 million hectares. Of this total, 17 million hectares were assessed as cultivable, comprising 8·6 million of farm land and fallow and 8·5 million waste. It is true that the bulk of the potentially cultivable area lies either in the outlying districts of Upper Burma, especially the northwest, or in the Tenasserim and Arakan coastal strips of Lower Burma (KTA 1953: 5-8). But the Burmese peasants have shown themselves enterprising and mobile in the past, provided that law and order prevailed and that they were given sufficient incentives and government assistance through major land reclamation work.

Probably the most important obstacle to faster expansion of agricultural production and area cultivated during the past two decades has been dacoity. In the less-secure areas incentives to expand production are reduced by demands of rebel 'taxation officers' for protection money. In districts where villagers have not come to terms with insurgents farmers have frequently ceased to cultivate fields too far from their villages. They must get back to the security of the village stockade before nightfall and can no longer sleep in field huts during peak cultivation times, as was the custom. Moreover, where outlying fields are cultivated in insecure districts, seed is often broadcast, to save time, rather than transplanted, thus reducing yields.

Total area sown to all crops, including double-cropping, averaged 7·8 million hectares during the five seasons 1936-7 to 1940-1 (Table 5.5), and it was not until 1961-2 that this area had been regained.[6] During the 1950s there was some shift of land from paddy to other more profitable crops, such as pulses, groundnuts, jute, and sugar, but even so the area under non-rice crops was by 1961-2 only 24 per cent larger than before World War II. Then in the following two seasons came a remarkable spurt in acreage extension, the total sown area increasing by nearly one million hectares, or 12 per cent, of which roughly 450,000 hectares were sown to paddy and 500,000 to other

[6] Rainfall and the annual flood are dependable in Lower Burma, where roughly 90 per cent of the cultivated area is sown to rice, and most rice in Upper Burma (where it occupies about one-third of the sown area, including double-cropping) is grown on irrigated or riverine land. Sown area is therefore a fairly accurate guide to farm planning for the rice sector, and also for other crops on *Kaing* (alluvial land seasonally inundated). Variations caused by unsuitable sowing conditions on *Ya* (dry upland crop land) are probably not very significant overall.

Table 5.5

SOWN AREA

('000 hectares)

	1936–7 to 1940–1	Per cent	1952–3	Per cent	1961–2	Per cent	1963–4	Per cent	1965–6ᵃ	Per cent
Paddy	5,193	67·0	4,181	63·3	4,597	59·2	5,048	57·8	5,014	57·1
Pulses	538	6·9	429	6·5	559	7·2	750	8·6	691	7·9
Sesame	567	7·3	537	8·1	619	8·0	652	7·5	809	9·2
Groundnuts	327	4·2	301	4·6	564	7·3	603	6·9	532	6·1
Millet and wheat	289ᵇ	3·7	288ᵇ	4·4	225	2·9	269	3·1	333	3·8
Cotton	183	2·4	139	2·1	191	2·5	273	3·1	229	2·6
Other cropsᶜ	659	8·5	727	11·0	1,009	13·0	1,133	13·0	1,166	13·3
Total	7,756	100·0	6,602	100·0	7,764	100·0	8,728	100·0	8,774	100
Double-cropped	493	6·4	n.a.	n.a.	532	6·9	768	8·8	876	10·0
Irrigated total	637	8·1	576	8·7	536	6·9	758	8·7	751	8·6
Of which private	139ᵈ	1·8	n.a.	n.a.	130	1·7	287	3·3	251	2·9

ᵃ Provisional.

ᵇ Millet and irrigated wheat.

ᶜ In 1965–6 the principal other crop areas were (in '000 hectares): sugar and palm 86; condiments and spices 92; fruit, vegetables, tea, and coffee 336; tobacco and narcotics 101; rubber 87; jute 29; and fodders 123.

ᵈ Data for 1938–9; source: Andrus 1947: 57.

Source: MNP 1951a–(1955: 10; 1964: 13); CSED 1952–(1955: 18); UBRC 1964–(1966–7: 7, 62–5).

crops. Almost one-quarter of this extra area was accounted for by an increase in double-cropping, largely supported by an extension of both private and public irrigation areas. Between 1963-4 and 1965-6 the expansion of paddy area halted and that under other crops slowed down markedly.

Yields per hectare for almost all crops other than rice, groundnuts, sesame, and pulses have declined since the late 1930s. Yields for paddy, at around 1,400 to 1,600 kg per hectare, compare favourably with those of Thailand, India, Pakistan, and the Philippines. But for most other crops, which currently occupy over two-fifths of Burma's cultivated area, yields are well below those of many other underdeveloped countries. Burmese yields per hectare expressed as a percentage of the corresponding yields for Thailand have been reported to be: maize 29; groundnuts 48; jute 40; sugar cane 71; tobacco 34; rubber 40 (UBRC 1964-(1965-6: 13, data undated, and in a variety of measures)).

Trends in rice production since the early 1950s will be found in Table 5.6. Ignoring the 1966-7 season, which was abnormally poor, it will be seen that the paddy crop averaged 7·78 million tons in the five seasons 1961-2 to 1965-6. This was 3 per cent more than the average of 1936-7 to 1940-1, the result of an 8 per cent increase in yield per hectare and a 5 per cent fall in the area cultivated. The slight rise in yields has occurred in spite of the fact that a higher proportion of paddy land is now sown broadcast (some 30 per cent as against roughly 10 per cent in the 1930s). Fewer large farmers can afford to pay for the seasonal labour necessary for transplanting, although transplanting techniques raise yields generally by nearly one-third. Increased productivity per hectare can probably be attributed as much to abandonment of production on marginal lands as to the application of additional or improved inputs. The latter consist largely of extra labour on small farms and superior quality seed.

Marketable Surplus of Farm Products

Because the farm population has increased more than food output over the past thirty years, the marketable surplus of the agricultural sector has declined. Moreover, a higher share of the surplus is being absorbed by the home market (as discussed below).

The reduction in the marketable surplus of the farm sector affects particularly foreign trade. In poor agricultural seasons, such as those of 1966-8, export earnings from farm products have been adequate only to keep the economy going at minimum levels and have contributed little or nothing to growth. This is the more important because exports of non-agricultural products (minerals, petroleum, and timber) have fallen very sharply since World War II. They contributed an average

Table 5.6

RICE: AREA, PRODUCTION, YIELDS, EXPORTS, DOMESTIC RETENTION

	1936-7 to 1940-41 average	1945-6	1950-1 to 1953-4 average	1954-5 to 1957-8 average	1958-9 to 1960-1 average	1961-2	1962-3	1963-4	1964-5	1965-6	1966-7
Sown area ('000 hectares)	5,193	2,691	3,980	4,145	4,287	4,680	4,837	5,048	5,109	5,014	4,989
Paddy production ('000 metric tons)	7,545	2,672	5,584	5,845	6,802	6,908	7,665	7,783	8,507	8,055	6,635
Yield per hectare (kg)	1,453	993	1,403	1,410	1,587	1,476	1,585	1,542	1,665	1,606	1,330
Rice production (milled '000 tons)[a]	4,902	1,737	3,630	3,799	4,421	4,490	4,982	5,059	5,530	5,236	4,313
Exports, milled rice ('000 tons)	3,249	n.a.[b]	1,271	1,785	1,773	1,872	1,646	1,519	1,330	1,129	559
Export % of total volume	66	—	35	47	40	42	33	30	24	22	13
Retained for domestic use[c] ('000 tons)	1,653	1,737	2,359	2,014	2,648	2,618	3,336	3,540	4,200	4,107	3,754
Retention per head of population per year (kg)	101	96	120	96	118	112	140	146	170	163	146

[a] Conversion ratio 65 per cent.

[b] Insignificant. Remainder of column calculated on assumption exports nil.

[c] Seed, consumption, waste and stock.

Source: CSED 1952-; UBRC 1964-; United Nations FAO (June 1967); Working Peoples' Daily, 27 Nov. 1967.

of only 3 per cent of gross domestic product in the ten fiscal years to 1963-4, compared to 15·4 per cent in 1938-9.

In the late 1930s nearly two-thirds of Burma's rice crop was exported. After World War II the proportion of output exported was gradually raised from about one-third in the early 1950s to a high point of 49 per cent in 1955-6. Since that date the share exported has been gradually whittled down to less than one-quarter (Table 5.6). This change was mainly due to the growth of population, but the trend towards shortage of export supplies has also been exacerbated by an apparent rise in per capita consumption since the early 1960s.

Although exports of other products have somewhat increased since the early 1950s, they have been unable to compensate for the drop in both the volume and unit value of rice exports. The result is that the share of total exports in gross domestic product has declined from 32·7 per cent in 1938-9 to an average of 23·8 per cent between 1950-1 and 1952-3 and to 16·2 per cent between 1959-60 and 1963-4 (MNP 1951b-(1956, 1964: 1A and V)). Since 1963-4 it has undoubtedly dropped further.

Government Planning and the Agricultural Sector

Burma is a poor country, at least in terms of conventional statistical measurement. Gross domestic product amounted to an average of K313 (US$65) per head in 1966-7 (CSED 1967-(1967: 97)). It has, however, devoted a surprisingly large proportion of its national income to reconstruction and development since the early 1950s. Between 1952-3 and 1963-4 an average of 18·5 per cent of gross domestic product was utilised for gross fixed capital formation (public and private) annually, one of the highest shares for any underdeveloped country in the ECAFE region. But the proportion of investment which has gone to the farm sector has been low, and the return to capital throughout the program has proven disappointing.

Despite the primary importance of traditional agriculture, all postwar governments have tended to neglect it in favour of other sectors. In 1947 Burmese leaders laid down in broad outline development plans for an economy in which the land and other key factors of production were to be nationalised. The main emphasis was on the need to shift the economy away from the pattern of primary production and export orientation established in colonial times, towards a more diversified and sophisticated economy in which local manufacturing would replace many imports and mechanised farming gradually supplant hand and bullock techniques. This emphasis on industrialisation, diversification, and greater self-sufficiency is the *leitmotiv* running through the succession of plans drawn up in socialist Burma since independence. It has affected the development of the agricultural sector directly, in the type of state farm programs initiated, and indirectly, in the share of national resources channelled to non-farm sectors.

Repeated government assurances to peasants that agriculture merited priority in development planning have never been translated into action. Investment in the traditional sector has been left largely to private enterprise. It seems to have been generally assumed that smallholder production could be expected to progress automatically, at least in the short run, without too much assistance from the state, once the peasant was relieved of capitalist exploitation by agrarian reforms, assuring security of land tenure, moderate rents, state farm credit facilities, and modest but guaranteed prices for his produce. Such state investment as has taken place to the direct benefit of the agricultural sector has tended to concentrate on capital-intensive projects, with benefits in the long term.

The most detailed post-war development plan was the Pyidawtha program of 1952-60 (ESB 1954). This program was drawn up largely on the advice of a group of American economists and engineers (KTA 1952, 1953). The plan envisaged an increase of 79 per cent in gross domestic product in real terms from the low level of 1951-2 to K7,000 million by 1959-60, at which point it would have been 31 per cent above that of 1938-9. This called for a total net investment expenditure of K7,500 million in prices of 1950-1, of which K4,020 million was to come from public net capital formation and K3,470 million from private net capital formation. The bulk of investment in agriculture, for which no guidelines were suggested, was expected to come from the private sector. The principal state investment planned of direct benefit to agriculture was for irrigation, which was allotted K370 million, 9 per cent of the proposed state total. Of the remaining K3,650 million of planned state capital expenditure, 37 per cent was allotted to transport and communications, 28 per cent to industry, power, and mining, and 26 per cent to government and welfare services.

Actual public capital expenditure on agriculture and irrigation is shown in Table 5.7. Over the fourteen years 1952-3 to 1965-6 it amounted to K848 million, an average of K61 million (US$13 million) a year, spending having been increased since 1962, when the Revolutionary Government assumed power.

Planning priorities seem to be based on two attitudes: first, the notion that the share of production value extracted from smallholder output under colonial capitalism in land rents, interest, trading and milling profits was so large that it could well be split between the cultivator and the community under socialism, leaving both more

TABLE 5.7

PUBLIC CAPITAL EXPENDITURE

(annual averages, million kyats)

	Agriculture and irrigation	Total	Agriculture as % of total
1952–3 to 1959–60	51·8	504·2	10·3
1960–1 to 1961–2	51·4	443·7	11·6
1962–3 to 1965–6	82·7	649·0	12·7

Source: MNP 1951a–; UBRC 1964–(1966–7).

prosperous than before; second, that the marginal product of investment in traditional farming was low compared to that in technically modern agriculture or in other economic activities.

Government policy of diverting the country's scarce investible funds to sectors other than traditional agriculture has not yet proven productive. Returns to capital spent under successive development plans have been disappointing, and little substantial diversification of the economic structure has been achieved in comparison with the colonial period. It was not until 1958-9 that gross domestic product in terms of constant prices was estimated to have passed the pre-war level. By 1963-4 (when the constant price series was suspended) gross domestic product was 26 per cent above the real level of 1938-9, while per capita output of the larger population was still only 85 per cent of pre-war.

The high capital:output ratio seems to be attributable to three main causes: much waste occurred when newly created capital was destroyed or under-utilised because of rebel activity; plans were over-ambitious both in terms of available domestic skills and capital, so that resources were spread too thinly to be effective; and the state programs allotted too large a share of funds to capital-intensive infrastructure investment designed to diversify the economy in the long run, compared to directly productive projects, especially in established primary sectors.

Agricultural production and price policies. Although the various post-war governments have from time to time announced plans and targets for agricultural production and for particular crops, at no time have mandatory crop programs been laid down which farmers must follow. The government's influence has been exercised mainly through its farm credit system or through price and purchase policies, and has been confined largely to rice, jute, groundnuts, and cotton, all of which it has bought in sufficient share to dominate the market. Ex-farm prices of paddy have been effectively controlled until very recently, but those of most other food crops have been allowed to fluctuate freely, with a resulting distortion of price relationships. Thus while wholesale prices of all cereals (mainly paddy) since the early 1950s have remained around three times their 1938-40 level, those of pulses and oilseeds have ranged from five to nine times pre-war levels and those of other foodstuffs from six to ten times (CSED 1952- (1964: 146)).

A central feature of Burma's economic policy, with important effects on trade, public finance, and agriculture, has been the government's control of rice exports. State trading was, for the policy-makers of 1948, both a socialist and nationalist doctrine and an inheritance of practices from the Burmese kings. It also arose out of the disruptions of war. Indeed, the State Agricultural Marketing Board (SAMB, now UBAMB) was established by the British in 1946 to handle the export of rice and other farm products.

Rice exports have been a monopoly of the state since independence[7] and SAMB has been looked to as a prime source of foreign exchange earnings to finance development. During the early 1950s SAMB made substantial profits for the state on foreign sales of rice. There are no data showing SAMB's contribution to the exchequer, part of which is made through income taxation and part merged with balances of other state boards. Levin (1960: 223) estimated SAMB profits at about K560 million (US$117 million) on exports of 1·2 million tons in 1952-3. Profit, net of all costs including milling, transport, and storage, was estimated at 54 per cent of the selling price. Walinsky (1962: 418) estimated that the Board's contribution to government finances averaged about K400 million during the three fiscal years 1952-3 to 1954-5, which would have been óver 7 per cent of gross domestic product in that period and at least 40 per cent of government revenue.

Since the early 1950s SAMB's net earnings on rice sales have been very greatly reduced, both because rice export prices have fallen and because purchase prices from the farmers have been gradually raised (Table 5.8). Taking price factors during the period into account, it seems probable that between 1961-2 and 1964-5 SAMB's net earnings on rice export sales were not much in excess of K100 million (US$22 million) per annum. Since that period profits have undoubtedly been much lower, because the volume of exports has declined sharply, so that the contribution of rice profits to the development program must now be slender.

Official paddy price policy has been seen as important in guaranteeing minimum prices to the growers; in holding down retail prices of

[7] Up to the end of 1962 a small share of exports was handled by private merchants under licence from the Board.

TABLE 5.8

OFFICIAL RICE PRICES

(kyats per metric ton[a])

| Period[b] | Official purchase prices | | Unit value of exports | Minimum purchase price as % of export price |
	Paddy	Rice, Equiv.[c]		
1950–1	137	211	545	38·6
1951–2	137	211	691	30·5
1952–3	137	211	836	25·2
1954	137	211	612	34·4
1955–8	137	211	455	46·3
1959–60 to 1960–61	137–49	211–29	407	51·7
1961–2 to 1962–3	144–51	222–32	467	47·5
1963–4	144–92	222–95	496	44·6
1964–5	149–96	229–302	485	47·2
1965–6	149–96	229–302	492	46·5
1966–7	163–211	251–325	564	44·5
1967–8	172–219	265–337	n.a.	

[a] Converted from data expressed in baskets of 46 lb.

[b] Fiscal year or calendar year, as stated.

[c] Based on a milling ratio of 65 per cent.

Source: MNP 1951a–(1951–64); CSED 1952–(1955, 1965); UBRC 1964–(1966–7: 102); DI 1962a–(15 Apr. 1967).

rice, and thus living costs for the mass of the people; and in acting as the principal means of taxing the farm sector, apart from customs duties and other indirect taxes.[8] The official paddy price was fixed at the equivalent of K137 per metric ton delivered up country from 1948 to 1958, compared to a Rangoon low of K25 and high of K125 between 1913 and 1940 (KTA 1953: 141). The Ne Win caretaker government of 1958-60 introduced in 1959 a range of K137-149 per ton for various qualities and delivery dates. Since this date there have been modest increases for ordinary grades and more generous ones for superior qualities, so that prices for the former were in 1967-8 26 per cent over their 1958 level and those for the latter 60 per cent above (Table 5.8).

Prior to 1966-7 farmers, particularly in the delta, had little opportunity to obtain prices better than official prices for their paddy, domestic supply being in more or less constant surplus. The rice short-

[8] Land revenue, which contributed 33 per cent of government revenue in 1938-9, has contributed an average of only 3 per cent since 1952-3.

ages which developed in 1966-8 enabled paddy farmers for the first time to sell substantial quantities on the black market.[9] For a brief time the government then permitted rice farmers to barter paddy against other goods, only to revert to the system of legal monopoly of paddy marketing by state agencies in May 1968.

Government policy of maintaining retail prices of rice at levels well below international prices meant that part of the margin between producer and international prices flowed to domestic purchasers of rice as well as to the Treasury through exports. In the early 1950s home sales of rice (excluding paddy farmers' repurchases of milled rice) seem to have absorbed about one-third of rice sold. The home market share of sales has been gradually rising and exceeded that of exports in 1966-8.

Domestic consumption was stimulated after 1964 when retail prices of rice were equalised throughout the Union at levels prevailing in the delta. This was said to have cost the state some K70 million in subsidies in 1964-5 (*The Guardian*, 23 June 1965).[10] Since 1967 the government has, like the Burmese kings of old, given priority to supplying the home market. To the extent that it succeeds in holding down retail prices by subsidy or by diverting export supplies for local sale, it transfers income to urban consumers and reduces both farmers' and state resources for investment and the country's foreign exchange earnings. The alternative of permitting a free market in grains to develop in the hope that it might stimulate production is thought to be not only unlikely of success, but dangerous politically.

Agrarian Reform

Successive governments have looked to agrarian reform as the principal means whereby the state could in the short run increase the well-being and raise the living standards of the peasants by assuring them of a larger net share of the gross ex-farm value of their crops than they had received in colonial times. The farm policies carried through by the Revolutionary Government which has ruled since March 1962 have not differed substantially in concept from those of the parliamentary or 'caretaker' governments in office from 1948 to February 1962.[11] What the revolutionary régime has done is to apply

[9] The diversion of supplies away from authorised buying centres was exacerbated during this period by rebel concentration in the main rice districts on harassing farmers prepared to sell to the state. Official deliveries could thus frequently be made only in armed convoys.

[10] Since mid-1965 official retail prices have been uniform throughout the country, but have ranged from the 1964 flat rate of K0·24 per kg for the cheapest grade of rice to K0·62 (U.S. 13 cents) for the best quality.

[11] This is a strong contrast to the revolutionary régime's policies in non-agricultural sectors (and in policies concerning the trading and processing of farm products) which have been far more radical in concept.

established agrarian policies more extensively and rigorously, giving
the farm sector ever stronger doses of egalitarian socialism.

The agrarian pattern inherited on independence was, as we have
seen, one of low-yielding smallholder production, not greatly changed
in its techniques from pre-colonial times, although much altered in its
ownership and marketing structure. There were no great estates where
powerful local landowners held entrenched rights and, although almost
half of the arable area was under tenancy, most owners of tenanted
land had acquired it fairly recently and lacked the political backing
to assert their rights to it. At the same time little change had occurred
in farm production units as such, since little consolidation of holdings
had taken place. Land properties consisted mainly of scattered plots
small enough to be worked as family enterprises, with the help often
of seasonal labour. Social and economic conditions in the colonial
period, as in feudal times, had inhibited the growth of a class of
entrepreneurial farmers able and willing to invest in commercial-
scale farming. The landlords of late colonial days were mainly non-
agriculturalists serving no entrepreneurial farming function save the
provision of capital. There were thus no technically modern farms
in Burma at the time of independence which could serve as models
for change. All these factors made plans to nationalise the land
more realisable, but made modernisation of farm practices more
difficult.

Neither wing of the government coalition, the Anti-Fascist Peoples'
Freedom League (AFPFL), wished to see the emergence of a class of
large-scale capitalist farmers. Opinion on the structure of farming to
be aimed at was divided between those who held that smallholding
was best suited to Burmese socialism, that the faults of smallholding
could be attributed to colonial and capitalistic mores and could be
remedied by the agrarian reforms of welfare socialism, and those who
held that Burmese farming could only be efficient and truly socialist
under collectivisation. The moderates predominated and an ill-defined
compromise policy was adopted. The short-term aim was for the state
to help the smallholders to prosperity by ensuring fair conditions and
assistance in modernisation. The long-term aim was to achieve 'a
thorough-going revolution in the structure of [the] agricultural
economy, and in the methods of producing and marketing agricultural
products' (ESB 1954: 42). This long-term objective was to be achieved
by organising smallholders in co-operatives or collective work units
large enough to be able to utilise large farm machinery and the mutual
purchase of farm inputs and marketing of produce. The radical wing
by and large acquiesced in this compromise because it considered that
the smaller and more equal the plot held by each peasant, the less

would be his ultimate resistance to collectivisation. Ownership of all land was, in the 1948 constitution, vested in the state, and the state given comprehensive rights to acquire land for distribution in any form and to regulate land tenure.[12]

Land nationalisation. A land nationalisation measure was passed in 1948, but proved impracticable and was replaced by the Land Nationalisation Act of 1953. This Act prohibited the mortgaging, selling, or subdivision of farm land without authorisation, and set the pattern for much of the agrarian reform program that was to follow. It limited the size of holdings for resident Burmese and provided that land in excess of these limits, together with all land owned by non-residents, should be resumed by the state for distribution either to individual farm families or to collective units. A total of 2·74 million hectares of land, 90 per cent of it paddy land, was expected to be nationalised under this Act, of which 2·1 million hectares was in Lower Burma and 0·64 million in Upper Burma (United Nations FAO 1962: 21). The limits of holdings to be exempted varied according to the type of land and owner. Working farmers who had been in possession of their land since January 1948 were permitted to retain 20 hectares of paddy land, 10 hectares of *Ya*,[13] 4 hectares of *Kaing*,[14] or 4 hectares of sugar cane land, provided that they worked it personally with the help of family or hired labour. Working farmers who had been in possession since June 1953, or non-agriculturalist families living in the village, were allowed 8, 4, 2·4, and 2 hectares respectively, and there were also allowances for other types of owner.

There were complex provisions for the distribution of expropriated land by elected village land committees to defined classes of applicants. The law clearly aimed to promote co-operative farming, in that those receiving land were required 'to undertake to join such agricultural organisations as might be formed under the Act'. Distribution was to be made in *tadontun* (the area which could be worked by one yoke of cattle) which was to be varied in the different districts according to population density and soil fertility, but which, in practice, rarely exceeded 4 hectares and was usually smaller. This legislation proved cumbrous and contentious to carry out. By the end of 1958 only 1·3 million hectares had been surveyed, of which half had been exempted and half resumed, and 0·6 million hectares had been distributed to

[12] Section 30(i) states, 'The State is the ultimate owner of all lands' and section 30(ii), 'Subject to the provisions of this constitution the State shall have the right to regulate, alter or abolish land tenures or resume possession of any land and distribute the same for collective or co-operative farming or to agricultural tenants'.

[13] Dry upland land, used mainly for crops other than rice.

[14] Alluvial land, used for cash cropping.

178,000 families. No land had been allotted to collective units. The program's slowness and uncertainties had in the meantime inhibited longer-term investment in farming and land maintenance work on the 1·4 million hectares likely to be nationalised (Walinsky 1962: 293-4). The Act was suspended by General Ne Win during his first caretaker period of office in 1958 and was not implemented further by the U Nu government which followed or by the Revolutionary Government of March 1962.

Tenancy and rent controls. The tenancy problem has been less widespread since independence than before, both because Chettyar landlords fled the country during the war and because of the subsequent nationalisation of land. Thus only one-third of the cultivated area was farmed by tenants in 1963, compared to almost half in the late 1930s. Moreover, most tenants of land owned by Chettyars paid no rent for it. From 1946 onwards a series of laws established standard rents for smallholders with less than 20 hectares.[15] Rents for the different types of land were tied to the state land revenue, which was itself frozen at pre-war rates in 1948.[16] For paddy land, a rent equal to the pre-war land revenue was prescribed in post-war legislation, while for land used for other (and more profitable) crops rents were fixed at two or three times the land revenue rates. Provided that these modest rents were paid, security of tenure was, at least nominally, secured to the sitting tenant. A 1963 decree restated these provisions and deprived landlords of their right to select tenants, the right being vested in Village Land Committees.[17] Legislation protecting all cultivators against seizure of their land for debt under civil suit was introduced under the parliamentary régime and extended in 1963 by the Revolutionary Government to cover all farmers' working property, such as cattle, tools, produce, etc., from attachment by private persons against debt settlement.[18] These laws helped to reduce the pitch of rents, but they did not effectively control them. Moreover, even the low controlled rents remained a significant factor in the agricultural economy, since rural incomes were so low.

In April 1965 all land rents were abolished by decree, so that all tenanted land was expropriated *de facto* in favour of sitting tenants

[15] The principal early legislation was the Tenancy Act of July 1946 and the Tenancy Standard Rent Act of 1950, as amended in 1951 and 1954.

[16] Land revenue, based on the imputed net produce value of land in colonial Burma as reassessed every twenty or thirty years, averaged K5·7 per hectare cultivated between 1936 and 1941.

[17] The Tenancy Law of 25 March 1963, as reported in *The Nation*, 29 March 1963.

[18] The Farmers' Rights Protection Law of 25 March 1963, as reported in ibid.

without a process of survey, exemption, distribution, and compensation.[19] According to official estimates this decree affected some 2·96 million hectares of land, which were leased by 1·1 million tenants for K13 million per annum. The decree was thus wider in its scope than the nationalisation law, affecting 200,000 more hectares than had been expected to be resumed under the 1953 Act. Including land already resumed, the 1953 Act and the 1965 decree together expropriated 3·52 million hectares of farm land, 44 per cent of the estimated cultivated area of 8 million hectares in mid-1963.[20] The 1965 decree was of particular consequence for the 2·07 million hectares owned by national landlords, ownership being roughly equally divided between non-agriculturalist owners living outside the village (1·1 million hectares) and village landlords (0·96 million). It is not known how the decree affected tenants farming holdings in excess of the *tadontun* established for each area, but it appears that their excess holdings were distributed to the landless.[21]

Even though land rents were modest, their abolition was an important and controversial measure, especially where it affected land owned by village landlords, the élite of the rural areas. In Upper Burma its repercussions must have been greater than those of the land nationalisation program, because most holdings in this region were small enough to have claimed exemption under the Act, whereas the share of farm land under tenancy was significant. Throughout the country it appears to have affected particularly tenants of very small plots, who traditionally looked to village landlords for the maintenance of bunds and private irrigation systems, frequently for the provision of seed, cattle, and ploughs, and sometimes for the drawing up of crop programs, the last especially in Upper Burma, where farm management is necessarily more varied and sophisticated because natural conditions are more hazardous. An official report (UBRC 1964-(1965-6: 12)) paints a gloomy picture of the disorganisation resulting from the ousting of the landlords, and suggests collectivisation as the remedy.

State agricultural credit. The state agricultural credit system, initiated on a limited scale by the parliamentary régime, has been

[19] The Tenancy Law Amending Law of April 1965, published with a statement of the Ministry of Agriculture in *The Guardian*, 7 April 1965. The decree contained no compensation provisions, but it was stated that hardship cases would be considered.

[20] Eight million hectares was given as the cultivated area in mid-1963 in the Ministry statement. It appears to be the net sown area, excluding fallow.

[21] There were in 1962-3 nearly a quarter of a million tenants with holdings of four hectares or more (UBRC 1964-(1965-6)).

built up in volume and scope by the Revolutionary Government.[22] The programs of both régimes have been based on the same general principles and have benefited mainly the rice farmers.

The demand for farm credit since independence has been concentrated almost entirely on seasonal and short-term loans to meet farmers' cultivation expenses, and sometimes living expenses also. Such working capital has been required because profitability of many farms —especially rice farms—was so low that savings were insufficient to carry the farmers through until their crops were sold. There has been little post-war demand for long-term loans since capital could no longer be used for land purchase, and little private long-term investment in land improvement or farm equipment has been undertaken. Because of the lack of demand for long-term credit, the total volume of private and public farm credit has probably been lower in relation to the value of agricultural production since independence than in the mid-1920s (Aye Hlaing 1958: 16).

Although the total demand for farm credit has probably been lower since independence than before, the proportion of this credit which had necessarily to be supplied by the state has been increased with each successive intervention by the state in the farm sector. Agrarian reforms prohibiting land mortgage, etc., have reduced the security which cultivators could offer to private lenders. Nationalisation of Chettyar holdings and state participation in rice trading and milling have progressively eliminated major commercial sources of funds. Interest rates on private farm loans have thus been even higher since independence than in the colonial period.[23]

The post-war state programs aimed primarily to supply a rising share of farmers' needs for seasonal loans and so reduce the burden of high interest rates on their cash incomes. Working on 1953 census data, which covered farms near the main towns only, Professor Aye Hlaing (1958: App. III) estimated that, for the 6·5 million hectares of land then under cultivation, a total volume of K350 million in loans should be adequate to meet current expenses of farmers in Burma Proper. If the average cultivator in the delta had been able to borrow the whole of his credit needs at the government interest rate of 7 per cent, instead of obtaining 70 per cent from commercial sources at interest rates ranging between 48 and 60 per cent per annum, his gross income would have been raised by 12 per cent and his net income by 25 to 30 per cent (Aye Hlaing 1958: 7).

[22] A fuller description of the post-war state agricultural credit programs in Burma is given in a recent article by the author (Richter 1968).

[23] Post-war legislation fixed the maximum interest rate on private agricultural loans at 18 per cent per annum, but controls have been completely ineffective.

During the period 1945-6 to 1953-4 the state lent out between K30 and K54 million per annum in farm credit, the higher figure representing roughly one-sixth of total farm credit from all sources in 1953-4. In the early 1950s the government had ample funds to extend the credit program, thanks to the boom in rice export earnings, but it was discouraged from doing so by the administrative difficulties involved and by the poor recovery record on loans made. Almost all funds were allotted in seasonal cultivation loans, issued partly through co-operatives and partly by government loans teams. In 1953 the State Agricultural Bank was created to handle a rising proportion of funds, through a hierarchy of village and district banks. The co-operative societies in 1957-8 lost their function as loan agencies. Over the decade to 1961-2 K681 million was lent out, of which 74 per cent was recovered, not allowing for interest or administrative costs.

After the *coup d'état* of 1962 the Revolutionary Government quadrupled the volume of loans, lending out K350 million in 1962-3. The following fiscal year it allotted K700 million for farm credit, in an attempt to meet the whole of demand for seasonal and short-term loans, although not more than half of this sum was actually issued. The 1963-4 program included harvest loans, and these, and the advance purchases of standing crops undertaken by the government between 1964-6, were intended to undercut the traditional *sabe-pe* (loan made and repaid in kind) and *pindaung* (forward purchase of standing crop) loan systems, under which some of the poorest of the peasants sell their crops forward at very substantial discounts.[24] In 1962-3 the recovery ratio on loans deteriorated so sharply that repayment conditions were tightened. From the 1965 winter crop season a system of collective security was introduced, under which the village tract became the primary unit of lending. Defaulting tracts were in 1966 excluded from qualifying for fresh loans until repayments of outstanding loans reached satisfactory levels. At the same time the State Agricultural Bank was made responsible for almost the whole of the credit program. Stricter accountancy seems to have brought finances into better balance through 1966-8, but the volume of loans issued was reduced to around K160 million per annum.

The state farm credit program was designed by the U Nu government largely on a welfare basis, and this stress on needs has been increased by the revolutionary régime. Roughly 80 per cent of all funds have been granted to paddy farmers, the poorest of the cultivators. There is little or no link with the ability of the recipient to use the

24 The state advance purchase system was confined to paddy and jute. It was dropped in November 1967 in favour of harvest loans because recovery was poor. The provision that recovery be made in kind from deliveries to state purchasing depots encouraged peasants to sell on the black market.

credit productively, nor is there any supervision of its use. Over 90 per cent of funds has been allotted in seasonal loans, issued on a flat rate area basis. Up to 1961-2 this flat rate was K20 per hectare for all crops. In 1962-3 it was raised to K62 per hectare for paddy and to a range of K25-309 for other crops, according to their estimated production costs. But rate calculations are not even based on the costs of using the best traditional farming techniques, and are wholly inadequate to finance new ones. The present K62 rate for paddy, for example, is said to be sufficient to meet the expenses of land preparation and the broadcasting of seed. It does not provide sufficient funds either to hire labour for the more productive line transplanting of rice or to purchase fertilisers or other inputs.

The loan system is seen primarily as performing a welfare role, not an investment function. This is demonstrated by the fact that larger farmers are excluded from qualifying and the loan limit for individual farmers is set low. Harvest loans are granted only to those who employ no labour. From 1964-5, monsoon cultivation loans have been confined to peasants working eight hectares or less save in special cases, thus setting, at the rate of K62 per hectare for paddy, a maximum of K500 per paddy farmer. Winter crop loans were limited to K500 per recipient from 1965-6 onwards.

Only a very small share of credit (since 1961-2 an average of 6 per cent) is tied to particular purposes. Most of this is allotted in kind, as, for instance, in farm inputs by institutions such as the Agricultural and Rural Development Corporation (ARDC) or in farm services, such as the hire of tractors and pumps to be paid for after the crops are sold. A minute fraction is granted to finance longer-term investment, such as the purchase of plough cattle, land improvement, fruit-growing, or rubber-planting.

Co-operative farming. The co-operative movement has without doubt proven the least successful sector of the agrarian reform program. The moderate wing of the AFPFL reformers assigned after independence a crucial theoretical role to co-operatives in raising the prosperity of the peasantry. The movement was expected to evolve more or less spontaneously from below, once social justice was achieved for the smallholder. This was because of the technical and financial benefits expected to flow from the increased scale and capital intensity of farm operations, and from producer diversification into the storage, simple processing, and marketing of farm products. By 1966 a network of over 12,000 'agricultural and multipurpose' co-operatives had been built up throughout the country (CSED 1952-(1965: IV, 148)) with some three-quarters of a million members.

In practice the role of the co-operatives seems to have been unimportant in the parliamentary period and to have been reduced to insignificance by the revolutionary régime. The motive for the creation of the societies was often formal, in that membership was a necessary qualification for land grants or state loans. They played a small active role in the allotment of state farm credit until 1958. They have a nominal role in organising some of the village works which are undertaken mutually by established custom. They are expected to help to channel farm produce to the state buying agencies and to handle a very small part of retail sales of state-controlled consumer goods. It seems clear that for the moment the government is relying on leadership from above to initiate technical change and investment in farming, and to control directly the marketing and processing of farm produce.

The agrarian reforms aimed to cure the ills of the smallholder sector, which were seen as arising out of an over-rapid commercialisation of farming, which exposed the peasantry to capitalistic exploitation. They aimed in the short run particularly to bring security to the rice farmers and labourers of the delta, whose conditions had been most adversely affected by the *laissez faire* policies of the colonial period, and to carry through for the peasant sector as a whole a 'fair shares for all' program.

In social justice much has been achieved by reform since independence, but it has almost all been in the general direction of aiding the lower strata of the cultivators, the landless tenants paying excessive rents, and smallholders, especially paddy-growers, paying high interest on loans to meet current expenses. To the extent that the reforms have in fact been implemented, both the security and net income of such poor farmers have been raised.

The remaining longer-term aims of the reforms have not yet been achieved. A viable co-operative system of farming has not been created and the share of individual farmers who are commercially oriented has declined. Much of the risk has been taken out of farming, but much of the profit and opportunity for enterprise also. The reforms have been of little benefit to many of the small owner-cultivators of mixed-crop farms, while for the few larger owners and tenants they have been detrimental. Thus the trend created by the reforms, at least in the short run, has generally been to raise the margin for consumption within the farm sector and reduce that for savings and investment. As yet they do not appear to have had any positive effect on production, and it is arguable that their impact has been negative.

It is, of course, possible that government policies towards the farm sector are laying the groundwork for future growth and development. An increased assurance for the majority of cultivators created by the agrarian reforms could in the longer run stimulate growth. Enlarged education and health services in the countryside should raise the efficiency of the rural work force. Direct benefits should eventually flow from government expenditure on long-term agricultural projects and institutional services. If marketing incentives are added, the reforms may yet prove to have been a necessary pre-condition to raising savings and productivity of the farm sector.

Structure and Productivity of the Farm Sector

The farm population of Burma is generally well fed by Southeast Asian standards, and adequately housed in traditional dwellings built with little cash outlay. But production for the market has been discouraged since the early 1930s. Since the land/man ratio and the size of farms is gradually diminishing without a commensurate rise in yields per hectare, the marketable surplus of the farm sector is being progressively eroded, and its contribution to capital formation reduced.

The agricultural share of national income is low and is probably diminishing. It is not proposed to discuss here Burmese estimates of national accounts; as in all semi-subsistence economies such estimates present difficulties. It is sufficient to state that the share of gross domestic product[25] attributed to agriculture, livestock, and fishery declined from 39 per cent in 1938-9 to 28 per cent in 1963-4[26] (MNP 1951b-(1956, 1964: 1A)). Recent estimates show a stagnation or reduction in the value of agricultural production since 1963-4, which was offset by a slight rise in those of other sectors (UBRC 1964-(1964-5; 1966-7: 3)). However debatable the statistics, there seems little doubt that the farm sector is making a declining contribution in real terms to economic growth.

This pattern of gradual decline can be seen from the productivity of farm labour, which occupied 65 per cent of the total work force in 1965, almost the same proportion as in 1931. In 1966 there were 58 per cent more workers on the land than in 1931, while agricultural production was estimated to have risen by 9 per cent comparing 1936-41 to 1962-6, and probably by some 13 per cent since 1931.[27] This would show, between 1931 and 1966, a reduction of some 29 per cent in the productivity of farm labour. Taking into account the fact that

[25] Calculated on the product (value added) method.

[26] Agriculture excludes rice milling or the processing of other farm products. Farm incomes are, of course, supplemented by earnings from other activities, such as petty trading, cottage industry, and construction.

[27] Data for the whole of the 1930s are available only for sown area and production of main crops of Burma Proper. These show a rise of some 4 per cent between 1930-1 and 1936-41. For the purpose of the above calculation, it has been assumed that production of all crops throughout Burma also rose by 4 per cent between 1930-1 and 1936-41.

the dependency ratio on farms has risen by more than the work force, a rough guess that the volume of the marketable surplus of the farm sector declined by almost one-third between 1931 and 1966 seems reasonable.

It is difficult to judge the relative importance of causative factors in the decline of farm productivity since independence. The impact of government policy, through agrarian reforms, farm prices, taxes, state investment, and current expenditures, has probably thus far been negative. Other social and economic factors have been significant. The rise in the rate of population growth has increased the dependency ratio within farm families. A lack of employment opportunities outside farming has discouraged urbanisation, so that underemployment in farm families has increased. Dacoity has reduced incentives to expand production. Incentives have been reduced by chronic shortages of consumer goods and the time-consuming inefficiency of the retail distribution system. The end result has been a reversion by many smallholders to the subsistence attitudes to farming of feudal times, partly stimulated by these difficulties and shortages, and partly by welfare government policies supporting the poorest of the peasants.

But the sluggishness of agricultural productivity is also part of the long secular decline, caused by lessening profitability resulting from deteriorating terms of trade, which has affected most sectors of the Burmese economy since the mid-1920s. The main spurt in the expansion of cultivation occurred in the period 1855 to 1910, when the increase in the area sown to rice in Lower Burma exceeded population growth by a wide margin (Table 5.1). A period followed to 1930 when the expansion of rice area was only fractionally faster than that of population in both Lower and Upper Burma. During most of the time from the opening of the Suez Canal to the depression of the late 1920s rice prices were either slowly rising or steady. From 1931 to 1941 the country's population rose by an average of 1·4 per cent annually; the area sown to rice remained static and paddy prices low. Then came the disruptions of World War II and the Karen rebellion of 1948-9. By 1952-3 85 per cent of pre-war rice land had been restored to cultivation. It is the exceptionally slow restoration of production from the early 1950s onwards which is significant, especially when set against increases in population, estimated to average 1·8 per cent per annum during the 1950s and 2·07 per cent annually between 1960 and 1966.

Price factors have been consistently unfavourable to most farmers over the past two decades and profit margins on most cash crops so slender that it is not surprising that leisure is preferred to field work and investment is minimised. The single-cropping of land to ordinary grade paddy or to any of the bulk field crops (excluding vegetables, tobacco, spices) for the market has been an activity only marginally economic in Burma at post-war price levels. Walinsky (1962: 42) estimated from 1953 census data that the annual receipts of the average four-hectare farm surveyed were K1,139 and expenses K716, leaving a profit of K423, of which the net cash income of those farmers reporting farm consumption as part of income was some K280. A delegate to the Kabaung Seminar of 1964 (DI 1962a-(7 Mar. 1964)) gave similar figures, applying to a four-hectare paddy farm in the Henzada district, annual income being put at K930, costs at K680, and profit at K250. The last figure is equivalent to US$52.[28] Official estimates of the value of production for the principal crops in the four seasons 1962-3 to 1965-6 show that the gross annual average value of production per hectare during this period ranged between K235 and K257 (US$49-54) for paddy, and between K282 and K346 (US$59-73) for all other crops combined (UBRC 1964-(1966-7: 62-5)). Against gross income must be set not only cash expenses for wages, interest, purchased seed, tools, manure, freight, etc., and amortisation of capital assets, but also the fact that *wunza* (home consumption, seed, and waste) greatly reduces the marketable surplus of smaller farms, which are in the great majority.

The various factors outlined above, dacoity, agrarian reform, lack of market incentives and of employment opportunities outside farming, have created a situation in which a cultivated area, not greatly enlarged since the colonial period, is shared amongst a much larger farm population. This farm population consists today largely of owner-cultivators or tenants secured on their holdings, who are disinclined or unable to invest in raising the productivity of their land.

The number of farms, 1·7 million in 1931, has since more than doubled. Agrarian reforms have not only increased the number of farms and reduced their overall average size, but have also ensured that there is a far greater equality in the size distribution of holdings. According to the 1954 census, less than one-quarter of all farmers cultivated holdings under 2 hectares, accounting for 6 per cent of the farm area surveyed. But the bulk of land was worked either in holdings of between 2 and 4 hectares (34 per cent of all farmers, 23 per cent of area, mainly in Upper Burma, mixed crops) or between 4 and 8 hectares (31 per cent of farmers, 38 per cent of area, some Upper Burma, but mainly Lower Burma, rice). The remaining one-third of the area was held in plots of over 8 hectares by 11 per cent of farmers (almost all Lower Burma, rice).

An estimate for 1964-5 (UBRC 1964-(1966-7: 6)) put the number of farms throughout the country at about 3·6 million and average

[28] This conversion into dollars has some relevance in this context, since many of the goods the peasant wants to purchase are either imported, or produced locally at prices at or above international levels.

TABLE 5.9

LAND CULTIVATED IN BURMA PROPER,[a] 1962–3

	No. of farmers ('000)	Hectarage cultivated[a] ('000)	Average holdings (hectares)	Average holdings		
				Paddy	Kaing (hectares)	Ya
Cultivating owners	1,806	3,841	2·12	2·35	1·13	2·18
Tenants	1,072	2,938	2·74	3·21	1·15	1·86
Total	2,878	6,779	2·36	n.a.	n.a.	n.a.

[a] Excludes double-cropping or fallow. Excludes Shan and Kayah states, Chin Special Division, Naga Hills and frontier areas (apart from Mayu district). Data do not show area worked in each farm size group.

Source: UBRC 1964–(1965–6: 5, 6, 12).

holdings at 2·2 hectares, but this estimate was apparently based on a blanket extension throughout the Union, including the states and frontier regions, of estimates for Burma Proper made in 1962-3 (Table 5.9), a decidedly dubious exercise.

Estimates for Burma Proper for 1962-3 (UBRC 1964-(1965-6: 6)), a date prior to the changes induced by the abolition of land rents, showed that 84 per cent of all holdings were less than 4 hectares, whereas 1954 census data showed 58 per cent of holdings in this size group.

As the size of holdings goes down, so does the proportion of farmers with holdings profitable enough, under present conditions and techniques, to produce margins for savings and investment. There are no statistics to show post-war private investment in the agricultural sector. On a limited scale there does seem to have been investment in private irrigation systems up to the early 1960s, a certain demand for pumps (which the government was unable to meet fully), and a sizeable investment in plough cattle, normally the cultivators' principal capital asset.[29] The number of plough cattle was increased from just over 3 million in 1952 to 3·7 million in 1965-6. But the disorganisation created by the agrarian reforms is illustrated by the fact that, although the number of plough cattle had by the mid-1960s been restored to pre-war levels in relation to the area cultivated, it had been halved in relation to the number of farms, there being 1·85 million yoke of cattle for 3·6 million farms in 1965-6, compared to 1·77 million yoke and 1·7 million farms in 1931. This means that many small farmers must now depend on custom ploughing or the hire of cattle or must combine to undertake ploughing co-operatively.

29 A pair of working bullocks costs around K500, and usually has a working capacity of about eight years.

Because cash incomes are so low, purchased farm inputs, including labour, are kept to a minimum and yields per hectare thereby reduced.[30] Very little chemical fertiliser is used, the total quantity imported averaging 31,000 tons per annum between 1961-2 and 1966-7.[31] Of this the bulk is used by rice farmers (MNP 1951a-(1962: 89)), mostly on nurseries producing high-grade strains. Green manuring and composting are little employed and rice stubble normally burnt off, because the nitrogen content of soil is too low to break ploughed-in straw down to humus. Cattle manure is, however, all used on fields and not burnt for fuel, as in India.

The use of chemical fertiliser has at no time since independence been an economic proposition for the farmer growing ordinary grades of paddy. The Knappen Tippetts Abbett consultants estimated (KTA 1953: 183) that the application of 112 kg of (imported) fertiliser per hectare at a cost of K84 would raise paddy yields by 30 per cent. Since the increased output would fetch K62 at official purchase prices, the return to the farmer would be negative. The ratio has become still more unfavourable to the farmer since that date.

Similarly, the reduction in farm size combined with stagnant productivity has reduced the marketable surplus. The demand of the farm family for basic foods remains roughly the same in present circumstances irrespective of the area cultivated, so that the share of output going to *wunza* increases almost in direct proportion as the size of the farm diminishes (not allowing for variation in seed usage). For example, *wunza* requirements for paddy farms in Lower Burma were estimated in 1956 at the equivalent of 1,565 kg per household, including single households (University of Rangoon 1957). Since the average of all-Burma paddy yields is just over 1,565 kg per hectare, *wunza* would absorb some 40 per cent of production of the average paddy farm.

POTENTIAL FOR GROWTH

Against this rather gloomy picture of declining surpluses may be set a number of more hopeful features. There does not seem to be any significant unsatisfied demand for grains and other staple foods (excluding proteins) in the farm sector, while that of the non-farm

[30] Seasonal labour hired for seven to ten months traditionally receives 2,000-3,000 kg of paddy plus board. But the trend since independence has been either to shorten the period to the rainy season only, paying 1,000-1,250 kg (Ne Win, in DI 1962a-(22 May 1963)), or to hire piece workers, principally for transplanting, by the day.

[31] Value of chemical fertiliser used in 1965-6 was K18 million, insecticides K4 million, quality seed K20 million, tractor services K34 million, the total of K76 million being equivalent to 3·3 per cent of value of all crops (UBRC 1964-(1966-7: 9-11)).

sector is probably small. Any increase in staple crop production should therefore produce an addition to exportable surpluses, unless affected by hoarding in the short run. The experience of rice shortage of 1966-8 must have proved a salutary shock to government planners, so that nominal may become actual priority for agriculture in state planning and expenditures. Recent opportunities to sell paddy at premiums over official prices may stimulate production from farmers.

Although the value of modern inputs used by farmers is still very limited in relation to the value of production, the state farm services are being gradually built up and some of the longer-term research and extension work may soon have a significant effect in raising yields. The Gyogon Research Institute and its thirteen affiliated experimental stations are carrying out basic research on the principal crops, and the Department of Agriculture is making more quality seed available through private foundation stock farmers.[32] Of the new rice strains, Rizal, developed from stock imported from the Los Baños Rice Research Institute, is said to be proving itself in field trials, with yields three to five times average (*Working People's Daily*, 20 Oct., 18 Nov. 1967).[33]

Construction of chemical fertiliser factories (originally planned in 1952) at length started in 1967. Two plants, based on natural gas from the Chauk oilfields, each with a capacity of 60,000 tons of urea per annum, are scheduled for completion by 1970, one being built at Kyunchaung with West German technical assistance and credit, and the other at Sale by a Japanese group. Survey work on the giant Mu River irrigation scheme, planned in 1953 to irrigate 445,000 hectares, began at the end of 1967 with assistance from the U.N. Special Fund. The Kyetmauktaung dam, built at the confluence of the Taungzin and Kyaukpon with Soviet aid to irrigate 20,000 hectares, was completed at the end of 1967, the Washawng dam at Myitkyina to irrigate 12,000 hectares in early 1968, and a number of smaller projects are under way.

More thrust may be put behind plans to extend the double-cropped and irrigated area, which it was proposed in the early 1950s to raise to about 2·4 million hectares (Trager 1957: 47). Double-cropping cannot be greatly extended without tractor ploughing, since the work involved is beyond the capacity of bullocks and it is estimated that, to plough

[32] In 1963-4, 288,000 tons of improved paddy seed were distributed (enough to plant over 400,000 hectares), 4,900 tons of groundnut seed, 2,700 of cotton seed, and some maize, beans, etc. (MNP 1951a-(1964: 135)), the total valued at K7·7 million (UBRC 1964-(1966-7: 11)).

[33] It is, of course, too early yet to judge whether such new strains will prove high-yielding and disease-free in extensive use. Moreover, no data are available on the quantity and price of fertiliser needed to induce high yields or on the sophistication of fertiliser usage and water management required.

2·4 million hectares, 20,000 tractors will be required (Laurence French
Publications 1946-: July 1963). Thus far the mechanisation program,
now controlled by the ARDC, has run into continuous trouble. Walinsky (1962: 288) estimated that the 130 tractors used for ploughing in
1956 each worked 500 hours, well below economic operation. By 1966-
7 the number of tractors had been built up to 4,511 and the area
ploughed to 276,000 hectares (*The Guardian*, 23 Sept. 1967). But the
number of working hours had dropped to 250 per machine per annum,
while state losses on custom ploughing amounted to K26 million in
1965-6 (UBRC 1964-(1966-7: 9)).

As already stressed, political and social factors have been and remain
major hindrances to economic development generally in Burma. Until
political solutions to racial disunity within the Union are achieved
and attitudes to non-Burmese enterprise and foreign participation are
modified, economic growth and effective resource allocation must
necessarily be limited.

Granted these constraints and the very limited domestic resources
of skill and capital at the country's disposal, there are theoretically a
number of directions in which the socialist economic planners could
aim. Let us dispose first of two alternatives which do not appear
practical in present circumstances. Socialisation of the farm sector
could become much more radical—mandatory programs of crop production and state procurement could be enforced, and collectivisation
(which has so far been confined to a few reclamation settlement
schemes) pushed through rigorously throughout the country. Although
this would have technical advantages, especially in eliminating
boundary lines to small farms to facilitate mechanical culture, it seems
most unlikely that the already grossly overloaded government machine
could handle such a program, even if it were politically feasible.

Nor does market socialism on the Yugoslav pattern appear likely to
be able to provide an answer for the farm sector while the economy
is in such low key. Such a solution, entailing a liberalisation of prices
of state enterprise, would be practicable only if the state had adequate
international reserves to finance a constant and ample supply of consumer and producer goods to the rural sector, either directly, through
imports, or indirectly, through allocations of additional foreign exchange to domestic manufacturing industry. Nor (save to the extent
that it can obtain higher export prices) can the state afford to offer
higher purchase prices for ordinary-grade paddy, since this would
further reduce state resources for investment, and might add to inflationary pressure by raising farmers' demand for consumer goods rather
than stimulating farm investment.

There would seem to be only two realistic choices. The first would
be the maintenance of present 'big planning' policies, which aim in the

long run to diversify the economy, build up local manufacturing industry, and create a technically modern agriculture; and in the short run to give, within the traditional farm sector, social objectives precedence over economic objectives, relying on socialist exhortations to stimulate growth, until such time as long-term social and economic investment bears fruit. Such policies have thus far spread resources of talent and capital so thin that no sector has progressed satisfactorily, but their reversal would not be popular politically.

The second would be to back-pedal on the industrialisation program and long-term agricultural projects until a larger surplus can be stimulated from the traditional farm sector by a variety of short-term pragmatic measures, and to reduce within the farm sector the great emphasis on egalitarianism which has characterised policy since the 1940s. While the general pattern of agrarian reform could be maintained, any additional resources could be highly concentrated on those areas or farmers, whether individuals or co-operative groups, that could demonstrate a capacity to use capital productively, resulting in a greater inequality of income between efficient and inefficient farmers than at present. Since the bulk of state resources flowing to the farm sector is expended in the farm credit program, it would seem logical to raise as rapidly as possible the proportion of credit which is allotted in kind, and to back this up with the provision of inputs and technical advice from state institutions. An obvious first step would seem to be to think out and then publicise a properly integrated policy on the fertiliser/paddy price ratio. When devising such a policy, it would appear more sensible to subsidise fertiliser use to raise production of rice than to subsidise retail rice prices and raise consumption.

Whichever of these general strategies for growth is pursued, the difficult choice still remains of deciding priorities between rice, alternative export crops, and import replacement crops such as oilseeds and fibres. With internationally traded rice supplying less than 6 per cent of world rice consumed, world trade is clearly vulnerable to the marginal changes in production which may well result from higher-yielding rice strains, and to switches by consumers to alternative grains. Nevertheless, in the immediate future demand prospects for Burmese rice appear fairly good, especially as food aid from the United States to India and other traditional Burmese markets seems likely to decline, while there is an income elasticity in these markets in demand for rice. But while pushing ahead with raising rice output and export surpluses, yields of other crops, both for export and import replacement, will clearly have to be raised if farm profitability is to be increased, the balance of payments position improved, and foreign trade diversified in the longer run.

6

Malaysia

E. K. Fisk

MALAYSIA is a relatively new state, and, as its composition has changed in the last few years, it may be advisable to outline its political and geographic composition. It is, in fact, a federation of all the territories in the Malaya-Borneo region that were in the early 1950s under British guidance or control, with two exceptions: the very small state of Brunei, which declined to come in to the federation; and the Republic of Singapore, which withdrew from the federation in 1965.

The Malaysia of 1968, therefore, comprises three main political components: West Malaysia, made up of the nine states and two settlements of the older federation of Malaya; the state of Sarawak, in Western Borneo; and the state of Sabah, which, prior to incorporation into Malaysia, had been known as British North Borneo.

The geographic, and even ethnic, links between the three major political components of Malaysia are not particularly strong, and are in fact looser in some respects than those which each component has with its immediate neighbours outside the federation. The states of Sarawak and Sabah have a very small length of common land frontier, but a very long one, in each case, with the Indonesian region of Kalimantan. They are both separated from West Malaysia by hundreds of miles of the South China Sea. West Malaysia, on the other hand, has a common frontier with Thailand and with Singapore, and is within 32 km (20 miles) or so of the Indonesian island of Sumatra at its nearest point, and even closer to the Rhio Archipelago to the south. Ethnically, the Malay majority of West Malaysia have more in common with the people of Rhio and West Sumatra than with most of the indigenous people of East Malaysia, though there is a substantial minority population of Malays in Sarawak.

However, in economic, social, and political structure, and in the institutions and attitudes on which this structure is based, East and West Malaysia have a great deal in common, and are very much closer to each other than they are to their Indonesian and Thai neighbours.

Table 6.1 sets out some features of the economic geography of Malaysia. From this, some fundamental differences between East and

Map 7 Malaysia

TABLE 6.1

BASIC DATA, 1966

Region	Population (million)	Land area sq. km	sq. miles	Gross national product (M$ million)
West Malaysia	8·42	131,300	50,700	8,058
West Malaya	6·96	53,600	20,700	
East Malaya	1·46	77,700	30,000	
East Malaysia	1·46	201,000	77,600	1,182
Sabah	0·58	76,000	29,400	531
Sarawak	0·89	125,000	48,300	651

Source: Derived from information supplied by Department of Statistics, Kuala Lumpur, December 1967; also from Malaysia 1965b: 3; Rueff 1963: 9.

West Malaysia are apparent. East Malaysia has 60 per cent of the total land area, but only 15 per cent of the population, and produces only 12·8 per cent of the gross national product. Population density in West Malaysia averaged 64 per sq. km (166 per sq. mile) in December 1966, but in East Malaysia the average density was only 7 per sq. km (19 per sq. mile). Nevertheless, in terms of gross national product per head, the difference is much less, averaging M$957 (US$313) in West Malaysia, as against M$810 (US$265) in East Malaysia in 1966.[1]

THE ROLE OF THE RURAL SECTOR

Before concentrating on the rural sector, some features of the Malaysian economy as a whole will be examined in order to show the role of rural industries in the economy, and the economic setting of the rural sector.

Table 6.2 shows the main structural features of the economy. Primary production is the source of 38 per cent of gross domestic product, compared with industry's mere 15 per cent. Services (government and commercial) are highly developed, providing nearly 47 per cent of gross domestic product. Comparable figures for earlier years are not available, as Malaysia dates only from late 1963, and in its present form only from 1965. However, figures for West Malaysia are available for 1960 to 1964, and these are summarised in Table 6.3.

[1] Within West Malaysia a somewhat similar distinction can be drawn between east and west sections of the peninsula. In 1965, 83 per cent of the population lived in the more developed western section, whilst the less developed eastern section (60 per cent of the land area) supported only 17 per cent of the population.

TABLE 6.2

GROSS DOMESTIC PRODUCT, 1965

(by industry of origin)

Industry	M$ million	Per cent
Rubber	1,095	14·0
Other agriculture, forestry, and fishing	1,250	15·9
Mining and quarrying	625	8·0
Primary production, total	2,970	37·9
Industry	1,215	15·5
Services	3,645	46·5
GDP at factor cost	7,830	100·0

Source: Malaysia 1965b: 62, Tt. 3–13.

The picture of the Malaysian economy presented by these figures is fairly clear. It is an economy based on primary industry, in which secondary industry is as yet poorly developed, and with a high proportion of tertiary (service) industries. This balance between sectors is, however, changing quite rapidly, with the relative importance of primary industry declining in relation to secondary and tertiary industries.

Within this pattern, the composition of sectoral incomes is significant. In the primary sector, there is a decreasing, but still heavy, dependence on rubber. The secular decline in the price of rubber has been an important factor in this change, but it is by no means the only one. The increasing diversification of rural production, in particular through the rapid development of the timber and oil palm industries, and the healthy increases in rice, tea, fruit, and fish production, have also been important.

In secondary industry, manufacturing has responded to various government incentives, but the increase in total value added, though significant, is on a small base. This development is spread over a considerable number of small industries, and, whilst processing of primary products is still the major activity in this sector, the extent and sophistication of the processing is increasing, and there is a significant extension into new types of manufacture for the domestic market.

The employment situation emphasises even more clearly the dependence of the Malaysian economy on primary production, and in particular on agriculture. Table 6.4 shows the position for West Malaysia in the years 1960 and 1965. By the latter year the proportion

TABLE 6.3

WEST MALAYSIA: GDP AT FACTOR COST (BY INDUSTRY OF ORIGIN, SELECTED YEARS)

Industry of origin	1960 M$ million	Per cent	1962 M$ million	Per cent	1964 M$ million	Per cent
Rubber	1,226	23·5	1,023	18·5	991	15·7
Other agriculture, forestry, and fishing	750	14·4	808	14·6	879	13·9
Mining and quarrying	306	5·9	391	7·1	502	8·0
Primary production, total	2,282	43·7	2,222	40·2	2,372	37·6
Manufacturing	453	8·7	472	8·5	612	9·7
Construction	158	3·0	240	4·3	289	4·6
Electricity, water, and sanitary services	70	1·3	86	1·6	112	1·8
Transport, storage, and communications	189	3·6	199	3·6	220	3·5
Wholesale and retail trade	817	15·7	900	16·3	1,034	16·4
Ownership of dwellings	245	4·7	264	4·8	291	4·6
Public administration and defence	339	6·5	342	6·2	428	6·8
Banking, insurance, and real estate	71	1·4	86	1·6	98	1·6
Other services	596	11·4	716	13·0	854	13·5
Total	5,220		5,527		6,310	

Source: Malaysia 1964: 19, T. II.

TABLE 6.4

WEST MALAYSIA: EMPLOYMENT

Industry	1960		1965	
	Thousand	*Per cent*	*Thousand*	*Per cent*
Agriculture, forestry, fishing	1,277	55·2	1,388	51·8
Mining and manufacturing	196	8·5	234	8·7
Construction, transport, utilities	150	6·5	210	7·8
Government services	200	8·7	257	9·6
Other trade and services	351	15·2	429	16·0
Unemployed (overt)	138	6·0	160	6·0
Total labour force	2,312		2,678	

Source: Based on figures given in Malaysia 1965b: 35.

of the total labour force engaged in agriculture had fallen, but it was still over 50 per cent, despite the fact that the proportion of gross domestic product derived from agriculture had fallen to somewhere under 30 per cent. Overt unemployment was moderately high for both years at 6 per cent, but had not increased, despite the substantial increase in the size of the total labour force over the period.[2] However, this unemployment figure is of limited significance; it excludes a great deal of rural unemployment that remains unregistered, and does not indicate the extent of under-employment which, particularly in the rural areas, is certainly substantial.

In export performance, the primary sector is completely dominant. The major Malaysian exports are rubber, timber, oil palm products, coconut products, pepper, and iron. This is a disturbingly short list, and although 'other' exports have increased faster than total exports over the period 1961-7, this has been on a small base, and in 1967 accounted for less than one-fifth of the total. The four major agricultural exports and timber accounted for 60 per cent of gross exports in 1961, and 55 per cent (provisional) in 1967. Taking all merchandise exports together, practically the whole originates in the rural sector, and the direct contribution of other sectors is confined largely to the processing of rural products, with a small credit for services to offset the much larger debit for imported services.

THE SETTING FOR ECONOMIC DEVELOPMENT

The economy of Malaysia has a number of difficulties to contend with in planning its economic growth. Population growth is about 3 per

[2] In 1967 there was an increase in overt unemployment to about 6·5 per cent (Tan 1968: 9).

cent per annum. The most important economic product, rubber, has been subject to a long, severe, and apparently irreversible decline in price. The key mineral resources are either being worked out (tin) or subjected to severe competition from larger and richer ore sources overseas (iron). The internal market is relatively small, and the wage level high for Southeast Asia, so that rapid development of large-scale manufacturing is difficult. There are racial and political problems that act as constraints on development policy; some of these will be discussed in detail below.

On the other hand, in most respects Malaysia is well endowed, and in the economic, political, and social setting for development it is more favourably placed than many developing countries. Pressure on land resources has not yet developed, and even in West Malaysia there are large tracts of virgin land still available for settlement. The economic infrastructure—roads, rail, and air services, telephone and postal communications, financial institutions, distribution and marketing services, a sound and stable currency—all essentials for economic development, are provided and maintained on a scale and to a standard very much higher than is found in other rural-based economies of South and East Asia. Malaysia is well-endowed with skills, both technical and managerial, and the education system can supply most of the trained manpower required for development at secondary, and even at tertiary levels. Savings are a limiting factor, particularly for capital formation in the public sector, but only in relation to the high rate of public investment planned. In the years 1965 and 1966 capital formation proceeded at 20 per cent and 19 per cent respectively of the gross national product. Considering the relatively high level of gross national product per head, this represents a healthy and substantial sum.

The balance of payments position has been reasonably sound and it is backed by considerable reserves. In recent years the balance on merchandise account has been uniformly favourable, with a net credit of M\$527 million and M\$554 million (US\$172·2 million and US\$181 million) in 1965 and 1966 respectively. Added to this were official grants received for modest sums (US\$36·3 million in 1966), but there has invariably been a substantial net deficit in other invisibles, leading to a net deficit on current account from 1961 to 1964 inclusive, with a small surplus (US\$18·2 million) in 1965 and a complete balance in 1966. Private capital inflows and official loan receipts financed the deficit on current account from 1961 to 1966, with the result that Malaysia's official reserves increased slightly (by US\$19 million) during that six-year period. Considering the high level of government development spending during the period, this was a formidable performance.

In December 1966 Malaysia's foreign reserves stood at M$2,625 million (US$858 million), sufficient to finance about nine months' imports at the 1966 level.

However, it seems clear that prospects for increasing Malaysia's foreign earnings from exports will be impeded by world market conditions for its main products, and, despite improvement in efficiency and quantity of production, Malaysia will have to rely to a greater extent than previously on foreign loans and grants, to sustain the development expenditure planned. It should be in a good position to do this, for, apart from military assistance during its battle against communist insurgency and during Indonesian confrontation, Malaysia has depended mainly on its own resources for government investment and current expenditure. The public debt of the central government was 32 per cent of the gross national product in 1966, and of this only about 17 per cent was external debt. Under these circumstances, and with its record of financial soundness and stability, Malaysia should be able to obtain the still modest foreign loans and grants required to complete its current five-year plan.

Finally, wages have been rising gradually over the last six or seven years, whereas retail prices have remained remarkably stable. Table 6.5 gives some indication of the order of wage increase in some selected rural occupations between 1959 and 1965. During this period the 'All Races' retail price index rose from 100·0 to 102·3, which shows not only that the wage increases were a real gain to the recipients, but

TABLE 6.5

WAGE RATES, SELECTED OCCUPATIONS, 1959 AND 1965

Occupation	1959 M$ per day	1965 M$ per day	Increase (%)
Rubber estate foremen	3·60	5·35	48
Rubber estate tappers	2·90	3·15	9
Coconut estate foremen	3·55	4·15	17
Coconut estate weeders	2·25	2·55	13
Oil palm estate foremen	3·70	4·25	15
Oil palm estate sprayers	3·05	3·70	21
Tea estate foremen	4·00	4·30	8
Tea estate pruners	2·90	3·20	10
Tin dredge foremen[a]	14·84	19·43	31
Tin dredge crew, semi-skilled	4·90	6·20	27
Bulldozer drivers	3·95	5·50	39

[a] Calculated from monthly rate.

Source: Derived from Federation of Malaya 1961a: 133–4; Malaysia 1967c: 136–7.

also the degree of stability sustained in the Malaysian economy during a period of intense government development activity.[3]

Politically and socially the setting for development is also favourable. There is a stable and strong central government backed by a competent civil service. Law and order is well maintained. There is some racial friction, mainly between people of Malay and Chinese races, but this seldom breaks into open disturbances and is firmly controlled. The language problem of a multi-racial society has been overcome, at least to the extent that virtually all citizens can communicate effectively with one another, and complete illiteracy is becoming rare. Social services are well developed, and a start has been made with population control through family planning. Compared with most developing countries in Asia, Africa, or the Pacific, Malaysia is well placed in these respects.

However, by no means everything in the Malaysian situation is favourable to economic development, and there are very serious problems that act as constraints on policy and modify the course of development.

THE RURAL DILEMMA

The economic importance of the rural sector has been indicated. Including mining, the rural sector is the source of 38 per cent of the GDP, over 50 per cent of gainful employment, and virtually the whole of Malaysia's exports. But the political importance of the rural sector, and its implications, are even more marked.

In West Malaysia, by and large, the urban population is predominantly Chinese, while the rural population is predominantly Malay.[4] In East Malaysia there is a somewhat similar pattern, except that the rural population comprises mainly indigenous Borneo races such as Dyaks and Melanus. The Malaysian constitution gives the rural electorates proportionately more representation in Parliament than the more densely-populated urban electorates. The result is not so much that the rural people rule Malaysia—for the people with the skills of politics and government come often from urban backgrounds—but those who rule Malaysia do so only with the consent of the rural people. No government could long continue in office, by democratic means, unless its policies in the main pleased the rural people. Moreover, as the majority of the rural people are Malays in West Malaysia,

[3] See Malaysia (1967c: 124). The index tended upwards more sharply in 1966, and by June 1967 had reached 108·5. An important factor was the 30 per cent increase in the price of rice between 1965 and June 1967.

[4] Chinese dominance in the main urban areas has been reduced somewhat in recent years, but is still very marked.

and Sea Dyaks, Land Dyaks, etc., in East Malaysia, it is these people in particular that the government of Malaysia must satisfy with its policies of development.

This causes serious planning problems for any Malaysian government. For there is another structural defect in the Malaysian economy which we have, as yet, hardly touched. Malaysia is, by Asian standards, a wealthy country. Per head, its gross national product is one of the highest in Asia—but this product, and the income derived from it, is very unevenly distributed.

Figures for this are difficult to obtain. Some light has been shed on the matter by the *Household Budget Survey* for the year 1957-8 (Federation of Malaya 1959a), for West Malaysia. This survey indicated the relative levels of consumption of Malay, Chinese, and Indian households in rural and in urban areas. For the three races combined, the average of all income groups was M$1,920 (US$627) per annum for rural households, and M$3,113 (US$1,017) for urban households. These figures are now more than ten years old, apply only to West Malaysia, and in any case were based on a small (0·25 per cent) sample of households. However, they are supported by other, more general indicators,[5] and although few would be so bold as to attempt a precise quantification, there can be little doubt that the difference is substantial. In other words, although the gross national product per head in 1966 averaged M$957 in West Malaysia and M$810 in East Malaysia, there is no doubt that the average income per head in the urban areas of both regions was substantially higher than this, and the average in rural areas lower.

Even this is only part of the story. The Malaysian rural economy can be further divided into two segments which to a large degree are distinct.

First, there is an advanced segment. Within this, rural industry is very efficient, highly capitalised, advanced in technology, and very productive. This is the segment of the estates and mines and the timber-logging firms, upon which the favourable situation of the Malaysian economy has largely been constructed. In 1966 it produced half of the rubber, most of the palm oil, virtually all the tin and timber, and all the iron. On the other hand, it employed only about 410,000, or 16 per cent of the available labour force. At least 90 per cent of these are employed as unskilled or virtually unskilled labour at the lowest end of the Malaysian wage structure.

Second, there is a backward segment. This comprises small, often fragmented holdings in which rice, rubber, coconuts, and fruits, in

5 For example, agriculture employs over 50 per cent of West Malaysia's work force, but produces less than 30 per cent of GDP.

the main, and to a lesser extent pineapples, pepper, vegetables, and jungle produce, are produced with little capital—though often with considerable skill—and in which under-employment and low levels of income (by Malaysian standards) are common. Here a large proportion of families have incomes that are probably less than M$300 per head, or less than one-third of the comfortable-sounding average for Malaysia as a whole.

This distribution of income may have some theoretical advantages in purely economic terms, as, for example, from the point of view of the level of voluntary savings and investment it makes possible. However, politically and socially it is clearly disadvantageous, and when one considers that the backward segment is largely indigenous and contains most of the voting population, whilst the advanced segment is still largely non-indigenous in ownership and control, it is clearly untenable politically. We shall shortly see just how restrictive this is.

First, however, let us look at the heavy dependence on imports, without which present levels of consumption could neither be sustained in the short run, nor improved in the longer run. This dependence is particularly important, from the viewpoint of development, for three categories of imports: capital goods, raw materials and components used for productive processes; petroleum products needed to keep the productive capital operating; and certain foods, other consumer goods not locally produced, and even some consumer durables.

To pay for these imports, Malaysia relies mainly on five major export commodities: rubber, tin, timber, palm oil, and coconut products. One of the important features of Malaysia's development program must therefore be, on one hand, to increase proceeds from these exports, and where possible to add new ones, and on the other hand, to substitute local production for imports.

MALAYSIAN DEVELOPMENT POLICY

In its simplest terms, therefore, Malaysian economic development is dependent on raising the productivity of the economy in export industries and in import replacement industries. The rural sector, encompassing agriculture, forestry, fishing, and mining, must play the major part in this process for many years to come, as it is in these spheres of economic activity that Malaysia is, so far, most favourably placed.

This is not to suggest that manufacturing industry should be neglected, but merely that it will develop from a much smaller base, and that for some time the opportunities seem likely to be limited, on the home front by the small local market, and on the export front by the relatively higher cost of some inputs compared with certain foreign producers. Whilst these difficulties may eventually be overcome in

some industries, it is clear that the main thrust of Malaysian development must be initially in the rural sector.

The next choice in development policy must be between expansion
of the advanced or the backward segments of rural production, or both.

From many points of view the best policy might appear to be to
devote all available resources to the expansion and development of the
advanced segment of the estates and mines, and to let it encroach
upon, and as far as possible absorb, the backward smallholder segment. This would provide not only the most rapid, but also the most
efficient and economic means of expanding production with the
limited resources available. It would be making the best use of the
skills and institutions, built up over the last half century or so, on
which Malaysia's prosperity has been structured, and in which it has
a unique advantage. One could visualise the efficiency, advanced technology, capital, and high productive capacity of the estate system
spreading rapidly, with minimal government support,[6] not only into
new land and new crops, but even into traditionally smallholder crops
such as paddy. It would seem plausible to expect that the return per
dollar of government development expenditure in support of the
advanced segment would be higher, in terms of increases in national
product, than for expenditure in support of the backward sector.

However, this path of development has not commended itself to
the government of Malaysia, and indeed, whatever the merits or otherwise of the purely economic arguments in its favour, it is questionable
whether such a policy could be consistently followed by any demo-
cratically based government where the majority of the electorate are,
or aspire to be, peasant farmers. This is the more so where, as in
Malaysia, the advanced segment is largely owned and controlled by
foreigners, or Malaysians of races other than that of the electoral
majority. For this reason the governments of Malaya, and subsequently
of Malaysia, have for many years pursued an agricultural policy in
which development effort and resources have been heavily concentrated
in support of the backward sector. This policy has become clearer and
more emphatic with the growth of responsibility to an electorate, but
it commenced in the colonial period, even before the introduction of
elections, and has been clearly in evidence in relation to rice production since before World War II.[7]

The development and implementation of this policy is of interest,
not only because it has shown a surprising measure of success, but also
because the factors responsible for this success have not been clearly

[6] To some considerable extent this took place in the early days of rubber planting; see Jackson (1968: Chs. 10 and 11 (in particular pp. 241-51)).

[7] The passing of the Malay Reservations Enactment in the F.M.S. in 1913 was
an early, though somewhat isolated step in this direction.

identified even in Malaysia itself. When this is done the prospect of their further exploitation and development, both in Malaysia and in other agriculturally-based economies of the region, is quite exciting.

In the whole field of rural enterprise, the scope for promotion of the small individual producer varies considerably. In mining it has diminished rapidly over the years as richer surface deposits have been worked out and industry has been forced to depend more and more on the extraction of deeper or lower-grade deposits of tin, or on lower-value ores such as iron, in which only medium- to large-scale production is economic and practical. In forestry there is scope for small capitalist and co-operative enterprise in timber areas close to existing roads and villages, but these are, in the main, less viable than the larger-scale companies with whom they have to compete. The rapid development of major new forest resources in remoter areas, as in East Malaysia, is often practicable only with the organisation, capital, machinery, and collection of skills that the larger companies can bring to bear. In fishing, the lot of the small individual or village-group fisherman has been improved by government assistance in mechanisation of fishing boats, in preservation (ice plants), and marketing, but it seems that the scope for further expansion of short-range in-shore fishing is now very limited. Further expansion of the industry may depend increasingly on more distant deep sea fishing, for which neither the equipment nor the way of life of the individual village fisherman is well suited.

In these industries, therefore, the scope for further development of the backward segment through the expansion of owner-operated small-scale production is quite small, and although it has been encouraged and supported where possible, the main thrust of government development effort has been to increase local participation, through local shareholding and employment (including employment at managerial and professional levels), without unduly limiting growth of the industry in the process.

AGRICULTURAL DEVELOPMENT POLICY

In agriculture the situation has been rather different, both in the size and political strength of the population of the backward component and in the scope for expansion by small individual producers. Land has not been a limiting factor, and the purely technical limitations on the effectiveness of the small individual producer seem less intractable. In agriculture, therefore, there is a combination of strong political incentive to develop through the small agriculturalist, and some hope that this might produce results if the right techniques can be developed.

Malaysian agricultural policy has developed along two main lines. First, towards the advanced estate sector, the policy has been one of intensive development. It has discouraged expansion into new land areas, but has encouraged and assisted the development and improvement of existing estate land through technological change and reinvestment of profits. In rubber, in particular, the implementation of this policy has not been particularly difficult with an industry already progressive and efficient, and has had considerable success.

Second, towards the backward smallholder sector, the policy has been to foster both *intensive* and *extensive* development. It has aimed to improve the efficiency of all smallholder production, and at the same time to expand considerably the area farmed by smallholders. A wide range of organisational and institutional innovations has been tried in implementation of this policy. Some have been unsuccessful; some have been discarded; many have been modified; but there have been successes, and from these a pattern is emerging that has not yet had the recognition it deserves. Once this pattern has been recognised, a large part of the long and costly process of groping towards smallholder development can be replaced by a more direct and economical approach, both in Malaysia and in other countries seeking to base their agricultural development on peasant smallholdings.

Basically the pattern is simple, although the forms through which it has to be implemented are complex. Success has come where the innovations have given the small peasant producer access to economies of scale. These economies, available to the large producer as internal economies, had to be provided as external economies to the peasant producer. Where the activities of the peasant producers have been organised or supplemented by government institutions to produce this effect, a substantial increase in the volume and efficiency of production has resulted; where the innovations have not produced external economies of scale, total production may have risen (as through the cultivation of new land), but the level of efficiency, and hence of economic returns to factors, has remained low. Let us consider how this has operated, first with rubber, and second with rice, in West Malaysia.

RUBBER

Rubber, still Malaysia's largest source of foreign earnings and largest industry, has faced increasing competition from synthetic substitutes since World War II, with the result that world prices, whilst still subject at times to considerable variation about the trend, have declined from an average of over US55 cents per lb in the boom year of 1951, to an average of under 18 cents per lb for the year 1967.[8] Moreover,

[8] These figures exaggerate the trend somewhat, as the high price in 1951 was due to the Korean War, but there is no question that there has been a secular downward trend over the period.

although one might hope that prices for natural rubber should now tend to level off somewhat, the likelihood of a return to prices of 25 cents or more, other than in a short-term fluctuation, seems at the moment of writing to be quite remote.

This trend in rubber prices has long been foreseen and the response of the rubber industry and the Malayan governments has been one of the success stories of Asian agriculture. The combination of research, a new technology, and reinvestment of profits, has led to the replanting of most estate rubber, and over 50 per cent of the smallholder rubber, in West Malaysia. This replanting, using new high-yielding clones, has greatly raised the yield of rubber produced per hectare in tapping, and substantially reduced the labour costs of its production. As a result, the replanted area has been better able to cope with the reduced prices than that planted with unselected trees. In addition, the increased output of rubber has, to a considerable extent, offset the effects of the lower price, and so has avoided an otherwise catastrophic reduction in the nation's earnings from its prime industry.

Fig. II illustrates the magnitude of this achievement. Line P is an index of the price for rubber from 1961 to 1967, constructed from the average prices of rubber f.o.b. in bales, Malaysian ports, with the 1961 average as base. Line Q is an index of the quantity of rubber produced annually in West Malaysia, with the quantity at the end of the period (1967) as base. Line V is the product of the two indices P and Q, and shows how the value of West Malaysian rubber production varied over the period.

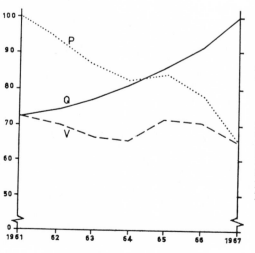

Fig. II   West Malaysia: indices of rubber price (P), rubber production (Q), and V, the product of index P and index Q

Year	P	Q	V
1961	100	72	72
1962	94	74	70
1963	87	77	67
1964	82	81	66
1965	84	86	72
1966	78	91	71
1967	66	100	66

Line P also shows what would have happened to West Malaysia's income from its major industry if production had not risen. In fact, had it not been for the success of the replanting program, it is quite likely that production would also have fallen substantially, not merely with the ageing of trees, but also with the collapse of some higher-cost producers who, without the reduction in costs derived from the increased efficiency of the high-yielding trees, might have been unable to continue at the low prices in 1966 and 1967. The effects on employment and on the general level of Malaysian economic activity, as well as on the balance of payments and on government revenue, would have been disastrous. The far-sighted action of the industry and government in the early 1950s, and the continued vigorous prosecution and improvement of these measures to the present day, have avoided, or at least greatly softened, what would otherwise have been a crippling blow to the standard of living and development prospects of Malaysia.

The earliest, and so far the greatest, successes with this rubber policy were achieved in the estate sector. By 1966 in West Malaysia 81 per cent of estate rubber area was under high-yielding clones, and the average production of rubber per hectare in the estate sector (including all immature stands not yet yielding) had risen from 550 kg (491 lb per acre) in 1956 to 712 kg (635 lb per acre) in 1966. Yields per hectare in tapping were estimated at 1,007 kg per hectare (898 lb per acre) for all types of estate rubber in 1966, and yields per hectare of high-yielding material were higher still, at 1,175 kg per hectare (1,048 lb per acre).

Amongst smallholders replanting proceeded more slowly, and was particularly slow to make a significant start. By 1966 about 59 per cent of smallholder rubber area was estimated to be under high-yielding material, and over the five years 1961-5 smallholder replanting (excluding new planting) averaged about 3·4 per cent of the total smallholder rubber area per annum.[9]

The reasons for this slower response in the smallholder sector are not difficult to understand. The commercial rubber tree was derived from a jungle species, and the planting and rearing of the original jungle tree in Malaysian conditions was not an exacting process. If the initial planting of the seeds could be combined with a catch crop during the first year or two, so that weed growth was controlled, the initial planting of unselected rubber did not make heavy calls on the capital or labour resources of the smallholder. Unselected seed could

[9] Barlow and Lim (1968), with the resources of the Rubber Research Institute of Malaysia, offer some up-to-date figures in this difficult area of smallholder statistics.

be obtained cheaply, sometimes for little more than the cost of gathering it from the ground. The cost of initial clearing could be offset by the value of the jungle produce and of the first two years' catch cropping. Thereafter, the newly-planted areas were commonly given little further attention until they had grown in their natural jungle-type environment to the point where they were tappable. Then narrow tracks were cleared from tree to tree and cropping commenced. This suited the rubber tree fairly well, and conformed in significant respects to the then favoured 'forestry method' of rubber cultivation. Moreover, labour and cash costs of maintenance were minimal, trees that died being replaced either by self-seedings, or by the growth of trees previously stunted by competition. In this way, when the annual yield of a reasonable estate was about 560 kg dry rubber content per hectare (500 lb per acre), the smallholder was able, with minimal special investment and little maintenance, to produce quite acceptable yields of 400 kg per hectare (350 lb per acre) and more, over a long period. There are, in fact, many areas of Malaysia and Southeast Asia where smallholder rubber of this type is still in production.

However, even where it is possible to sustain this level of output, falling rubber prices, together with rising levels of expectations regarding incomes, have made it an uneconomic proposition for most small producers. For example, with the rubber price about US17 cents per lb, a family of four living off the proceeds of a two-hectare (five-acre) rubber farm of the old type would have a gross income per head of about M$200 per annum. The same area under high-yielding clones would yield three to four times this amount with very little extra labour and other costs for cropping. For a small producer, therefore, the effective application of the new technology, combined perhaps with a small increase in the area of land he crops, can raise him from poverty to a reasonable standard of living.

The adoption of the new technology has been much more difficult for the small producer than for the estate. With unselected plants the disadvantages of the small-scale producer were less marked, and were to a significant extent offset by the ease with which the pattern of production and work fitted in to his preferred way of life. The high-yielding clones are more exacting as regards planting, rearing, and maintenance. For good results they require good land preparation, proper spacing, careful maintenance, and sophisticated tapping techniques. In addition, the use of fertiliser, fungicides, and other chemical preparations at the right time, of the right type, and in the right quantities, is often necessary. Other refined techniques, such as budgrafting, are of importance, requiring skills and special materials that the individual smallholder is normally unable to provide from his own resources.

Economies of scale are thus very much more important with the new rubber technology, and the small producer would find it difficult to take advantage of the new technology at all, unless the economies of scale could, to a considerable extent, be made accessible to him. This can only be done in the form of external economies, deriving from co-operation between smallholders, or from other external institutional innovations. It is in the creation of institutions which have achieved this that success in the Malaysian development of smallholder rubber has been found.

This applies not only to the improvement of existing holdings, but also to the establishment of new smallholdings. Here again success has accompanied schemes and projects that have made available substantial economies of scale. In this, the institution of the Federal Land Development Authority has been markedly successful. The device of the nucleus estate has also shown promise, though mainly with another crop (palm oil).

Space will not permit a full examination of all the institutions tried, with varying degrees of success or failure, over the years. However, in illustration of the theme, it will be possible to outline the working of some of the successful ones.

First, let us consider the institutions gathered under the Malayan Rubber Fund Board. These are financed by a cess on all rubber produced and exported from West Malaysia. A cess has been collected by government and used for this purpose since before World War II. For West Malaysia, this is a form of investment in research and publicity to which all producers, large and small, are forced to contribute. As research, in particular, is a field of activity in which economies of scale are most pronounced, this is of potential advantage to all producers, provided they are able effectively to utilise the innovations thus made available.

The institutions under this group fall into three categories. Two are directed primarily at the consumer, covering scientific and technological research into the compounding, processing and properties of natural rubber on the one hand, and information and publicity on the use of natural rubber on the other (Malaysia 1967d: 416). These serve all the producers indirectly by sustaining consumer demand for their product. The other category is directed at the producer, and covers research into all aspects of rubber-growing and production. The institution primarily concerned is the Rubber Research Institute of Malaysia, which is one of the most effective and successful institutions of its kind in the world today. However, this degree of effectiveness, particularly as regards the small peasant producer, was not rapidly or easily achieved. In the earlier days, from its foundation in 1926 up to 1950 or thereabouts, considerable difficulty was experienced in channelling the results of the research to the producer evenly and fairly, and the Institute was at times strongly criticised for concentrating unduly on estate producers to the detriment of the smallholder. This was the result of greater difficulties in assisting the small producers, compared with the large estates.

The estates, as a result of the size and form of their organisation, enjoyed considerable advantages of scale. Communication of the results of research to them was a relatively simple matter. They numbered less than 2,500 and most were managed by educated men with some technical training in agriculture. Much could be achieved quite simply by articles in the journals of the planters' associations, and by letters and circulars. Visits by technical officers could be readily made by a relatively small staff.

With smallholders, communication was more difficult. For one thing, the large number of rubber smallholders (probably not far off 200,000) was itself a serious barrier. For another, the level of education of the average smallholder was not high, and communication of technical information by normal publishing and circular letters was almost ineffective. Even today this remains a difficulty, though it has been greatly reduced by methods of communication designed specifically to fit the way of life, needs, and capacities of large numbers of smallholders. The Smallholders Advisory Service operated by the Institute has greatly expanded, and in 1966 employed over two hundred trained advisory officers. These officers make use of village meetings, co-operatives, and other village-level institutions to convey much of their advice by word of mouth: they supplement this by film shows, posters, models, and other aids, and by arranging demonstrations wherever possible. This expansion, and in particular, this development of special organisation and techniques for communication, has itself enabled the Rubber Research Institute to extend to the smallholder external economies of skills and knowledge that compensate to a considerable extent for their deficiency of internal economies in this particular respect.

This development has involved much trial and error over the years, and is still proceeding; but the methods developed and the lessons learned merit wider understanding and publicity, and should not have to be learned again, whether in other contexts within Malaysia, or in other countries faced with similar types of problems.

But this alone would have achieved little, for the new technology demanded facilities other than access to knowledge and skills. The small-scale operator was at a disadvantage in the production of high-yielding budwood or seeds, in the purchase of fertiliser by small lots, in access to credit and capital, and even in the maintenance of standards of weeding and other aspects of husbandry that had been unnecessary in his previous experience. The financial problems of the really small-scale producer were particularly intractable, for apart from materials that had to be purchased for money, new planting or replanting with the new technology involved a very large input of labour in the first year or two, and then a quite significant, though smaller, labour input for a further five years. Moreover, no return in cash or kind would be available for six or seven years.

The large estate had the same problem, but by replanting about 3 per cent of its area each year the burden, in relation to the resources and total earnings of the estate, was relatively manageable. For the small producer, depending for his livelihood on 1·5 or 2 hectares, the minimum area technically practicable for replanting is likely to be nearer 40 per cent than 3 per cent, which in terms both of the proportionate call on the labour resources of the smallholding, and of the income forgone, is more than many smallholders can afford.

The progress made by the Rubber Research Institute has, therefore, depended also on the successful introduction of complementary institutional innovations, quite independent of the Rubber Research Institute, that have made available to the small producer other essential external economies. Of these, the various smallholder replanting schemes and smallholder new planting schemes are of particular importance. The first of these schemes was initiated in 1952.

These schemes are under the control of the Rubber Industry (Replanting) Board. A flat-rate replanting cess of M$0·1 per kg (4½ cents per lb) is collected by government on all Malaysian rubber exported, and the proceeds are directed into two funds administered by the Board, fund A for estates, and fund B for smallholdings. The money is divided between the two funds in proportion to the rubber produced by estates and smallholdings respectively in the year of collection.

The moneys in fund A are returned to estates on proof of the quantity of rubber produced, and no further control is exercised over the funds. The replanting policies of the larger estates have in general been adequate and government has been content to leave the matter in the hands of the companies concerned. With some of the smaller estates, however, especially those operated as family concerns, replanting and adoption of the new technology has lagged seriously. This remains one of the few disappointments in the otherwise excellent record of the rubber estate sector.

The moneys in fund B, supplemented at times by grants from general revenue, have been used to help the smallholder take advantage of the new technology both in replanting and in new planting. The schemes provide direct financial assistance in the cost of the material and labour required for effective replanting. The original replanting scheme provided for payments totalling M$980 per hectare (US$130 per acre) to smallholders successfully replanting up to one-third of their holding to specified standards.[10] By 1965 this grant had been increased, and under certain circumstances could be as high as M$1,966 per hectare.

This financial assistance has been supplemented by other services, the importance of which greatly exceeds their cost. The replanting proposal made by the smallholder is carefully examined and criticised by expert staff, and the financial grants, which are paid in instalments over the whole replanting and maintenance period, are made subject to the fulfilment of certain conditions specified by the Rubber Replanting Board. These conditions cover the proper preparation of the land and the correct procedures of planting and maintenance, and are strictly enforced. In addition, the Board supplies high-yielding planting materials from official nurseries, and undertakes bulk buying, distribution, and supply of approved fertilisers. These services, in close co-operation with those provided by the Rubber Research Institute, provide external economies to the smallholder in the field of capital, management, supply of materials, and skills, that make up, to a considerable extent, for the internal economies available only to the large-scale producer.

New planting has been assisted similarly through the Replanting Board, though this has been restricted to the landless and to the small smallholder. This financial assistance has been channelled in several ways, with varying degrees of success, and the differences between these, and in the results achieved, are of great interest.

The most generally successful new rubber-planting in smallholdings has been in schemes directly sponsored and controlled by the Federal Land Development Authority.[11] This planting was also by far the most expensive in terms of use of public resources, but this was a feature of the political, rather than the economic, requirements of the schemes. Were it politically practicable in the particular circumstances affecting new land settlement in West Malaysia, it would, in the writer's opinion, be financially possible for the schemes to be fully self-supporting in the long run.

The features responsible for success in the Land Development Authority schemes are again the provision of external economies, but here they go farther, and are more obviously related to the economies of scale of the estate type of production. In these schemes large blocks

[10] Where the holding was less than 6 hectares (15 acres), assistance was granted for up to 2 hectares (5 acres), irrespective of the proportion.

[11] This is not to suggest that FLDA schemes have not met with difficulties or disappointments. Other less controlled and organised schemes, however, have been considerably less successful on the whole.

of land are selected, comparable in size with quite large estates, but divided into contiguous rubber areas, contiguous dwelling areas, and areas for fruit or other crops. The land is cleared and planted by contract under the direction of a trained staff provided by the Land Development Authority, very much on estate lines. Houses are built by contract; roads, community facilities, and amenities are provided either by contract or by the Public Works Department; provision is made for the bulk handling and processing of the rubber through a factory, usually under some form of communal ownership and control. The settlers are moved in after the houses are built, and during the establishment of the rubber, and even during the maintenance period prior to tapping, the work of the settlers is very largely directed by the management staff of the scheme, enabling the smallholders as a group to enjoy many of the economies of scale that would normally be available only to large producers with a pool of wage-labour. The direct planting costs, including wages and allowances to the settlers during the period prior to tapping, together with some of the indirect costs such as the cost of dwelling houses, and an interest charge, are debited to a loan account to be recovered from the settlers.[12] The costs of the Land Development Authority administration, including management, staff housing and transport, construction of roads in the agricultural areas of the schemes, and other amenities not provided by other government departments, are met from government grants to the Land Development Authority. Other costs, such as main road access, schools, and medical clinics, normally considered the function of other government departments, are provided by, and charged to, those departments in the normal way.[13]

New planting for smallholders has been sponsored under other schemes.[14] These schemes have involved very large areas, but in general have been less successful both in the proportion of new land effectively planted, and in the standard of planting achieved. It is notable that in these schemes there was in general no special institution to provide direct management and control; administration and supervision were dependent for the most part on the normal government administrative institutions in the areas concerned. Whilst the scale and standard of planning, co-ordination, and supervision so provided varied widely according to the resources available in the localities concerned, each

[12] Most settlers become eligible for a grant under Smallholders' New Planting Scheme No. 1 or No. 2, in which case the grant is credited to their loan account, thus reducing the amount to be recovered after their rubber comes into bearing.

[13] For a detailed examination and exposition of the financial aspects of FLDA schemes see Singh (1965).

[14] Many of these schemes were also financed by grants under Smallholders' New Planting Scheme No. 1 or No. 2.

new planter within such schemes tended to move at his own pace, in his own way, as his other work and preferences allowed. Communal works and services were difficult to organise and even more difficult to sustain. Under these circumstances the dis-economies and impediments of very small-scale operation were still much in evidence, and in many important respects the institutions providing external economies to counterbalance these had insufficient resources for the task. The result has been a wide range of success and failure in the large number of schemes initiated.[15]

RICE

With rice cultivation there have been significant differences both in the form of government policy and in the mode of its implementation. One important difference from rubber is that there is no 'advanced' estate sector of large-scale rice producers; rice is solely a smallholder crop in Malaysia. Another difference is that government assistance has tended to be channelled more directly through government departments rather than through independent, or semi-independent, institutions outside the normal government structure, though the latter have played some part. Third, the motivation for rice production has seldom been purely financial, either for the smallholder or for government. For the smallholder an important factor at all times has been the attachment of Malays to the cultivation of rice as a part of their way of life, so that social and political motivation augmented the economic inducements. As the financial returns to factor inputs in rice culti-vation have rarely been as great as in other forms of farming in Malaysia and have often been considerably lower,[16] this non-economic motivation has been important. For government, the policy of assist-ance to rice-producers has been heavily influenced by two political

[15] In 1962 I recommended the establishment of a Land Rehabilitation Autho-rity to deal with problems arising from fragmentation of holdings, multiple owner-ship, neglected holdings, and replanting of very small holdings (Silcock and Fisk 1963: 187). In 1966 the Malaysian Parliament passed the National Land Rehabilita-tion and Consolidation Authority (Incorporation) Act in terms that appear to be based on the lines suggested, but which may not, in fact, go quite far enough to be effective. It is understood that the first function of this Authority is to be the rehabilitation of some of the group settlement and fringe alienation schemes that have gone wrong. This is a particularly difficult task, and it would be a great pity if the Authority foundered as a result, and were unable to fulfil its important economic role of bringing some of the critical Land Development Authority types of economies of scale to the wide range of smallholdings that need them outside these faltering schemes.

[16] See, for example, Bauer (1948: 60-3). Only in the last few years does the new rice technology, combined with increasing prices for rice and lower prices for rubber, appear to be changing this relationship.

H

motives, that of providing special assistance to Malays in their traditional economic enterprise on the land, and that of winning self-
sufficiency in the staple diet of the people.[17]

Despite these differences, there is clearly detectable in the measures
that have been successful a pattern very similar to that found in
rubber. Even in an industry where large-scale producers are absent,
economies of scale have been vital to the success of the small man, and
the institutions that have provided access to these have been the most
effective instruments of policy.

After rubber, rice is the most widely planted crop in Malaysia.
Figures of area and production are subject to some uncertainties, particularly as regards East Malaysia, but for West Malaysia at least, the
general orders of magnitude are not in doubt. In the 1966-7 season the
area planted to rice was roughly as follows:

	'000 hectares	'000 acres
West Malaysia		
Wet paddy, main-season crop	300	760
Wet paddy, off-season crop	40	100
Dry paddy	16	40
East Malaysia (rough estimates)		
Wet paddy	50	125
Dry paddy	75	185

As the off-season crop is a second crop from land already used for a
main-season crop, the area utilised for rice in Malaysia would be something over 450,000 hectares (1·1 million acres), producing probably
700,000 tons rice equivalent, or a little less.

Average yields per hectare (acre) mean very little in a situation
such as this, where types of land and types of cultivation vary from
hit-or-miss dry-land cultivation on semi-exhausted soils, to double-
cropping on high quality soils with first class water control. Under
the former conditions, a yield of half a ton of rice to the hectare might
be considered a good result, whereas under the latter, a yield of five
tons to the hectare might be disappointing. Even changes in average
yields over the years are of little significance where land is not generally a scarce factor, and dry-land rice can be cultivated as a catch crop
when opening up land for other purposes.

To evaluate the progress made with rice cultivation in Malaysia, it
is necessary first to distinguish three main types of cultivation. First,
there is dry-land rice cultivation which, in the main, is practised either
as a form of shifting or long-fallow 'slash and burn' cultivation in the

[17] For a useful review of the development of government policy and motivation
towards rice cultivation see Ding (1963: Chs. III and IV).

more primitive regions, or as a catch crop in the earlier stages of cultivation of other crops, such as rubber. In some areas continuous re-cultivation with dry rice is practised with the aid of fertiliser, but problems of weed growth, disease, and pests are such that this is not, and does not seem likely to become, an important source of rice in Malaysia. The 'slash and burn' type of cultivation is still practised quite widely in East Malaysia, but is unrewarding in comparison with wet rice cultivation in terms of returns to factor inputs, and is beset with other problems of land use, such as erosion; it must be considered in most respects an unsatisfactory stop-gap that should be replaced by better methods as soon as practicable.[18] Dry rice as a catch crop may perhaps have some continuing role to play on a small scale, but seems unlikely to be an important source of major food supplies. For these reasons, dry rice cultivation will be omitted from consideration in the rest of this chapter.

Second, there is wet-land rice cultivation undertaken in areas naturally flooded, and with little water control other than that available from locally improvised ditches and small earth dams. Much of the world's rice is grown under such conditions, and Malaysia is no exception. In Malaysia, some rice areas in this category are highly productive, but others are not, and there is a considerable proportion of 'marginal' land that is effectively cultivated only in favourable seasons or when special circumstances demand it.[19] In addition, there are considerable areas partially irrigated under what the Malaysian government describes as 'sub-standard' irrigation schemes. If these are included, almost 230,000 hectares (560,000 acres) of Malaysia's wet paddy land would have been in this second category in 1966.

Third, there is wet-land rice cultivation undertaken in large-scale irrigation areas designed, constructed, and maintained by government. The first such scheme in Malaysia was the Krian irrigation scheme, commenced in 1895. By 1966 there were in West Malaysia some 130,000 hectares (322,000 acres) included in government irrigation schemes in full operation during that year, and on about 70,000 hectares (170,000 acres) of this, irrigation services were available for double-cropping.

Rice-growing, particularly with irrigation, is a form of cultivation in which economies of scale are quite crucial. An individual who has a plot of naturally irrigated land can plant it to rice, and with proper care and reasonable weather the rice will grow satisfactorily. However,

[18] A brilliant account of the economics of dry-land rice cultivation in Sarawak is found in Freeman (1955).

[19] Rice imports, on which Malaysia has to date relied for a substantial proportion of its requirements, have failed it on several occasions, notably during World War II, and earlier in 1918 and 1920.

as it begins to ripen all the birds for miles around will descend on it, and the grain left to be harvested would be a meagre return for the grower. If village people co-operate to build a simple brushwood dam and extend the cultivated area to 20 or 40 hectares (50 or 100 acres), the bird population feeding in the area will not greatly increase, but the effect of their depredations on the harvest per hectare, and per grower, will be greatly reduced. Moreover, twenty-five or more growers working together or in shifts can be much more effective in controlling birds, rats, and flooding than can one grower on his own. The effect of these and many other simple economies of scale are well known to the Malaysian rice-growers, and they have long developed village institutions and traditions that enable them to work together on village-scale schemes of this type.

However, beyond village size, this type of organisation is difficult to operate and manage, and a more elaborate institutional approach becomes necessary. Larger irrigation schemes have even larger economies, and the areas that can be effectively served with high-class water control, levelling, access roads, etc., can thereby be greatly extended. But this requires a very high order of planning, skills, organisation, and large capital investment. This type of institution can only be provided by governments, or by large-scale private enterprise. As rice production was in general less profitable than sugar and coffee up to about 1908, and than rubber and coconuts for most of the period since then, large-scale private enterprise has not been much attracted to the industry, and on the few occasions when it has displayed an interest, it has been discouraged by the inclination of government to give preference to Malay smallholders in the alienation of the suitable land.[20] Consequently, only government has been in the field in Malaysia.

Apart from the Krian irrigation area completed in 1906, little progress with the encouragement of rice production in West Malaysia was made until the late 1930s. This was largely due to the preoccupation of both government and people with rubber. However, a serious rice shortage in 1918, when the Indian government restricted rice exports from Burma, and another in 1920, when Thai rice exports were suspended as a result of drought, again raised self-sufficiency in rice as a policy objective. The collapse of rubber prices, and thus of the main source of external income, further underlined the vulnerability of Malaya's food supplies in the early 1930s. It was in this context that the Malayan government in 1932, in the depths of the depression, created a new major institution for the stimulation of rice production, the Drainage and Irrigation Department. This was the prelude to

[20] Some instances in which larger-scale enterprises were discouraged are mentioned in Ding (1963: 16, 34).

more active and effective support for the rice industry, and to heavier public investment in large-scale irrigation schemes.

Since that time government policy has developed and has been refined, but the main line has been consistent and sustained. This was based on the advice of Dr H. A. Timpany, Director of Agriculture, who in 1930 advocated two main lines of policy:[21] first, the extension of good rice-growing areas by means of irrigation; second, improvement of the types of rice grown and the use of better agricultural practices. To these main broad lines of action others have since been added, and the development of these additions is of particular interest.

The large irrigation schemes were exceedingly complex, requiring highly skilled technical surveys, planning and design, advanced large-scale engineering, land administration, provision of roads, schools, services, and amenities for the settlement of a substantial population of farmers, followed by a complex organisation for selecting, moving, settling, and advising the settlers. In all of these aspects, the effect was to bring to the small rice farmer scale economies that would have been entirely beyond his reach as an individual on his own, or as a member of a traditional village-level co-operative group. This, however, is fairly obvious. What is more interesting is the degree and manner in which government intervention developed further round this core.

First, the grouping of large numbers of rice farmers, in contiguous areas of land served with similar water control, experiencing the same seasons, with known soil and climatic conditions, with good road access and services, greatly facilitated fruitful extension work and local adaptive research. The Department of Agriculture was thus able more effectively to pursue the second line of action (improvement of rice types grown, and improvement of agricultural practices) in the irrigation areas, since it was possible to operate the research and extension service in each area as though it were one large estate. Similarly facilitated was the supply of special inputs, such as selected seed and fertiliser, for not only could the quantities and types required be more easily determined, but the times and places of delivery were more concentrated.

Credit[22]

From time to time other institutional innovations were added. One requirement of great importance, which cannot yet be said to be satisfactorily met, is the requirement for credit. Most small farmers need

[21] Cited by Ding (1963: 25n.).

[22] I have included this discussion of credit in the section on rice because it is the only major smallholder crop in Malaysia with marked seasonal variations in income, and thus with an acute need for seasonal credit.

credit at some time. The needs are of three kinds. The first is long-term credit to meet the investment costs of opening up and settling new land. This is a need that the Federal Land Development Authority meets so effectively in its large-scale settlement schemes. For the rice farmer settling in a new irrigation scheme this need has been met only in part, but to that extent on generous terms. The costs of the basic improvements to the land, including all the main irrigation works, are met by government, and although the farmer is subsequently required to meet certain annual charges, these charges are insufficient, normally, to meet the costs of operation and maintenance of those works and services. In many areas also, a settlement grant is made to the settler to assist with the initial requirements of tools and materials.

However, there remains a long and difficult period of investment (usually in the form of labour) before the whole rice holding in a new irrigation area is brought into production. It is not uncommon to find considerable proportions of controlled-irrigation rice areas still uncultivated seven or more years after initial settlement. This is a waste of government investment, as well as a loss to the individual planters, and is again a problem of access to scale economies. Further intervention on the part of government, by undertaking large-scale mechanised preparation of the holdings, completing the clearing, levelling, bunding, and initial cultivation of the whole area on behalf of the settler before handing over responsibility to him, and recovering the costs from the greatly increased crop over the first few years of occupation, would add considerably to the returns on the total investment, both to government and to the individual planter. The economics of this have been discussed in detail elsewhere (Fisk 1964: Ch. VI). The individual rice-producer, because of the small scale of his individual operation, urgently needs the external economies of large-scale mechanical operations and of large-scale managed credit that would be provided by such intervention.

The second type of credit need is the short-term seasonal one. The income flows of small rice-producers tend to be very 'lumpy', the main amounts coming in once or twice a year after the rice harvests, while expenses, both of cultivation and of living, involve an outflow over most of the year. This need has commonly been met by credit from shopkeepers, repayable at harvest in rice at a very heavily discounted rate. This has been very expensive indeed to the rice-grower, basically because the cost of administering small unsecured loans is itself high, and losses are also substantial. Moreover, the monopsonistic situation which is an essential element in such credit (it often provides the only effective inducement to repayment) also provides the opportunity for additional monopsonistic gains to the shopkeeper concerned.

Many efforts have been made in Malaysia, as in other countries, to improve the facilities for seasonal credit, and most have been relatively unsuccessful. Rural banks have never provided an adequate answer, and never can. The reason is that the high cost of administering rural loans derives from the number and small scale of the borrowers, and the economies of scale can be expected only from aggregation and organisation on the borrowing side of the transaction. In fact, the costs of lending are lower for the small-scale lender who lives nearby, and who knows and can bring personal pressures upon the borrower. The larger, more impersonal banking institution lacks these advantages. Loans made in the 1950s by the Rural and Industrial Development Authority (RIDA), in the form of credit for tractor operations, were very expensive to collect and had a high rate of non-payment for this very reason. The individual small contractor, for whose operations the RIDA tractor schemes were designed to pave the way, has also had problems in this matter, but they have proved less difficult.

Several institutional approaches to this problem have shown some signs of success. It is significant that all enable the rice farmers to borrow in groups, as large-scale borrowers, thus reducing the costs of loan administration and collection. At the same time they retain, or replace, the personal and other constraints necessary to enforce repayment.

The co-operative movement has had some success through the use of joint liability and local representative management, which has gradually built up the creditworthiness of those societies that have persisted long enough, but, despite many decades of operation with government support, it has provided adequate seasonal credit for only a small proportion of the rice-growers in Malaysia. The movement tends to emphasise its missionary and teaching role, and concerns itself with the improvement of the attitudes and principles of its members in business. This educational role is inevitably slow of fulfilment, and to the extent that the satisfaction of the immediate credit needs of the farmers is dependent on that fulfilment, co-operative societies have failed to provide a sufficiently expeditious answer for most growers.

Two other approaches are now being tried in Malaysia. One is based on Farmers' Associations, which, in Taiwan at least, have proved a most effective institution for bringing farmers together to enable them to enjoy economies of scale in agricultural credit. Another is the Federal Agricultural Marketing Authority (FAMA), established late in 1965. This Authority, though at present concerned only with marketing and processing, has powers to require individuals or areas to market their produce through the Authority. It could therefore readily be placed in a position where it could, at the grower's request, deduct debt repayments and interest charges from the proceeds of sales, and pay them on his behalf to a co-operative, to the relevant Farmers' Association, to a bank, or even to a loans division of FAMA itself. This Authority, either by itself or in co-operation with other credit institutions, if successful in its main function as a marketing authority, could also enable small farmers to enjoy economies of scale in the field of credit.

The third type of credit need is more intractable. It arises from exceptional cash requirements such as for sickness, marriages, births and deaths, or for pilgrimages to Mecca. These needs have traditionally been met by recourse to the pawn shop, the money-lender, or the local middleman, and have frequently led to the mortgaging and ultimate loss of ownership of the land.

An interesting approach to one part of this problem was the institution of the Muslim Pilgrims' Savings Corporation, which commenced operations in 1964, and by 1968 was said to have over 7,000 members saving for a pilgrimage to Mecca. Unfortunately, very few of its members are full-time farmers.[23] The difficulty and cost of collecting small subscriptions in rural areas are certainly partially responsible for this situation, and here again there would seem to be a great opportunity for collaboration with FAMA to extend the effectiveness of such credit institutions to the rural people.

Other Institutions Affecting Rice Production

Other fields in which government intervention has been, or may be, of benefit to the rice-grower are bulk buying, distribution, processing, and marketing. Bulk buying has at times been arranged by the Department of Agriculture, and by co-operative societies, and can be undertaken by the Farmers' Associations or by a combination of such institutions. It is most common in the purchase of fertilisers, seeds, and pesticides, but can be, and is occasionally, extended to consumer requirements through co-operative stores. However, bulk buying and distribution, at best, require a close link with credit, and thus with marketing, to avoid excessive loan administration costs. This could be achieved better if there were a formal relationship for the purpose between the Farmers' Associations and/or the co-operatives on the one hand, and FAMA .

Processing has long been a point of intervention by government in some areas, and the first government rice mills were set up in the state of Perak in 1919. Though this has not been widely extended, the

[23] My information is based on verbal communications by Professor Ungku A. Aziz of the University of Malaysia. It was largely on the basis of suggestions made by Professor Aziz that the Corporation was created.

system of licensing and the willingness of government to intervene with its own rice mills if necessary, has undoubtedly reduced, but by no means eliminated, the areas where monopsonistic advantages can be exploited by large-scale buyers. For this reason, the widening operations of FAMA, whether in a monopoly position (as was instituted in the Tanjong Karang district of Selangor) or merely in competition with private dealers and processors, should, if skilfully and efficiently run, offer further external economies to the rice-grower. The fact that such institutional intervention may deny the grower a certain freedom of choice regarding the place and manner of sale of his produce will be unpalatable to some, but some control is unavoidable in the provision of these particular economies of scale. Provided FAMA is efficiently and honestly run, the cash value to the grower of the economies extended to him should more than compensate for the degree of regimentation involved—particularly if the credit advantages mentioned above are also thereby made available to him.[24]

Another measure for the encouragement of rice-growing, unrelated to external economies, but nevertheless very important in the long-term development of the West Malaysian rice industry to date, has been government price support for locally-produced rice. This commenced in 1919, when the government guaranteed a minimum price of 13 cents per *gantang*[25] for local rice (in the husk) delivered at the government mill. Since then, the subsidisation of paddy prices by this device has been at times an important factor in the economics of rice production in West Malaysia, not only because of the price advantage when world prices were low, but also through the removal of one risk in rice-growing.

Table 6.6 shows rice production in recent years in West Malaysia. Of particular significance is the steady increase in the three-year averages over the whole period, and the increasing part played by double-cropping since 1960. These figures indicate the regular returns earned by government intervention in favour of rice-growing. However, there are indications that two other factors, complementing the work that has already been done, may now be producing an extra response in production that will be in the nature of a breakthrough. First, there is the rapid advance now being made in the production of high-yielding strains of *indica* rice species that respond handsomely to nitrogenous fertilisers. This, together with adequate water control, facilities

[24] Many other advantageous measures, such as irrigation control and the consequent control of the timing of planting and harvesting operations, involve a similar degree of regimentation.

[25] A *gantang* is a volumetric measure and equals one gallon. A *gantang* of paddy weighs approximately 5·6 lb, whereas a *gantang* of milled rice weighs approximately 8 lb.

TABLE 6.6

WET-LAND RICE PRODUCTION IN WEST MALAYSIA

(metric tons of paddy)

Season	Main season crop	3-year average	Off-season crop	3-year average	Total wet-land crop	3-year average
1952–53	584,312		2,375		586,687	
1953–54	545,908	555,403	5,038	4,999	550,946	560,402
1954–55	535,990		7,586		543,576	
1955–56	553,704		3,271		556,975	
1956–57	635,252	612,481	7,143	5,519	642,395	618,001
1957–58	648,489		6,144		654,633	
1958–59	571,297		7,326		578,623	
1959–60	733,654	701,029	11,225	13,766	744,879	714,795
1960–61	798,136		22,749		820,885	
1961–62	739,083		33,845		772,928	
1962–63	792,133	740,729	48,267	45,142	840,400	785,872
1963–64	690,972		53,315		744,287	
1964–65	807,892		62,119		870,011	
1965–66	789,941	781,510	101,958	94,655	891,899	876,166
1966–67	746,698		119,891		866,589	

Source: Derived from Narkswasdi 1968: 25, T. 10.

for double-cropping, and credit arrangements for the purchase of fertiliser, is making possible a more rapid and more substantial increase in the returns to factor inputs for the rice farmer in the controlled-irrigation areas. Second, there has recently been a tendency for rice prices to increase, whilst the prices of other crops in West Malaysia, and in particular rubber and coconuts, have declined steeply. The result has been a sharp increase in the absolute and relative profitability of rice-farming, and consequently in the economic incentives to raise production. It remains to be seen to what extent this can be sustained, and much will depend on the effectiveness of FAMA and the other institutions discussed above. However, for the first time, the dream of self-sufficiency in rice for Malaysia appears to have come within reach.[26]

NUCLEAR ESTATES

The establishment of nuclear estates, and the development of small-holdings around them, is another means of giving small producers access to economies of scale, especially economies of processing, skills,

[26] An exceptionally good economic survey of rice production in Selangor and Malacca has become available recently; see Narkswasdi and Selvadurai (1967a, b, and c). See also the excellent overall report on the rice economy of West Malaysia, Narkswasdi (1968).

and marketing. In Malaysia, as a deliberate instrument of policy, this has been important only in smallholder palm oil production.

Malaysian palm oil production has expanded very rapidly and by 1968 Malaysia had become the world's largest exporter of this product. In 1967 production was estimated at 210,000 tons of oil and 48,000 tons of kernels.

Smallholder participation had for many years been impeded by the very large throughput necessary to make the processing of the fruit economically possible, and by the need for the processing plant to be near to the growing palms. For this reason, even though established under the Federal Land Development Authority, the earlier oil palm smallholdings were sited in areas close to estate producers, to economise both in the technical skills, and in the large and costly processing plant. More recently, with its own great resources, and with the experience and technical know-how gained through its participation in the nuclear schemes, the Authority has been able to undertake unit smallholder oil palm schemes of sufficient size to provide all the necessary services itself.

Although this method is not widely used in Malaysia, it has been successful and is used elsewhere (e.g. Africa and New Guinea). It is of particular importance in countries which lack the resources to rely wholly on the more costly alternative methods available.

CONCLUSION

In many developing countries the political and social situation demands that agricultural development should be effected more through the expansion and improvement of small peasant farms than through large-scale estate production. On the other hand, economies of scale are important in rural production, and without access to such economies the peasant producer loses heavily in efficiency and in returns for his efforts. The rise in the level of expectations of the rural people is such that low returns to factor inputs are no longer acceptable to the peasant producer as a substitute for the economies of scale enjoyed by the large-scale estate producer.

The experience of Malaysia, and of Malaya before it, has covered a very wide range of institutional innovations undertaken by government in pursuit of its policy of rural development through the smallholder segment of its economy. The effectiveness of these institutions has varied greatly, from failure, through indifference, to success. In this chapter[27] an attempt has been made to show that a major factor

[27] I expressed these views briefly in a paper incorporated in the Review Report on the *Asian Agricultural Survey* by the Consultative Committee (of which I was a member); see Asian Development Bank (1968: 121-2).

in the success or otherwise of many of these attempts has been the external economies provided to the small producer, and that recognition of the importance of this factor should prove valuable in further planning, both in Malaysia itself, and in the many other countries that are similarly faced with the need for agricultural development through the small-scale peasant type of producer.[28]

Within Malaysia itself this is, of course, only a part of the development problem, albeit an important one. Malaysia has many of the institutions necessary for such development, and the infrastructure is already well developed.[29] However, it has yet to make some of the most important of the institutions (e.g. FAMA, Farmers' Associations, and the Land Rehabilitation Authority) work effectively. Some extension of powers and functions (e.g. in credit) is still required, and many strong and telling attacks from vested interests have yet to be weathered and defeated. There is a shortage of the top-level management and business skills necessary for the key institutions to operate, and it seems clear that these key institutions can only be adequately manned at the expense of some less essential institutions already in operation. In the meantime, a large part of the rural population, especially that part in the old-established unirrigated smallholdings, is facing increasing unemployment and under-employment, and their already modest levels of income are being eroded by population pressure and low prices. The lesson of the Malaysian experience is one that should be of use and encouragement to many other developing countries. It would be tragic if Malaysia failed to make effective use of these opportunities itself.

[28] In Fiji, for example, sugar growing has been organised on a smallholder basis very successfully, despite the highly competitive nature of the world market for sugar. For a very interesting account of how this was done, see J. C. Potts, 'An Outline of the Successful Development of the Small Farm System in the Fiji Sugar Industry', *The Fiji Society Transactions and Proceedings*, Vol. 9, pp. 26-38. In this case, the external economies were provided by the operations of a large commercial firm.

[29] No mention has been made in this chapter of the system of the Operations Room and the Red Book, with which so many official visitors to Malaysia have been impressed in recent years. This is because that system is primarily a device for increasing the efficacy of the government administrative machinery in getting things done. As such, its discussion belongs in a book on public administration, rather than rural economics. A good description of the system is available in Ferguson (1965).

The Philippines

Richard Hooley and *Vernon W. Ruttan*

IN this chapter we review the role of agriculture in the economic development of the Philippines during this century. We then turn to a review of the historical growth of output and productivity of the major agricultural commodities—rice, maize, sugar, and coconut—and to an analysis of the environmental, technical, and institutional factors affecting their growth. Finally, consideration is given to some of the policy issues relevant for the design of a strategy for agricultural development in the Philippines.

LONG-TERM GROWTH OF THE PHILIPPINE ECONOMY[1]

When discussing economic growth in the Philippines, economists have typically concentrated on the recent past. Golay's work (1961) is a thorough coverage of development after 1946. Goodstein's (1963) study of the growth of real product extends our knowledge back to 1938, but the four decades prior to World War II which are essential for constructing broad secular movements, are still by and large unaccounted for.

Although there are a number of other studies which provide some insight into Philippine economic development before World War II, many of these views are mutually inconsistent.[2] Furthermore, none of the previous investigators has approached the last six decades as a unit and examined the evidence on a quantitative basis.

Growth of Aggregate Output

Over the past six decades, total output rose about $6\frac{1}{2}$ times, or at a compound rate of slightly more than 3 per cent per annum (Table 7.1).[3] There was considerable variation among sub-periods, with the

[1] The material in this section is discussed in greater detail in Hooley (1966).

[2] E.g. see Hartendorp (1958), Taylor (1964: 87-8), Mitchell (1942: Ch. 11) and Hayden (1945: 7-9).

[3] For a discussion of the methodology employed in constructing the long-term trend in output presented in this section see Hooley (1966: Ch. 4, 4-9).

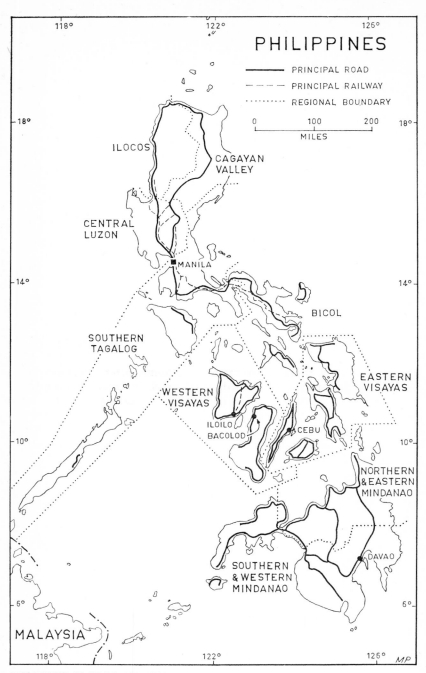

PHILIPPINES

PRINCIPAL ROAD
PRINCIPAL RAILWAY
REGIONAL BOUNDARY

0   100   200
MILES

ILOCOS

CAGAYAN
VALLEY

CENTRAL
LUZON

MANILA

BICOL

SOUTHERN
TAGALOG

EASTERN
VISAYAS

WESTERN
VISAYAS

ILOILO
BACOLOD

CEBU

NORTHERN
& EASTERN
MINDANAO

SOUTHERN
& WESTERN
MINDANAO

DAVAO

MALAYSIA

DEPARTMENT OF HUMAN GEOGRAPHY, A.N.U.

Map 8   The Philippines

Table 7.1

GROSS VALUE ADDED IN AGRICULTURE AND NON-AGRICULTURE FOR SELECTED YEARS, 1902–61

(constant 1939 prices)

Year	Output			Agricultural in total (%)	Annual percentage increases (compound)		
	Agricultural[a] (m. pesos)	Non-agricultural[b] (m. pesos)	Total (m. pesos)		Agricultural	Non-agricultural	Total
1902	134·0	109·8	243·8	55·0	—	—	—
1918	294·5	193·0	487·5	60·4	5·1	3·6	4·4
1928	313·5	270·2	583·7	53·7	0·6	3·4	1·8
1938	328·0	375·9	703·9	46·6	0·4	3·3	1·9
1948	328·7	340·8	669·5	49·1	0	−1·0	−0·5
1961	537·5	1063·2	1600·7	33·6	3·8	9·1	6·9
1938–61	—	—	—	—	2·2	4·6	3·6
1902–61	—	—	—	—	2·4	3·9	3·2

a Includes crop and livestock production.

b Excludes government and private sector services.

Source: Censuses of 1902, 1918, 1938, 1948, and 1961. For methods of constructing individual series, see Hooley 1966.

most rapid rates of growth achieved during the first and last periods (1902-18 and 1948-61). The very high growth rates achieved during the last period are, however, somewhat difficult to interpret, owing to the fact that the period covers post-war reconstruction. Therefore the growth rate for 1938-61 is shown separately, and this is obviously much lower (3·6 per cent). On this basis, the highest rate of growth was achieved during the 1902-18 period.

The growth rate of agriculture is particularly interesting. Over the whole period, the increase in agricultural output was about fourfold, compared with a nearly tenfold increase in non-agricultural output. The rate of growth of agricultural output was highest over the 1902-18 period, and was particularly low during the inter-war period. Since the first period, the non-agricultural sector has been the major contributor to increases in the rate of growth of output for the economy as a whole.

The fast rate of increase in agricultural output between 1902 and 1918 was partly due to a sudden increase in demand for Philippine products. The primary cause was the opening of the United States market to Philippine products, which began in 1902, and culminated in 1909 with the Payne-Aldrich Tariff Act permitting the duty-free entry (under a system of quotas) of Philippine products into the United States.[4]

The growth of non-agricultural output followed a somewhat different pattern. For the period as a whole, the average annual rate of growth of non-agricultural output was 3·9 per cent, half as much again as that of agriculture. The decline from an early peak of 3·6 per cent per annum in 1902-18 probably reflected the connection between industrial and agricultural output.[5]

The very high growth rate of industrial output in the last subperiod (1948-61) has been due partly to the policy of import substitution followed since 1949. However, the growth of non-agricultural output before World War II was also substantial, and so post-war growth in industry appears more like a continuation of past trends than the emergence of a new industrial system *ab initio*. In fact, the situation is more complex.

Industrial growth in the Philippines seems to have three major roots: first, import substitution (e.g. the shoe and glass bottle industries); second, growth in domestic demand; and third, growth in exports, although the share of non-agricultural output in exports is

---

[4] For a full discussion of this and other U.S. laws affecting Philippine exports to the U.S., see Abelarde (1947).

[5] This linkage can be both through intermediate demand (e.g. when sugar exports require milling) or through the impact of agriculture on final demand for industrial products.

small, being concentrated on unprocessed minerals (assuming log exports are agricultural). On balance, it appears that the most important one of the three has been the growth of final domestic demand, which means that the rate of growth in the agricultural sector represents a significant determinant of the potential rate of growth of output in the non-agricultural sector.

Growth of Per Capita Output

Total per capita commodity output grew by an average of 1·0 per cent per annum over the entire 1902-61 period (Table 7.2). While non-agriculture contributed at a rate of 1·6 per cent per annum, per capita agricultural output rose by a meagre 0·1 per cent per annum.

TABLE 7.2

GROSS VALUE ADDED PER CAPITA, AGRICULTURE AND
NON-AGRICULTURE FOR SELECTED YEARS, 1902–61

(constant 1939 prices)

Year	Agriculture (pesos)	Non-agriculture (pesos)	Total (pesos)	Compound rates of growth (%)		
				Agriculture	Non-agriculture	Total
1902	17·6	14·4	32·0			
1918	28·9	18·9	47·8	3·2	1·7	2·5
1928	25·1	21·6	46·7	−1·4	1·3	−0·1
1938	20·8	23·8	44·6	−1·9	1·0	−0·5
1948	16·4	16·8	33·2	−2·4	−3·4	−2·9
1961	19·1	37·0	56·1	1·2	6·2	4·1
1938–61				−0·4	1·9	1·0
1902–61				0·1	1·6	1·0

Source: Table 7.1; population estimates from Bureau of Census and Statistics 1947.

Had the growth of total output matched that of non-agricultural output, income per capita would have increased threefold rather than a mere 1·7 times between 1902 and 1961. As it was, the stagnant character of agricultural output per capita over the period as a whole reduced the nation's average annual rate of growth by nearly half.

Examination of output growth for the several sub-periods suggests that the growth rates of the two sectors are closely related, for the agricultural and non-agricultural series moved in the same direction in all sub-periods. This close relationship may simply reflect autocorrelation in the two series, arising mainly from the impact of conditions of world trade. It may also arise in part from changes in aggregate demand (transmitted through agriculture) for the products of industry, and vice versa. To the extent that the final demand hypothesis is true, then stagnation in agricultural output has retarded total national growth even more than appears from the simple calculations shown earlier. Conversely, an acceleration in agricultural output growth would also accelerate the rate of industrial growth, and therefore would have multiple effects on the growth of output of the economy as a whole.

Growth in Trade

In an economy like that of the Philippines, trends in foreign trade are obviously of great importance for the growth and composition of output.

Foreign trade (in real terms) was a fairly stable fraction of total output over the period as a whole (Table 7.3). It grew faster than

TABLE 7.3

FOREIGN TRADE FOR SELECTED YEARS, 1902–61

(constant 1939 prices)

Year	Imports (m. pesos)	Exports (m. pesos)	Imports + Exports ÷ 2 (m. pesos)	As % of total output[a]	As % of agricultural output[a]
1902	70·4	67·5	69·0	28·4	51·5
1918	112·7	154·5	158·9	32·6	53·9
1938	215·6	239·0	227·3	32·3	69·3
1948	236·8	125·3	181·1	29·9	55·1
1961	427·7	349·7	388·7	26·6	72·4

[a] The reader should be cautioned against drawing inferences from the data with respect to the magnitude of the share of trade in national income or product. In order to arrive at such a comparison, additions must be made to output as shown to derive total product. The data in this table are, however, suitable for making inferences concerning the trends.

Source: Bureau of Census and Statistics 1947; Central Bank of the Philippines (1950–); Table 7.2.

either total output or agricultural output between 1902 and 1918, partly because of the rise in the relative importance of export crops in total crop production, and particularly to the increased volume of sugar exports to the United States. During the inter-war years, the growth of foreign trade was much more rapid than agricultural output and about equal to that of total output. In this period the decline in foreign demand for some agricultural products during the great depression was offset by a rise for other products, such as minerals and logs.

During the last decade foreign trade has grown much more rapidly than agricultural output, but more slowly than total output. This confirms what one would intuitively hypothesise: that the period of exchange controls (1949-61) resulted in a modest reorientation of the economy away from its traditional commercial-centricism. In manufacturing, import substitution became the conscious policy of government. This resulted in a decline in imports of finished products which was, however, largely offset by increased importations of semi-finished products (Sicat 1966).

The data also make clear the relatively satisfactory record of export expansion during the period—a growth more than fivefold since 1902, as against agriculture's fourfold growth. Obviously this strong performance in the growth of primary exports has had a great deal to do with the country's ability to expand output in the non-agricultural sector (using capital-intensive techniques, as will be shown below).

For the future, it seems clear that rapid growth in both total and per capita agricultural output will represent an essential element in any strategy to achieve rapid economic growth in the Philippine economy. Rapid growth of total output is necessary in order to keep the terms of trade from turning against industry, to provide for continued growth of export earnings, and to earn foreign exchange for importing food.[6] Rapid growth of per capita output—or labour productivity—in the agricultural sector could contribute further to total economic growth by providing a mass market for the products of the industrial sector.

An important consideration in the design of a strategy for more rapid agricultural growth is the extent to which the barriers to growth in the past, and in the immediate future, stem from the supply or demand sides. It is possible, for example, that productivity growth resulting from technical change has not been sufficient to offset diminishing returns to labour and capital in agriculture and the other resource sectors. It is also possible that growth in domestic and foreign demand has not permitted resource industries to expand in line with technical capacities.

In the following discussion we attempt to determine the significance of those limitations, stemming primarily from the supply side, for agricultural growth in the Philippines.

SOURCES OF OUTPUT GROWTH: RESOURCE USE AND PRODUCTIVITY

The shift from a situation in which growth of agricultural output is accounted for primarily by changes in traditional inputs—land, labour, and traditional forms of capital—to one where the source is increasingly new inputs, which embody technical changes resulting in rapid

[6] Rice imports rose rapidly during the mid-1960s. In 1968, however, production and consumption were roughly in balance.

growth of total and partial productivity, represents an important turning-point in a nation's agricultural development. The relatively greater elasticity of supply of the newer technical inputs releases agricultural output growth from the classical constraints imposed by diminishing returns. In this section we examine the sources of growth in aggregate agricultural output, and in output of the major subsistence commodities—rice and maize—and the major commercial (export) crops—coconut and sugar.

Total and Partial Productivity

Estimates of changes in total and partial productivity for Philippine agriculture have been prepared by Hooley (1966) for benchmark years in the period 1902-61 (Table 7.4) and by Lawas (1968: 76, 82) for 1948-61 (Table 7.5). Although both authors emphasise the provisional nature of their results, the broad trends indicated by their estimates are basically the same. Total productivity has declined continuously throughout this

TABLE 7.4

TOTAL AND PARTIAL PRODUCTIVITY INDICES FOR PHILIPPINE
AGRICULTURE, SELECTED YEARS, 1902–1961

	Relative factor shares	Output and input indices				
		1902	1918	1938	1948	1961
Output		100	224	251	254	410
Inputs:						
Land	0·4277	100	186	313	286	603
Labour	0·4691	100	161	256	243	350
Machinery	0·0119	100	187	300	231	595
Animals	0·0913	100	267	457	301	605
All inputs combined[a]		100	181	299	266	484
Productivity measures:						
Output/land		100	120	80	89	68
Output/labour		100	139	98	105	117
Output/machinery		100	120	84	110	69
Output/animals		100	84	55	84	68
Total productivity		100	124	84	95	85

[a] Inputs have been combined on the basis of values expressed in constant prices. Farm land was valued at the average assessed value per hectare, as reported in the 1938 real property census; labour is at the rate of P1·100 per day, 200 work days per year. Valuation procedures for animals and machinery are described in Hooley (1966). The resulting weights for 1961 are roughly the same as those implicit in studies of operating farms. See Quintana 1965: 9–56, Tt. 9–14.

Source: Hooley 1966.

TABLE 7.5

CHANGES IN RESOURCE LEVELS AND FARM OUTPUT, 1948–60

	Relative factor shares	Per cent change, 1948–60		Per cent of output growth accounted for by variable
		Total	Weighted annual[a]	
Non-irrigated land	0·2491	94·56	1·40	41·18
Irrigated land	0·0823	57·37	0·32	9·41
Labour services	0·3908	16·47	0·49	14·41
Farm equipment services	0·0082	233·22	0·09	2·65
Farm building services	0·1801	14·16	0·20	5·88
Current expenses	0·0906	97·93	0·52	15·29
Total resource inputs			3·02	88·82
Technology		4·56	0·38	11·18
Farm output		49·54	3·40	100·00

[a] Obtained by multiplying the annual compound rates of change in the variables by their respective productivity coefficients.

Source: Lawas 1968: 76, 82.

century.[7] Labour productivity has either been constant, or has risen slightly in the past six decades, depending on the degree of adjustment that one is inclined to make for the drought in the 1902-3 period, while output per unit of land area declined continuously.

An interesting light is thrown on the decline of land productivity if, as Lawas has done, one distinguishes between irrigated and non-irrigated land. Lawas shows a small rise in total productivity during the post-war period, while Hooley shows a small decline. The difference can be accounted for largely by the fact that additional land area added to production was largely non-irrigated, and these additions receive a greater weight in Hooley's calculations. In fact this situation has persisted throughout the entire period since 1902 (Table 7.6). Irrigated land accounted for about one-quarter of total cultivated area at the end of the Spanish period, and then proceeded to decline sharply in the early years of the American period. This decline was largely arrested, but never reversed, during the remainder of the American period, and then continued after World War II.[8] The general

[7] The rise in productivity from 1902-18 shown in Hooley's data is ignored because we know that 1902-3 was a period of extreme drought, whereas 1917-18 were years of above average rainfall; see Coronas (1920: 114).

[8] There are interesting implications in this heretofore unnoticed fact, concerning the difference between the approach of Spanish and American colonial policies to Philippine agriculture, which might provide fruitful directions for further research. It seems apparent that American policy in the Philippines was also less development-oriented than Japanese policy in Taiwan during the same period; see Hsieh and Ruttan (1967).

TABLE 7.6

SHARE OF IRRIGATED IN TOTAL CULTIVATED AREA, 1902–60

Year	Irrigated area ('000 hectares)	Cultivated area ('000 hectares)	Proportion irrigated (%)
1902	320·4	1,298	24·7
1918	324·6	2,415	13·4
1939	523·0	3,953	13·2
1948	400·1	3,711	10·8
1960	620·5	7,595	8·2

Source: The estimate for 1902 was obtained by deducting hectares irrigated between 1902 and 1918 for each year, as shown in the annual reports of the Bureau of Public Works, and summarised in Bureau of Commerce and Industry 1929. Cultivated area is from the Department of Agriculture and Natural Resources. Other data are from Bureau of Census and Statistics 1918–.

conclusion is that productivity in Philippine agriculture declined steadily during the twentieth century, and this decline is due largely or entirely to the deterioration in the quality of land inputs.

It is possible, of course, that one or more individual commodity sectors may experience rapid productivity growth even though the agricultural sector as a whole remains unprogressive. Available data do not permit an analysis of changes in total productivity on a

TABLE 7.7

UTILISATION OF CULTIVATED LAND, 1948, 1955, 1962, AND 1966

Crop	1948 (%)	1955 (%)	1962 (%)	1966 (%)
Total cultivated area	100·0	100·0	100·0	100·0
Food crops	71·0	76·0	76·0	73·0
Rice (*palay*)	43·5	41·3	40·2	37·5
Maize	17·8	21·6	25·5	25·4
Fruits	4·2	5·5	5·0	4·6
Root crops	3·5	4·2	3·3	3·2
Other food crops	2·0	3·4	2·7	2·4
Commercial crops	29·0	24·0	23·3	27·0
Coconut	20·1	15·4	16·2	19·4
Sugar cane	1·8	4·2	3·2	3·8
Abaca	6·1	3·4	2·3	2·3
Other commercial crops	1·0	1·0	1·6	1·4

Source: Bureau of Agricultural Economics; Department of Agriculture and Natural Resources.

TABLE 7.8

CHANGES IN RICE PRODUCTION, AREA, AND YIELD, 1908–9 TO 1962–4

Period	Output ('000 metric tons, paddy)	Area ('000 hectares harvested)	Yield (metric tons/hectares harvested)
Amount			
1908–09 to 1909–10	798	1,174[a]	0·68[a]
1925–26 to 1926–27	2,140	1,781[a]	1·20[a]
1952–53 to 1953–54	3,163	2,650[a]	1·19[a]
1962–63 to 1963–64	3,905	3,124	1·25
Annual rate of change (%)			
1908–10 to 1925–27	6·0	2·5[a]	3·4[a]
1925–27 to 1952–54	1·5	1·5[a]	0·0[a]
1952–54 to 1962–64	2·1	1·7	0·5
1908–10 to 1962–64	3·0	1·8[a]	1·1[a]

[a] On an area planted basis. In crop year 1953–54, the Philippines area and yield data were shifted from an area planted to an area harvested basis.

Source: (1909–10 to 1924–25) Bureau of Commerce and Industry Statistical Bulletins, No. 8 and earlier issues; (1925–26 to 1952–53) Department of Agriculture and Natural Resources 1955–; (1953–54 to 1958–59) Department of Agriculture and Natural Resources 1958–59 and earlier issues; (1959–60 to 1963–64) Department of Agriculture and Natural Resources 1964.

commodity-by-commodity basis, but it is possible to analyse the extent to which output growth has been associated with changes in land productivity over time for a few crops, and to isolate some of the factors associated with differences in land productivity for the major commodities—rice, corn, sugar, and coconuts. These four crops account for a very high proportion of total agricultural land use (Table 7.7),[9] while total crop production itself increased in total agricultural output from 63 per cent in 1946-50 to around 75 per cent in 1961-5.

Rice

Production of paddy almost quadrupled between 1908-9 and the mid-1960s (Table 7.8). This represents an annual rate of increase of approximately 3 per cent per annum. During this same period the area devoted to rice increased from approximately 1·2 to 3·1 million hectares and the average yield from 680 to 1,250 kg per hectare.

[9] For discussion of other commodities see Wernstedt and Spencer (1967: Ch. 6). Other chapters contain an excellent introduction to the physical, cultural, economic, and regional dimensions of Philippine agricultural and natural resource use.

The rate of growth in output and productivity per hectare has varied sharply over the period since 1902 (Table 7.8). The most striking feature of the growth in production, however, is the long-term stability of the national average yield since the mid-1920s. Almost the entire increase in output since this time has been accounted for by increases in the area cultivated.

The stability of the national average yield in spite of the introduction of new varieties, increased use of fertiliser, and other changes in cultural practices, is a feature which requires comment. A likely explanation is that it represents a statistical fiction. It is possibly the result of combining an increasing area of low-yielding upland and rain-fed rice with irrigated areas where yields have risen, for, at the regional level, yields have typically declined where the area planted has expanded, and have risen where the area planted has declined or remained constant.

Variations in regional yields may reflect differences in environmental conditions of production, for example the proportions of irrigated, rain-fed, or upland rice or the proportions grown during the wet and dry seasons.[10] They may also reflect regional differences in technology employed, for example in cultural practices, varieties, and use of technical inputs such as fertiliser, insecticides, etc. The level of technology itself may reflect differences in economic incentives, such as factor and product prices; in institutional organisation (e.g. land tenure, credit, and marketing organisation), and social and cultural differences affecting the rate of adoption of innovations.

In the Philippines, rice is produced under many situations. Each province grows some rice in the wet season and some in the dry season. In each season some rice is grown under irrigated, rain-fed, and upland conditions. Regional or national average yields differ depending on season (wet or dry) and water treatment (irrigated, rain-fed, or upland).[11] Thus the average rice yield in each province

[10] Irrigated rice is typically grown in fields where water can be impounded by bunds or dykes and where water can be delivered to the field from surface storage, stream diversion, or wells. Rain-fed rice is grown in similar bunded fields but without access to water from surface storage, stream diversion, or wells. Upland rice is grown in fields where water is not impounded. Production of rain-fed and upland rice is typically confined to the wet season in Southeast Asia. In areas where seasonal differences are not too pronounced two crops of rain-fed or upland rice per annum are sometimes obtained. Typically, however, two or more crops per annum are obtained only where irrigation is available from surface storage or wells.

[11] Yield is measured in terms of kg of *palay* or paddy (i.e. rough rice) per hectare per season. Thus if both a wet- and a dry-season crop is grown on the same hectare it is counted as two hectares, and the average yield is the total production for both seasons divided by two.

or region and in the Philippines as a whole is determined by the yield obtained under different production conditions and the percentage of the total area on which different production practices are employed.

The non-irrigated or rain-fed first crop (wet season) accounts for the largest share of rice area in almost all regions. Irrigated first crop areas are substantial only in a few regions, such as Central Luzon, Bicol, and Southern and Western Mindanao. The area devoted to the irrigated and rain-fed second crop (dry season) rice is relatively small in all regions. The proportion of upland to lowland rice is relatively large in a few regions, such as Southern Tagalog and Southern and Western Mindanao.

Table 7.9 represents an attempt to estimate the effects of season (wet or dry) and water use (irrigated, rain-fed, or upland) on regional average yields. Standardised yields are calculated which are the average yields that would have been reported for the region if the distribution of rice area by season and water supply had been the same as the national average. The only year for which sufficient data are available to make this calculation is 1960-1.

TABLE 7.9

EFFECT OF DIFFERENCES IN REGIONAL PRODUCTION PATTERNS ON REGIONAL AVERAGE YIELD OF PADDY, PHILIPPINES, 1960–1

Region	Actual yield		Standardised yield[a]	
	(tons/ha.)	(index)	(tons/ha.)	(index)
Philippines	1·159	100·0	1·159	100·0
Ilocos	1·278	110·3	1·201	103·6
Cagayan Valley	1·087	93·8	1·174	101·3
Central Luzon	1·574	135·8	1·382	119·2
Southern Tagalog	1·049	90·5	1·143	98·6
Bicol	1·025	88·4	1·013	87·4
Eastern Visayas	0·891	76·9	0·891	76·9
Western Visayas	1·263	109·0	1·289	111·2
N. & E. Mindanao	0·847	73·1	0·851	73·4
S. & W. Mindanao	1·127	97·2	1·176	101·5

[a] In order to obtain standardised yields, regional yields of paddy from first crop, irrigated, non-irrigated; second crop, irrigated, non-irrigated; and upland areas are weighted by the national average distribution for the five categories.

Standardisation for the differences among regional yields (identified in the preceding paragraph) reduces the coefficient of variation for yields among regions in the Philippines by about 20 per cent (from 0·20 to 0·16). It seems reasonable to expect that if data were available to permit standardisation for differences in water control and season among provinces within regions and among villages within provinces the coefficient of variation for the standardisation yields would be even lower.

Source: Department of Agriculture and Natural Resources 1964.

In Central Luzon, for example, the actual average yield in 1960-1 was 1·574 metric tons of paddy per hectare, almost 36 per cent above the national average. The 1960-1 standardised yield was 1·382 metric tons, only 19 per cent above the national average. This means that almost half of the difference between the actual average yield and the average national yield is explained by the relatively favourable area distribution with respect to season and water treatment, and the rest by actual yield differences under similar environmental conditions. In the Ilocos region, about three-fifths of the margin of actual yield over the national average yield results primarily from the favourable area distribution.

In the Cagayan Valley, Southern and Western Mindanao, and Southern Tagalog regions, the relatively high proportion of upland area accounts for the below-average yield obtained in each.

In Western Visayas there is a close agreement between the actual and standardised yield, which implies that the higher-than-average yields are primarily the result of factors other than area distribution. In Bicol, Eastern Visayas, and Northern and Eastern Mindanao yields are still low compared with the national average, after allowing for area distribution, again apparently due to other factors.

The limited area of rice irrigated in both the wet and the dry seasons represents a major barrier to increased production and higher yields in most regions. Even in Central Luzon, where yields are relatively high, a shift of one hectare from production of one crop of rain-fed rice to production of irrigated rice during both the wet and dry seasons would add almost 2·37 tons to the total production, assuming the cultural practices of 1961. This would represent a 121 per cent increase in rice production per hectare per annum.[12]

The conclusion that environmental influences, primarily irrigation and water control, represent a dominant factor accounting for differences in output per hectare among the rice-producing regions of the Philippines, is reinforced by comparisons with other Southeast Asian

[12] Rice crop:

	metric tons per hectare
Wet-season irrigated	1·97
Dry-season irrigated	1·81
Total irrigated	3·78
Wet-season rain-fed	1·41
Increase from irrigation	2·37

This is clearly a conservative estimate of the increased output that would accompany irrigation. The dry-season yield reflects a situation where there is inadequate water throughout the dry season. Experimental evidence from the International Rice Research Institute and elsewhere indicates that with adequate irrigation water the dry-season yield for the typical varieties utilised by Philippine farmers should exceed the wet-season yield by 25 to 50 per cent.

countries. Hsieh and Ruttan (1967) found, for example, that rice yield differences among regions in Thailand are also primarily associated with differences in the environmental factors. Rice yields in the areas in which irrigation is most highly developed, Central Luzon and central Thailand, are essentially the same. They are also similar in the frontier regions of the two countries, where the area planted is expanding rapidly under rain-fed and upland conditions.

Maize

The area devoted to maize is second only to that devoted to rice. Production has risen from less than 200,000 metric tons in 1902 to 1,300,000 metric tons in the mid-1960s (Table 7.10). This represents an annual rate of change of 3 per cent per annum, the same as for rice. During this same period, the area devoted to maize rose from $0 \cdot 1$ million to approximately $2 \cdot 0$ million hectares. In the 1960s, output per hectare was slightly lower than during the early years of the century.

The two most striking features of the long-term growth of maize output and productivity are a very rapid growth in area and production since the early 1950s and a decline in the national average yield after the late 1920s. Between the mid-1920s and the early 1950s the area planted increased more rapidly than production. As in the case of rice, expansion of planted area has been the major source of output growth.

TABLE 7.10

PRODUCTION, AREA, AND YIELD OF MAIZE, 1909–11 TO 1962–4

Period	Production ('000 metric tons)	Area ('000 ha.)	Yield (metric tons/ha.)
1909–11	262·0	411·1	0·59
1928–30	355·7	516·0	0·69
1952–54	745·2	1,110·6	0·67
1962–64	1,282·8	1,923·5	0·67
Annual rate of change (%)			
1909–11 to 1928–30	1·6	1·2	0·8
1928–30 to 1952–54	3·1	3·2	−0·1
1952–54 to 1962–64	5·6	5·6	0·0
1909–11 to 1962–64	3·0	3·0	0·2

Source: (1909–10 to 1924–25) Bureau of Commerce and Industry Statistical Bulletins, No. 8 and earlier issues; (1925–26 to 1952–53) Department of Agriculture and Natural Resources 1955–(1955, 1956); (1953–54 to 1958–59) Department of Agriculture and Natural Resources 1958–59 and earlier issues; (1959–60 to 1963–64) Department of Agriculture and Natural Resources 1964 and earlier issues.

The relative stability of the national average yield since the recovery from World War II tends to mask substantial variations in yield trends at the regional and provincial level. Maize production is much more heavily concentrated geographically than rice production. Prior to the 1950s it was heavily concentrated in the Eastern and Western Visayas. During the 1950s it expanded rapidly in Northern and Eastern and in Southern and Western Mindanao. In contrast to rice, maize yields tend to be positively rather than negatively related to changes in area planted. The lowest yields have been obtained in the traditional production areas and the highest yields in the areas where production has been expanding most rapidly. The net effect of the shift in maize acreage toward the higher-yielding provinces and regions has been a smaller decline in the national average yield than if the acreage distribution had remained unchanged (Recto 1965: 81-7).

Environmental conditions and cultural methods for growing maize vary widely among provinces and regions. In most regions it is possible to grow three crops per annum; however, the proportion of land area devoted to each crop varies regionally. There has been a general tendency for an increase in the proportion of total area accounted for by the second and third cropping seasons.[13] In general, the first (wet) season yields are the highest and the third season yields are the lowest, although, in some instances, the availability of irrigation may raise yields during the dry season.

When the effects of variations in regional cropping patterns on average regional yields per hectare are taken into account (Table 7.11),[14] it is apparent that the relatively high average yields in the Cagayan Valley and the Western Visayas are due to the seasonal distribution. The relatively high yields in Southern and Western Mindanao and the relatively low yields in the Eastern Visayas are not, however, simply explained by the seasonal planting pattern.

Variations in productivity per hectare appear to be even more closely tied to environmental factors, both over time and among regions, for maize than for rice. In contrast to rice, maize yield has declined in the traditional centres of production. For rice, the rise in productivity in the traditional production areas appears to be related to investment in land development and irrigation, and to the complementariness of these investments with the new inputs—seeds, fertilisers,

[13] The first cropping season is the wet-season crop. The timing of the seasons also differs widely. In Negros Occidental, for example, the first (wet) season crop is planted in March and April, whereas in Bohol, Batangas, and Cagayan it is planted in May and June.

[14] The effects of irrigation, which were taken into account for rice, could not be estimated for maize.

TABLE 7.11

EFFECTS OF DIFFERENCES IN REGIONAL CROPPING PATTERNS ON REGIONAL AVERAGE YIELDS OF MAIZE, 1959–60

Region	Actual yields (metric tons/ha.)	(index)	Standardised yields[a] (metric tons/ha.)	(index)
Philippines	0·756	100·0	0·704	100·0
Ilocos	0·749	99·0	0·743	98·3
Cagayan Valley	0·857	113·3	0·820	108·4
Central Luzon	0·729	96·4	0·727	96·2
Southern Tagalog	0·728	96·2	0·729	96·4
Bicol	0·652	86·2	0·652	86·2
Eastern Visayas	0·564	74·7	0·564	74·6
Western Visayas	0·839	110·9	0·611	80·8
N. & E. Mindanao	0·824	109·0	0·821	108·5
S. & W. Mindanao	0·864	114·2	0·865	114·4

[a] Actual regional yields are weighted by the distribution of national maize acreages between first, second, and third cropping seasons. The weights were computed by dividing the area devoted nationally to maize in each cropping season by the total acreage. The standardised regional yields obtained are those which would have occurred if the regional distribution of acreage in each of the three cropping seasons had been the same as the national acreage distribution.

chemicals, and others—that have been introduced during the last two decades, whereas for maize, no new technology capable of reversing the decline in yields has become available to the small-scale cultivator.

The relatively high maize yields occur primarily in the frontier areas that have been recently opened up for cultivation. These are the same areas in which rice yields are the lowest. Unless new technology becomes available in maize production, it seems likely that these areas will also experience a decline in yields over time. Rice yields, on the other hand, could be expected to rise moderately, even in the absence of new technology, as labour-intensive investment in land development increases the proportion of irrigated and rain-fed area relative to the proportion of crop under upland conditions.

Coconuts[15]

The Philippines is the world's largest producer of coconuts and coconut products. It ranks first in the production and export of copra, coconut oil, copra meal and cake, and desiccated coconut. Throughout the post-war period coconut products have been the major export, exceeding earnings both from logs and sugar products.

[15] This section draws heavily on Nyberg (1966) and Hicks (1967).

TABLE 7.12

PRODUCTION, AREA, AND YIELD OF COCONUTS, 1910–11 TO 1963–5

Period	Production (million nuts)	Area ('000 ha.)	Trees Total (million)	Bearing (million)	Yield (nuts/bearing)
1910–11 to 1911–12	1,003·2	226·9	43·9	n.a.	n.a.
1919–20 to 1920–21	1,528·6	407·5	82·0	45·0	33·5
1935–36 to 1936–37	3,064·9	635·0	120·1	89·4	34·0
1946–47 to 1947–48	4,347·1	960·0	139·8	107·8	40·0
1957–58 to 1958–59	6,007·5	1,000·8	165·8	128·0	47·0
1963–64 to 1964–65	7,137·1	1,543·8	236·5	188·3	37·5
Annual rate of change (%)					
1910–12 to 1919–21	4·8	6·7	7·2	—	—
1919–21 to 1935–37	4·4	2·8	2·4	4·4	0·1
1935–37 to 1946–48	3·2	3·8	1·4	1·7	1·5
1946–48 to 1957–59	3·0	0·4	1·6	1·6	1·5
1957–59 to 1963–65	2·9	7·5	6·1	6·6	−3·7
1919–21 to 1963–65	3·6	3·1	2·4	2·4	0·3

Source: Computed from data assembled by Nyberg 1966.

Since 1910-11, the earliest year for which data are available, both the area planted to coconuts and total output have risen sevenfold (Table 7.12). Rates of growth in output, area, trees, and productivity have varied sharply over time. The area devoted to coconuts, and the number of trees planted, rose rapidly between 1910 and 1912, and at a modest rate between 1919-21 and 1935-37. Productivity per tree remained essentially unchanged during this period. Production, area, trees, and productivity declined sharply during World War II. From the end of the war to the late 1950s, area planted remained relatively stable, tree numbers rose modestly, and nut production and productivity rose sharply. Since the late 1950s, area and tree numbers have increased sharply and productivity has declined.

The problem of productivity measurement for tree crops such as coconuts is more complicated than for annual crops. For crops which require several years to reach maturity, output per unit area is not a useful measure, because it is influenced by the percentage of trees in the area which have reached bearing age. The number of nuts per bearing tree is a somewhat better measure, although this measure may also be biased, owing to variations in the number of bearing trees per hectare or variations in the average age of the bearing tree.

Three factors appear to be involved in the rise in productivity per tree since the mid-1930s: the southward shift in the geographical distribution of coconut production; differences in tree density among areas; and differences in the age distribution of bearing trees. In addition, yields in the Bicol region have apparently been reduced by the incidence of *cadang-cadang* disease amongst bearing trees.

Coconuts are grown in all the provinces of the country. The most important regions are southern Tagalog, Bicol, Eastern Visayas, and southern and western Mindanao. The seven provinces of Laguna, Quezon, Camarines Sur, Leyte, Samar, Davao, and Cotabato have historically accounted for close to half the total number of trees. There has been both an absolute and a relative decline in the number of trees in the Luzon area and a rise in Mindanao. Hicks (1967: 176) has estimated that the rise in the national average yield due to this geographic shift amounted to 2·2 nuts per tree, or approximately 20 per cent of the yield increase between 1918 and 1960.

It is more difficult to give a precise estimate of the effects of changing tree density on yield. Nut production per tree is generally higher in the Mindanao area, where numbers of trees per hectare are lower than in the Eastern Visayas and Tagalog regions (calculated from Bureau of Census and Statistics 1960).

There is also a tendency for nuts per hectare of bearing trees to be higher in the Mindanao area, where density of planting is lower. The relationship between tree density and yield is complicated partly by the fact that the new plantings have typically been on larger units than in the traditional Luzon, Bicol, and Visayas areas and partly because practically no new plantings are being made on tenant-operated farms (Nyberg 1966: 13-14).

The effect of the third factor affecting productivity, average age of bearing trees, is also difficult to measure quantitatively. A coconut tree begins to yield after 5-8 years and full production is reached in 15-20 years.[16] It is apparent, however, that the yield increases between the late 1930s and the late 1940s, and again between the late 1940s and the late 1950s, were occurring during periods of relatively slow growth in the number of trees of bearing age—periods when the average age per bearing tree was rising. Between the early 1920s and the mid-1930s, when the number of bearing trees was expanding at an annual rate of 4·1 per cent per annum, the average yield declined.

[16] The maximum number of nuts per tree by age is approximately as follows:

Bearing life	Nuts per palm
Fifth year	10
Sixth year	40
Seventh year	60
Eighth year	80
Ninth year	100

The above data are reported by Hicks (1967), quoting from Barrett (1946: 11).

Yield has declined again since 1957-8 when the annual growth of bearing-age trees rose to 5·7 per cent.

It seems apparent that a substantial share of the changes of coconut yields since the early 1920s can be accounted for by changes in area distribution and by the changes in the age distribution of the coconut trees. We classify these as environmental rather than technical changes. After allowing for these two factors, it seems apparent that very little real productivity gain, associated with technical and institutional changes, remains to be accounted for.

Sugar

The area planted to sugar has been considerably less than the area devoted to the other crops discussed above. Even after the increase which followed the reallocation, by the United States, of part of the Cuban quota to the Philippines as a result of the Sugar Act of 1962, less than 300,000 hectares were devoted to sugar. It has, however, represented a major source of Philippine export earnings in recent years. During the five-year period 1957-61 it was the second largest dollar earner, and contributed 26 per cent of the value of all Philippine exports. In 1961 it became the largest dollar earner and accounted for close to 30 per cent of export earnings.

A second important contrast with the other three crops is that, while the production and area of sugar has remained relatively constant, excluding the depression and war years, yield has risen significantly. Much of this increase has occurred during the post-war period (Table 7.13).

TABLE 7.13

LONG-TERM TRENDS IN SUGAR PRODUCTION, AREA, AND YIELD,
1935–6 TO 1966–7

Period	Production ('000 metric tons)	Area (ha.)	Yield (kg/ha.)
1935–36 to 1936–37	953	254,340	3,743
1951–52 to 1952–53	1,003	198,884	5,049
1960–61 to 1961–62	1,392	213,279	6,522
1965–66 to 1966–67	1,481	292,816	5,114
Annual rate of change (%)			
1935–37 to 1951–53	0·3	−1·5	1·9
1951–53 to 1960–62	3·3	0·8	2·9
1960–62 to 1965–67	1·2	6·5	−4·7

Source: (1935–37) Ilag 1964: 24, 26; (1951–67) J. C. Atienza, Philippine Sugar Association, personal communication, 1967.

The sugar industry also differs from the others discussed above in that it is produced primarily as a plantation crop, rather than by subsistence or peasant producers. Production is concentrated heavily in the Eastern and Western Visayas, Central Luzon, and Southern Tagalog regions.

A fourth important contrast is the tight controls exerted over production, marketing, and export. Exports are almost entirely to the United States, and changes in total production are dominated by United States sugar policy decisions affecting the size of the Philippine quota.[17]

The most striking feature in the growth of Philippine sugar production is the rapid increase in productivity between the early 1950s and the early 1960s and the decline in productivity since the early 1960s (Table 7.13). The rise in productivity during the 1950s was even more rapid in some of the major milling districts. A recent study by Ilag (1964)[18] in the Victorias milling district of northern Negros in the Western Visayas casts considerable light on the sources of productivity growth.

Research on sugar production in the Philippines is conducted both by the Philippine Sugar Institute, a semi-autonomous public corporation financed by funds from the sugar industry, and by several individual sugar-processing firms. Since the mid-1950s the breeding of new varieties has been primarily under the direction of the former. In spite of increased investment in research, sugar varieties of Javanese

[17] The U.S. government has played a major role in the Philippine sugar industry. In 1902 the U.S. Congress admitted Philippine raw sugar at 75 per cent of the duty of 1·685 cents per lb. The Payne-Aldrich Act of 1909 established free trade between the Philippines and the U.S. On 5 October 1913 the Underwood-Simmons Tariff Law abolished the 300,000-ton limit of sugar imports from the Philippines which was set by the Payne-Aldrich Act. By 1920 19 sugar centrals were in operation. By 1930 there were 40 centrals, which further increased to 46 in 1934, when the highest pre-war production record of 1,565,405 short tons was made.

The duty-free entry of Philippine sugar to the U.S. was limited to 800,000 long tons of raw and 50,000 tons of refined sugar each year by the Philippine Independence Law of 24 March 1934. The Jones-Costigan Act of 9 May 1934 and the Sugar Act of 1937 allocated larger quotas to the Philippines than the duty-free limitation set by the Independence Act. From 1934 to 1961 Philippine exports to the U.S. were limited to 952,000 long tons. In 1962 the quota was increased by 68,000 tons, as provided in the 1962 amendments to the U.S. Sugar Act. From the latter part of 1960 to date, non-quota purchase allocations had been granted to the Philippines owing to the banning of Cuban sugar from the U.S. market. This has greatly encouraged increased production.

After the expected termination of the Laurel-Langley agreement in 1974, the Philippines will export sugar to the U.S. only at the full duty of ₱3·50 per picul (of 63·25 kg), based on the exchange rate of ₱4 to US$1.

[18] This study represented a resurvey of an area previously studied in 1957-8; see Caintic et al. (1959).

I

or Hawaiian origin continue to account for a large share of total output. Much of the productivity growth has been due to the ability to transfer improved varieties adapted to Philippine conditions from other locations in the tropics and sub-tropics.

TABLE 7.14

YIELD CHANGES IN THE VICTORIAS MILL DISTRICT AND THE PHILIPPINES, 1947–8 TO 1966–7

| | | Yield (kg sugar per hectare) | |
Year	Victorias	Philippines	Ratio
1947–48	4,908	5,047	0·9724
1948–49	5,230	5,625	0·9298
1949–50	5,488	4,856	1·1301
1950–51	5,470	5,488	0·9968
1951–52	5,329	5,181	1·0284
1952–53	5,998	4,916	1·2201
1953–54	6,159	5,899	1·0441
1954–55	5,230	5,697	0·9181
1955–56	6,128	5,880	1·0423
1956–57	5,512	5,827	0·9460
1957–58	7,670	6,807	1·1267
1958–59	7,409	7,080	1·0465
1959–60	7,470	6,710	1·1133
1960–61	6,667	6,268	1·0636
1961–62	8,080	6,777	1·1924
1962–63	6,436	6,312	1·0195
1963–64	7,228	5,962	1·2124
1964–65	5,949	4,755	1·2510
1965–66	6,684	4,711	1·4188
1966–67	—	5,416	—

Source: (1947–48) Philippine Sugar Association 1964: 173; (1948–63) J. C. Atienza, Philippine Sugar Association, personal communication, 1967.

The Victorias Milling Company has been one of the leading processing firms in conducting research in recent years. Ilag indicates that the increase in yield between 1957-8 and 1961-2 (Table 7.14) in the Victorias district was due to the following six causes: change in varieties grown; increases in number of farms using fertiliser and lime and in the rates of application of fertiliser and lime; more effective land preparation associated with a shift from animal to mechanical power; improvements in planting materials and techniques; more effective cultivation and weed control; and use of more efficient post-harvest handling and transportation.[19] Ilag's analysis of the experience in the Victorias milling district provides rather clear support for the proposition that aggregate productivity growth in the sugar industry during the 1950s reflected changes at the micro-level rather than shifts in location or changes in environmental factors. Furthermore, these technical changes seem to be the product of investment in research and development and in more productive capital and current inputs.

The decline in productivity per hectare since the early 1960s is apparently a direct consequence of the rapid expansion in the area devoted to sugar production. As in the case of rice and coconuts, rapid expansion of area has been associated with a decline in yield. The decline in yield appeared first in a decline in the yield of sugar per ton of cane, then in a decline in the tons of cane per hectare. As the new areas brought into production are more fully developed for sugar cane production, and as available technology becomes fully employed, it seems likely that sugar yield will again resume its upward trend.

PRICE RESPONSE AND RESOURCE USE

In the modern industrial societies of the West, the notion that changes in relative prices play an important role in the modification of producer and consumer behaviour is firmly grounded in economic logic and is confirmed by observation and analysis. In most of such societies, whether they are relatively free from state control or are rigorously socialistic, prices play an important role in guiding agricultural production decisions and in directing the flow of commodities through the market.

In the past, a number of arguments have been advanced for anticipating that peasant producers would be unresponsive or would respond perversely to price incentives.[20] This argument is particularly significant for the formulation of growth policy in developing countries since production policy (primarily investment in research leading to new inputs and in plant and equipment to produce new technical inputs), and price policy (exerted primarily through import substitution policies and through direct market intervention), represent the major instruments for influencing the rate of growth of total agricultural output, and the allocation of resources to the production of different commodities.

[19] 'The research department played an important role in improving the productivity of the district. From 1957/58 until 1961/62, 180 experiments were conducted on fertiliser, 102 on variety, 68 on variety and fertilisers, 38 on liming and organic matter, and 124 others' (Ilag 1964: 102).

[20] For a review of the literature on price response see Krishna (1967: 497-540).

If a nation is to formulate a constructive agricultural price policy, an attempt must be made to obtain reliable information regarding first, the effects of relative changes of individual crop or livestock prices on resource use and production; second, the effects of changes in the prices of agricultural commodities relative to the prices of manufactured inputs and/or consumption goods purchased by peasants; and third, the effect of price changes on the marketed surplus of food crops in countries where a large part of the output is retained for home consumption.

In the Philippines we are able to examine the implications, first of relative price changes between food and commercial (export) crops, and second, of changes in the respective prices of rice and of maize, relative to other commodities.

Price Response of Commercial and Food Crops

Since independence in 1945, Philippine development policy has been oriented to the achievement of rapid economic growth. It has been dominated by two important value judgments concerning how growth should occur.

First, industrialisation should replace the primary export sector—sugar, copra, logs, and minerals—as the leading sector and basic generator of economic growth.

Second, the nation should become self-sufficient in basic foodstuffs —particularly rice and maize. The periodic 'rice crisis' should be eliminated.

The policies designed to achieve the growth objectives included an import substitution policy implemented through over-valuation of the peso in relation to other currencies, and a complex set of tariffs, import and exchange licences, and others. The over-valuation of the exchange rate, which held the peso/dollar exchange rate at ₱2/US$1, lasted until 1960, when steps were taken to free the exchange rate. These steps were completed in January 1962. Since that time the peso/dollar exchange rate has remained at approximately ₱4/US$1.

The effect of over-valuation and of the devaluation on income flows was first discussed by Legarda (1962: 18-28). The impact of the relative commodity price effects of exchange control and the devaluation on resource use as between commercial and food crops has been examined in greater detail by Treadgold and Hooley (1967). It is clear from this analysis that land use has responded to changes in the relative prices of commercial and food crops (Fig. III).

The higher peso price of the dollar since de-control appears to have contributed significantly to a change in relative agricultural prices. These in turn have induced a significant response in terms of resource allocation and in the relative rates of growth of the two alternative categories of output. These categories are food crops, prices of which (particularly in the case of cereals) are not in general affected directly by changes in exchange rates and international prices, and commercial crops, of which an important proportion are exported, and prices of which are more directly influenced by changes in exchange rates and international prices.

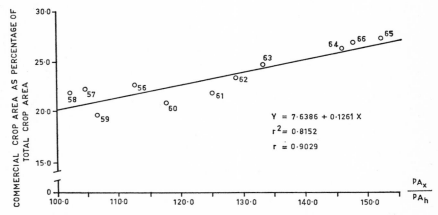

Fig. III   Relationship between land utilisation and relative prices in agriculture

Notes: Crop area data refer to area harvested (except for tree crops). Data are on a crop year basis; i.e. year t refers to the 12 months ending 30 June, year t. (Source: Department of Agriculture and Natural Resources; Bureau of Agricultural Economics.)

PA_x = price index of agricultural export products
PA_h = price index of agricultural products for home consumption
Both indices have bases of 1955 = 100.

Source: *Central Bank News Digest*, Vol. 18, No. 51, 1966.

The price ratio plotted against the area data for any crop year t is the average of the price ratios for the three previous calendar years t-3, t-2, and t-1. In the chart the figures alongside the plotted points identify crop years to which the relevant ratios relate.

Fig III suggests that the agricultural sector is influenced in the allocation of land resources between food and commercial crops by the relative prices which have prevailed for these two commodity groups in the recent past. The data imply that, for the period 1956-66, some 82 per cent of the variability in the proportion of total crop area devoted to commercial crops can be 'explained' by variability in the ratio of wholesale prices of agricultural products of an exportable nature to wholesale prices of agricultural products produced primarily for home consumption (largely food). The regression coefficient is easily significant at the 1 per cent level of confidence.

It is apparent from Fig. III that the de-control period brought about a change in relative prices in favour of commercial or exportable crops and that this evoked a response in resource allocation—namely, a relative diversion of land away from food crops into commercial crops. A rapid increase in the area devoted to commercial crops did in fact take place after crop year 1960, with a corresponding stagnation and decline up to 1964 in the area devoted to food crops.

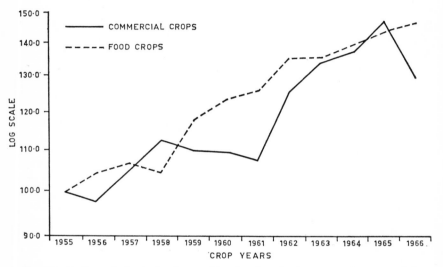

Fig. IV   Indices of physical volume of production, food crops, and commercial crops, 1955-66

Note: The two indices are base-weighted arithmetic averages of quantum relatives. The weights are based on the 1955 production values.

Source: The indices were computed directly from and value data supplied by the Department of Agriculture and Natural Resources.

These trends were reflected in terms of output, after making allowance for changes in yields (Fig. IV). Between 1955 and 1960 commercial crop output expanded at an average annual rate of only 1·9 per cent while food crop output rose at a rate of 4·4 per cent. Between 1960 and 1965 the commercial crop rate of output growth jumped dramatically to an average of 6·1 per cent per annum, while for food crops it fell away to an average of 3·1 per cent per annum.

The rapid expansion in commercial crop output explains much of the impressive export performance both during and after de-control.[21]

Price Response of Rice and Maize

Rice and maize occupy a dominant position among the agricultural commodities produced for domestic consumption. Together they account for approximately 5 of the 8 million hectares of cropland harvested annually. The traditional export crops—primarily coconuts, sugar, abaca, and tobacco—account for another 2 million hectares, and minor crops such as bananas and root crops account for approximately 1 million hectares.

In addition to absorbing a major share of the resources devoted to agricultural production, rice and maize price policy represents a major area of administrative and legislative concern. The President and the Legislature are under continuous pressure from producers to raise support levels, and from consumers to increase imports and to supply rice at subsidised prices.

Policy-makers and planners have typically assumed that expansion of rice and maize production in the Philippines is due primarily to new land brought into production as a result of transmigration to frontier areas, to intensification of land use by a shift to multiple cropping as a result of public investment in drainage and irrigation, and to technological changes, such as the introduction of new varieties. It is also generally assumed that the response of peasant producers of subsistence crops such as rice and maize to changes in the prices of these two products relative to other prices or to each other is low enough to implement consumer-oriented price and trade policies without substantially affecting aggregate production of the rice and maize.

There is little doubt that the transmigration of peasant families to the Cagayan Valley and to Mindanao was an important factor in the continued expansion of cultivated area during the first decade and a half following World War II, when the area planted to rice and maize expanded rapidly while the prices of both crops declined absolutely.

Studies by Mangahas, Recto, and Ruttan (1966a: 1-27; 1966b: 685-703)[22] have documented the response of resource use on rice and

[21] That part of the expansion which was contributed by sugar was to some extent independent of relative price movements. Prior to de-control, sugar exports to the U.S. were already pressing on the quota limits. The additional non-quota allocation of 176,426 short tons granted by the U.S., following its suspension of imports from Cuba near the beginning of the Philippine de-control period, was by itself an incentive for expansion.

[22] For further elaboration of the policy implications see Barker (1966: 260-76).

maize production to changes in the prices relative to each other and to other commodities.

Estimates of the price elasticities of hectarage were derived from statistical supply functions.[23]

For rice, the short-run elasticities, computed for the simple models, typically fall in the $0 \cdot 1$-$0 \cdot 3$ range. However, in the two regions with the largest relative shares of irrigated land (Central Luzon and Bicol), elasticities in the $0 \cdot 4$-$0 \cdot 6$ range were obtained. The elasticity in Western Visayas, the third-largest producing region, may also be in the neighbourhood of $0 \cdot 6$. Supply elasticities in the Cagayan Valley and in Northern and Eastern Mindanao could not be shown to be positive by any reasonable significance criteria.

For maize, short-run elasticities from simple models and from distributed lag models are in relatively close agreement. In Ilocos the supply elasticities were not significantly different from zero. In the Cagayan Valley, Bicol, and Southern and Western Mindanao regions, the maize supply elasticities fell in the $0 \cdot 1$-$0 \cdot 3$ range. Higher hectarage supply elasticities were obtained in Southern Tagalog, Eastern Visayas, and Western Visayas. The Southern Tagalog region is a major centre of commercial livestock production for the Manila market. The two Visayan regions are the major areas where maize is directly used as food. Supply elasticities in Central Luzon and in Northern and Eastern Mindanao could not be shown to be positive by any reasonable significance criteria.

It appears, therefore, that the supply elasticity for maize is highest in areas dominated by strong commercial markets, while for rice it is highest in areas with strong commercial markets and/or relatively extensive irrigation development. It seems reasonable to hypothesise that as the areas of recent settlement become more intensively developed, particularly in terms of irrigation and market facilities, peasant farmers in these regions will also become more responsive to price changes in allocating the area planted among alternative crops.

The short-run hectarage response elasticities for rice are usually larger than those for maize. The fact that marketing ratios for rice are greater than those for maize lends further support to the hypothesis that the price response of output in the Philippines is greater for

[23] See Table 7.16. It should be noted that the significance levels used are quite high. Although the available time series are relatively short and known to contain imprecisions, the seriousness of policy issues involving rice and maize in the Philippines requires that these data be analysed carefully for any economically important relationships that they may reveal. Thus stability of the regression coefficients over the several trials and tests is given relatively heavy weight, and the obtaining of standard errors indicating statistical significance of the regression coefficients at conventional levels $(0 \cdot 01, 0 \cdot 05,$ or even $0 \cdot 10)$ is given less weight than might be desirable under other circumstances.

crops produced under conditions where farmers are relatively market-oriented.

The price-elasticity estimates for the Philippines are similar to estimates obtained for the same crops and for other subsistence crops in other Asian countries. In those regions where production of rice and maize is highly market-oriented, Philippine rice and maize producers are at least as responsive to price changes as producers of commercial crops in India and Pakistan.

The price elasticity of hectarage is a lower limit to the price elasticity of output, provided that the price elasticity of yield is non-negative.[24] An estimate of a minimum value for the elasticity of the marketed surplus (Table 7.15) can be obtained by disregarding the price elasticity of home consumption (for which no estimates are available but which is hypothesised to be negative) and using the area (hectarage) elasticity as a lower limit on the estimate for output elasticity. The elasticity estimates resulting from the use of this procedure appear to be biased downward relative to the 'true' elasticity of the market surplus, except under conditions where the producers' income elasticity of demand is well above unity.

Conclusions and Implications

The data examined in this section indicate that Philippine farmers are reasonably responsive to changes in the relative prices of subsistence crops and commercial crops, even in the short run. These data imply that changes in relative prices are effective in determining the allocation of land among the several agricultural commodities. It seems quite clear, for example, that although the total area devoted to rice and maize before 1959-60 was expanding as a result of autonomous forces associated with transmigration and the opening up of new areas to cultivation, the declining price of rice relative to maize during the period before 1959-60 was associated with the relatively more rapid increase in maize area. In recent years, the decrease in area devoted to rice and maize and the increase in commercial crop area is apparently related to the rapid rise in the relative prices of sugar and copra.

While rice and maize prices have apparently been fairly efficient in resource allocation, there is little evidence to indicate that they have

[24] This follows from the definition that output equals hectarage multiplied by yield. The short-run elasticity of the marketed surplus with respect to expected prices is given by the formula

$$E_{M_t P_t}{}^* = E_{Q_t P_t}{}^* \cdot \frac{Q_t}{M_t} + E_{C_t P_t}{}^* \cdot 1 - \frac{Q_t}{M_t}$$

where $E_{Q_t P_t}{}^*$ and $E_{C_t P_t}{}^*$ are the price elasticities of output and of home consumption respectively.

been important incentives for the purchase of yield-increasing technical inputs such as fertilisers, insecticides, and herbicides. Efforts to identify a measurable yield response to price have not been successful.

TABLE 7.15

ESTIMATED SHORT-RUN PRICE ELASTICITIES OF RICE AND MAIZE HECTARAGE AND OF MARKETED SURPLUS IN POST-WAR PHILIPPINES

Region	Price elasticity of hectarage[a]	Average marketed proportion[b]	Price elasticity of marketed surplus (low estimates)[a]
Rice			
Ilocos (2)[c]	0·11 to 0·23**[d]	0·37	0·30 to 0·62
Cagayan Valley	neg.[e]	0·40	neg.
Central Luzon (2)	0·13 to 0·55**	0·65	0·20 to 0·85
S. Tagalog (2)	0·19 to 0·64**	0·50	0·38 to 1·28
Bicol (2)	0·38** to 0·41**	0·49	0·78 to 0·84
E. Visayas (1 & 2)	0·15 to 0·35	0·43	0·34 to 0·81
W. Visayas (2)	0·09 to 0·58	0·51	0·18 to 0·96
N. & E. Mindanao (2)	0·21 to 0·22	0·54	0·39 to 0·41
S. & W. Mindanao (2)	0·25 to 0·34	0·44	0·57 to 0·77
Maize			
Ilocos (1, 2, & 3)	0·04 to 0·08	0·24	0·17 to 0·33
Cagayan Valley (2)	0·17	0·36	0·47
Central Luzon	neg.	0·26	neg.
S. Tagalog (1, 2, & 3)	0·30 to 0·60	0·34	0·88 to 1·76
Bicol (1 & 3)	0·16 to 0·29	0·38	0·42 to 0·76
E. Visayas (2)	0·50** to 0·67**	0·27	1·85 to 2·48
W. Visayas (1 & 2)	0·42* to 0·49	0·19	2·21 to 2·58
N. & E. Mindanao	neg.	0·40	neg.
S. & W. Mindanao (1 & 2)	0·11** to 0·13	0·38	0·28 to 0·34

[a] The elasticities of hectarage and of marketed surplus are with respect to expected product price in the first-trial regressions and with respect to expected relative product price in second- and third-trial regressions.

[b] Rice 1959–60, maize 1954–55.

[c] The figures in parentheses after each region refer to the regression trials from which the price elasticity ranges were taken.

[d] One asterisk indicates that the coefficient is significant at the 40 per cent level; two asterisks, at the 20 per cent level.

[e] All estimates of price coefficients are negative.

LAND AND LABOUR[25]

Over the last half-century agricultural development has taken place through a process in which new land has been brought under cultiva-

[25] This section draws heavily on three sources: Fonollera (1966), Lawas (1968), and Golay and Goodstein (1967).

tion and the rural population redistributed to the margin of existing
cultivated areas and to the frontier areas of land colonisation. This
process has permitted a substantial increase in agricultural exports
while maintaining the per capita area planted to cereals and per
capita production of cereals approximately unchanged (Fig. V).

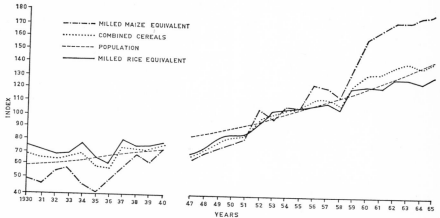

Fig. V   Production of cereals and population growth, Philippines, 1930-65
(indices, 1951-5 = 100)
Source: Golay and Goodstein 1967: 18.

Both the food crop sector and the export crop sector in Philippine
agriculture are, with the exception of sugar cane (and livestock) production, dominated by small-scale producing units (Table 7.16). The
average size of farms in 1960 was 3·6 hectares, of which 2·6 were

TABLE 7.16

SIZE OF FARM BY CROP CATEGORY, 1960

Type of farm	Number of farms ('000)	Area in farms ('000 ha.)	Average size of farms (ha.)	Proportion of total area		
				Farms of 10 ha. or less (%)	Farms of 10–199 ha. (%)	Farms larger than 200 ha. (%)
All farms	2,166·2	7,772·5	3·59	67	25·0	8·2
Rice (*palay*)	1,041·9	3,112·1	2·99	80	18·2	1·8
Maize	378·8	949·3	2·50	80	19·2	1·0
Sugar cane	17·8	249·4	14·01	20	36·8	43·3
Abaca	36·0	209·0	5·81	51	40·7	7·8
Tobacco	22·9	38·4	1·68	92	8·1	—
Coconut	440·3	1,938·6	4·40	62	34·8	3·4

Source: Bureau of Census and Statistics 1960.

under cultivation of annual and perennial crops.[26] The large-scale nature of sugar cane farming is notable, where 43·3 per cent of the total area cultivated was in farms of over 200 hectares.

There are indications that the traditional pattern of Philippine agricultural development—expansion of cultivated area through the extension of small-scale production units into frontier areas—is no longer as effective a source of agricultural output growth as in the past. Cultivated area per person in the total population, which rose almost continuously from the turn of the century, has declined sharply

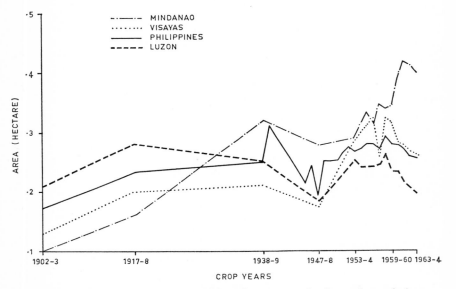

Fig. VI   Long-term trend in area cultivated per person in the total population, Philippines, 1903-64

since the late 1950s (Fig. VI). Cultivated area per person in the agricultural labour force has also declined in recent years. This reversal of trend reflects both a decline in the rate of expansion in the area cultivated and an increase in the rate of population growth.

Attempts to project the implications of current and anticipated rates of population growth for the distribution of labour force among

26 Size of farm, as reported here, refers to the size of the operating unit rather than the ownership unit. In rice production, where tenancy represents an important form of farm organisation, the average size of the ownership unit would, in most regions, be substantially larger than the average size of the unit operated by an individual tenant.

the agricultural and non-agricultural sectors imply that the prospect of reducing the proportion of the population dependent on agriculture is remote and that the size of the agricultural labour force can be expected to continue to rise for the next several decades (Ruprecht 1966: 289-312). The studies by Fonollera (1966: 145-270) and Lawas (1968: 93-124) imply a continued decline in cultivated land area per farm worker, even under optimistic assumptions regarding expansion of cultivated area.

These projections imply that it will be necessary for the Philippines to shift from a system of agricultural development based primarily on expansion of land and labour inputs with little change in output per unit of land or labour, to a system in which productivity growth, particularly output per unit of land area, accounts for an increasing share of output growth—a shift similar to that which occurred in Taiwan during the middle and late 1920s and accelerated after World War II.[27]

In some sectors, labour-intensive capital investment in land development for irrigated rice production, and in the extension of the land area devoted to coconut production, may represent an important element in this shift. Institutional modifications in the land tenure, marketing, and credit systems which increase incentives for the more intensive application of labour and capital inputs can also play an important role in some cases.[28] By and large, however, it seems likely that rapid growth in productivity will depend on investment in new technology which sharply increases yield potentials. It also seems likely that the realisation of yield potentials from any new technology will require substantial investment in three things: the production of inputs which farmers purchase from the non-farm sector, such as fertiliser and other farm chemicals; irrigation and drainage to permit continuous land use throughout the year; and rural infrastructure, particularly feeder road and highway development, in order to increase the real price to the farmer.

SUMMARY AND CONCLUSIONS

The results of our review of Philippine agricultural development during the twentieth century can be summarised in a series of generalisations.

Between 1902 and 1961, total commodity output grew by 3·2 per cent per annum in real terms. Non-agricultural output grew by 3·9 per cent per annum, while agricultural output increased by only 2·4

[27] See Ch. 2 above; also Hsieh and Ruttan (1967).

[28] For recent discussion of Philippine land tenure issues, see Ruttan (1965: 92-119; 1966: 42-63); Estanislao (1965: 120-4); and Morrow (1966: 380-5). For a review of the literature on agricultural factor, product, and credit markets in the Philippines, see Ruttan (in press).

per cent per annum. After accounting for population, this amounts to 1 per cent annual increase in aggregate commodity output per head. However, this fairly satisfactory aggregate record masks an annual per capita rate of growth of 1·6 per cent for non-agriculture and 0·1 per cent for agriculture.

Growth in non-agricultural development has been closely related to growth in the agricultural sector throughout this century. Agriculture and the other resource sectors have been the dominant sources of foreign exchange earnings to support imports of both consumption and investment goods. Furthermore, the products of an industrial sector dominated by import substitution industries have been closely related to growth in the purchasing power of the rural population. In the future it seems likely that growth of the industrial sector will be dampened unless the rate of growth in per capita income of the rural population rises above recent levels.

There has been little change in agricultural productivity over the past six decades. Output expansion is largely accounted for by expansion of traditional inputs. Total productivity has remained about the same; output per worker has remained the same or has increased very modestly. Output per land area has declined. If irrigated and non-irrigated land are separated, and each considered an independent input, most of the decline in output per land area can be accounted for. Thus, on the basis of available aggregate data, the stagnant character of Philippine agricultural yields is rather closely associated with the sharp decline in the quality of land inputs that have been added since 1902. This is partly a case of classical diminishing returns setting in as land area expands. It also seems to reflect decreased attention to investment in irrigation since the end of the Spanish period.

This general conclusion concerning the lack of productivity growth is substantiated for important specific commodities.

In rice, advances were made in the traditional areas through improved seeds, increased use of chemical inputs, and better farming practices. In the newly-settled areas, however, irrigation, and hence also the complementary technological improvements, have lagged far behind. The net result is that productivity gains in the traditional areas have been about equally offset by declines in the new areas, and nation-wide rice yields have remained essentially unchanged for at least fifty years.

In maize the nation-wide trend in yields is clearly downward. Between 1902-3 and 1959-60, average yields declined by about one-third. Even more than for rice, maize productivity appears to be closely tied to environmental factors. Also, in contrast with rice, maize yields have declined in the traditional production areas. The quality of land inputs related to maize has declined, and—again in contrast to rice—no new technology capable of reversing the decline in productivity has become available to the small-scale cultivator. In its absence, the decline will probably continue, as the relatively high yields in the frontier areas decrease.

Products derived from coconuts, such as copra, coconut oil, meal cake, and desiccated coconut, constitute a leading group of export commodities for the Philippines. Like many other crops in Philippine agriculture, expansion of output can be attributed largely to the expansion of traditional inputs. Productivity (measured as nuts per tree) has varied considerably over time, but there has been only a modest upward long-term trend since the mid-1930s. Environmental factors, such as a shift of production toward higher-bearing regions and variation in the age distribution of trees, account for most of this gain.

In contrast to other crops, area under sugar has remained constant over the past thirty-five years. Output has risen by 50 per cent, reflecting an equal gain in productivity. The gains in productivity have resulted from technological factors—improvements in sugar varieties, increased use of fertilisers and pesticides, increased use of mechanical power (in place of animals), improvements in planting materials and techniques, and improvements in post-harvest handling and transportation. Technological development has centred in the plantations, but has also 'spilled over' into the small farms. Technological change has been the product of both investment and research in development of higher-quality capital inputs, and in new current inputs.

Philippine farmers are, by and large, highly responsive to economic incentives. In the past, substantial shifts in land resources among crops have directly followed changes in product price ratios. Price incentives have not, however, had any apparent impact on productivity growth in the two major crops—rice and maize—which have been studied intensively. Lack of an observed yield response appears to be related to the fact that, for most Philippine agricultural commodities, the technical basis for a response to higher levels of inputs, other than the traditional factor of land and labour, has not yet been established. Until rice varieties with a high response to fertilisers are available, for example, it is not reasonable to expect that yields will respond to a change in the rice/fertiliser price ratio.[29]

It appears likely that the 1960-70 decade may represent the 'closing of the land frontier' in Philippine agricultural development. With the

[29] New rice varieties characterised by a substantially greater yield response to higher levels of fertiliser application became available to Philippine farmers in limited acres during the 1967 dry season. It is estimated that over 10 per cent of total rice hectarage in the Philippines was planted to these improved varieties during the 1968 wet season (see Barker 1968: T. 6).

possible exception of coconut and a few minor commodities, it seems unlikely that expansion in cultivated land area can continue under the traditional pattern of labour-intensive land development by peasant cultivators. Capital-intensive irrigation, drainage, and transportation investments appear necessary either to expand the area under cultivation or to intensify production on existing areas of cultivation.

Substantial investments in research and development of the agricultural supply sector also represent important elements in any future effort to achieve rapid gains in agricultural productivity and output. Growth of productivity in the sugar industry has clearly been related to a substantial research program emphasising development of new varieties, application of new technical inputs, processing technology, and more effective management of sugar plantations and centrals. It also seems apparent that the substantial increase in investment in rice research since the early 1960s is creating a new technology which should enable substantial gains in productivity and output.

As yet, however, the government of the Philippines is not effectively organised to carry out the essential investments in research and development, in land and water resource development, and in related areas of infrastructure and institutional development, necessary to achieve a rapid transition to a situation in which increases in output are primarily a product of higher yield per unit area. Responsibility for agricultural development decisions is highly fragmented. The Secretary of the Department of Agriculture and Natural Resources, while having the broad responsibility for agricultural development, does not have responsibility for such major areas as agricultural price policy and irrigation development and administration. Only part of the agricultural research and extension activities are under his direction.

The increase in demand for agricultural output generated by growth of population and income in the domestic market and the demand for foreign exchange earnings indicate the importance of rapid agricultural development during the next decade. A substantial redirection of investment flows and administrative talent into the agricultural sector would seem necessary if a rate of agricultural development consistent with a rapid growth in the total economy is to be achieved.

8

Indonesia

D. H. Penny

IN the last two hundred years or so Indonesia's agricultural economy
has grown substantially, in line with its population, but it has not
developed. Real income per head has not risen. In this period there
have been many changes in the general economic framework within
which farmers, rural merchants, exporters, and other decision-makers
in agriculture operate. The changes have included an increase in the
proportion of agricultural production traded on markets at home
and overseas, and increased monetisation of wage and rental payments,
but the changes have not led to any widespread upsurge in development.

Boeke (1953) has aptly described the process of growth-without-
development as 'static expansion'. Most students of Indonesia have
seen the process of growth-without-development as the result of the
impact of European colonial penetration;[1] but agriculture in the post-
independence period (after 1945) has also been characterised by
growth-without-development. An explanation of the persistence of
static expansion and an assessment of the possibilities for development
in the future will be given below. Consideration must first be given,
however, to the environment, physical and economic, within which
farmers and others make their decisions about the use of resources.

THE RESOURCE BASE

Indonesia straddles the equator and, like most parts of the hot, wet
tropics, has poor soils (Gourou 1953). It differs from the very lightly
populated Congo and Amazon basins only in those areas where
volcanic activity periodically renews soil fertility (i.e. in Java, Bali,
Lombok, and in parts of Sumatra and Sulawesi). Annual rainfall is
high in most parts of the country—1500 mm and more—but irrigation
is nonetheless necessary, particularly where there is a pronounced
dry season, as there is in much of Java, in Bali, and in Nusa Tenggara.

[1] See, for example, Wertheim (1956), Selosoemardjan (1962), and Geertz (1963).

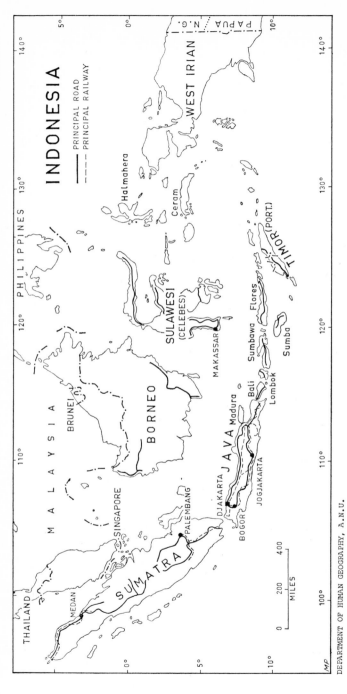

Map 9 Indonesia

DEPARTMENT OF HUMAN GEOGRAPHY, A.N.U.

Continuous cropping without irrigation and fertilisers is possible only in those areas with volcanic soils.[2] It is in these areas that 85 per cent of Indonesia's people are concentrated.

Elsewhere the population density is very low and, outside Java, shifting agriculture is widespread. In regions of non-volcanic soils, four main farming systems are in use: irrigated wet rice, perennial tree crops, 'Chinese agriculture', and shifting cultivation. Only the first two can provide the basis for a large-scale expansion of agricultural output. Each, however, has specific disadvantages that make agricultural development difficult: the first is expensive, land use tends to become specific, and agricultural mechanisation is hampered; the second for the most part depends on overseas markets for its produce, and these are limited; the third, involving as it does the slow build-up of soil humus through heavy organic fertilisation, is expensive, and suited only to the growing of vegetables and other crops of high value; and the fourth does not permit intensive land use.

It is possible, of course, that researchers will one day find other economic ways of using the soil and water resources of the equatorial regions, but for the present, Indonesian farmers are compelled to use one or another of the four farming systems mentioned.

Indonesia is, moreover, a country consisting of many islands, and transportation within and between the islands is little-developed. It is often easier to trade on the world market than it is on the domestic market, because Indonesia's national economy is as yet quite fragmented. For example, the price of rice in July 1967 was Rp5 per kg in Makassar, the capital of South Sulawesi, Rp11 in Djakarta, and Rp32 in Medan, the capital of North Sumatra.

Population and Land

Indonesia has the dubious distinction of having the highest population density in the world. The highest population densities are found in Java, and in particular Central Java (Table 8.1). In some agricultural areas of Java, population density already exceeds 2,000 persons per sq km.

Intensive land use has long been a feature of the river valleys of Central Java, and to a lesser extent, of East Java. The great monuments of Prambanan and Borobudur, both near Jogjakarta, are eloquent testimony to the high productivity of these lands a millennium and more ago. It is not known exactly why the kingdoms that erected these great monuments declined, but the remnants of their complex

2 For good descriptions of the difference between sedentary agriculture and shifting cultivation in Indonesia see the works of Pelzer (1945) and Geertz (1963).

TABLE 8.1

DISTRIBUTION OF POPULATION, 1961

Region	Total (million)	Density/sq. km
Djakarta Special Region[a]	3·0	5,152
West Java	17·6	380
Central Java[b]	20·6	553
East Java[c]	21·8	455
All Java[c]	63·0	477
Outer Islands	34·0	19
All Indonesia[d]	97·0	51

[a] The capital city area.

[b] Includes Jogjakarta Special Area.

[c] Includes Madura.

[d] At the end of 1967 total population was estimated to be 110 million. The estimated annual rates of increase are 2·24 per cent and 2·41 per cent for Java and the Outer Islands respectively.

Source: Biro Pusat Statistik 1961.

social organisation and of their productive, indeed sophisticated, agricultural system remain to this day. The difference between then and now, or between the first population count (1815) and now, is simply increased population density. Population figures for Java and Madura since 1815 are given in Table 8.2.

It was not until the end of the nineteenth century that Java 'filled up'. Before then, Javanese could find unoccupied land to cultivate reasonably close to their home villages. Now only some 23 per cent of the whole island is forested, hills too steep to stand upright on are cultivated, erosion has accelerated, destructive floods have become more common, and the capacity of existing irrigation systems has declined.

Adjustments that have occurred in the agricultural economy of Java to accommodate the increase in population have included the following:

First, an increase in the frequency of double-cropping. In the period 1955 to 1964, the index of double-cropping rose from 104 to 126. Without additional irrigation works, however, it is unlikely that double-cropping will become more widespread.

Second, a drastic decline in the area under export crops, with a corresponding increase in area under food crops. In the late 1920s over half of Indonesia's exports came from Java; by 1940 the propor-

TABLE 8.2

POPULATION OF JAVA (INCLUDING MADURA), 1815–1967

Year	Total (million)	Annual rate of increase (%)
1815	4·5	
1845	9·4	2·5
1860	12·5	1·9
1880	19·5	2·2
1890	23·6	1·9
1900	28·4	1·9
1920	34·4	1·0
1930	41·7	1·9
1961	63·1	1·4
1967	72·0[a]	2·2

[a] Estimate.

Source: (pre-1860) Soedigdo (1965); (post-1860) Biro Pusat Statistik.

tion had fallen to 45 per cent, while in 1958 the proportion was only 13 per cent.

Third, an absolute decline in the number of buffalo, cattle, and horses, which has likewise freed land for food-growing.

Fourth, a decline in the proportion of basic food needs that can be met by rice. Rice is clearly the preferred basic food, but yields fewer calories per hectare per annum than maize or cassava, the other important annuals. Between 1850 and 1900 annual rice production per head in Java was 106 kg; between 1900 and 1940 it was 96 kg; and by 1960-7 it had fallen further to 73 kg. The area of arable land per head in Java fell from 0·16 hectares in 1955 to 0·12 hectares in 1964.

There is already a large literature on the population problem in Java. The two major recent studies using primary data are those of Bailey (1961) and Timmer (1961). Both these men worked in the Jogjakarta Special Area in Middle Java. Other major studies have used secondary data, and include the writings of Iso and Soedarsono (1960), Geertz (1963), and Napitupulu (1968). The conclusions drawn by all these writers are deeply pessimistic, foreseeing no possibility of improvement in agricultural prosperity in Java until the birth-rate has been drastically reduced.

Java is fortunately not the whole of Indonesia, and there is as yet no other part of the country that has so acute a population problem. But the problem does exist elsewhere, as the bare hills of North Tapanuli (North Sumatra), Flores (Nusa Tenggara), and yet other places, amply testify.

AGRICULTURE IN THE INDONESIAN ECONOMY

The predominant position of agriculture in the Indonesian economy and the lack of progress in the manufacturing sector are clear (Tables 8.3 and 8.4).

The data show that Java has a somewhat smaller proportion of its work force engaged in agricultural production than the less densely populated Outer Islands. It also has a rather larger proportion of its work force engaged in manufacturing. But these data are misleading indicators of relative prosperity and of prospects for development, for

TABLE 8.3

NET DOMESTIC PRODUCT BY INDUSTRIAL ORIGIN, SELECTED YEARS

Sector	1939 (%)	1960 (%)	1965 (%)
Agriculture	61[a]	52	52
Manufacturing	15	12	12
Mining		3	3
Others[b]	23	33	33

[a] Mining included in agriculture.

[b] Includes government. The share of government was 6·7 per cent of the total in 1939, 7·9 per cent in 1953, and 13·7 per cent in 1958 (see Paauw 1961). The ECAFE figures show, however, that the share of government was only 4 per cent and 2 per cent in 1960 and 1965 respectively. Another source, Biro Pusat Statistik (1966a), gives 6·6 per cent as the sectoral share for government in 1958, or about half the figure cited by Paauw. Data discrepancies such as these are common. See also the appendix to this chapter.

Source: (for 1939) Paauw 1959: 206; (for 1960 and 1965) ECAFE 1966.

TABLE 8.4

PERCENTAGE INDUSTRIAL DISTRIBUTION OF EMPLOYED, 1930 AND 1961

| | Java[a] | | All Indonesia | |
Sector	1930	1961	1930	1961
Agriculture	71·2	69·3	73·9	73·3
Manufacturing	12·8	7·0	11·5	5·8
Mining	0·8	0·1	0·9	0·3
All others[b]	15·7	23·6	14·7	20·6
Total	100·0	100·0	100·0	100·0

[a] Includes Madura.

[b] Tertiary sector: trades, professions, government, etc.

Source: Adapted from Jones 1966: T. 2.

industrial output in Java has failed to increase, and the proportion of industrial workers in the total population has fallen.[3]

The growth of the tertiary sector in Java reflects in part the substantial increase in the number of government employees (since 1950), but the greater part of the rise is due to the increase in the number of persons engaged in petty trade: most of the latter are people who can no longer gain a livelihood in agriculture.[4]

A high proportion of exports originate from the agricultural sector: in the 1920s and 1930s some 70-85 per cent of all exports were agricultural products, and since 1950 agricultural exports have been about 55 to 65 per cent of the total. Almost the whole of the remainder are also primary products, minerals, timber, and petroleum.[5] Exports have, however, declined from 30 or 35 per cent of national income in the late 1920s, to about 25 per cent in the late 1930s, to some 15 per cent in the early 1950s, and to about 10 per cent in 1967.

Agricultural products are almost all exported unprocessed, many smallholder products going to Singapore and Malaya for upgrading and reshipment through the entrepôt trade (Richter 1966: 31-6) and others going direct to consumer countries for manufacture.

One of the major imports is an agricultural commodity, rice. Between 1955 and 1964, the value of rice imports ranged from 7 to 15 per cent of total imports, and often totalled more than one million tons per annum (Table 8.5). This is not a recent phenomenon, however, for between 1911 and 1941 rice imports were never lower than 109,000 tons per annum, and normally ranged between 250,000 and 500,000 tons.

Rice is by far the most important single commodity in the economy, and its production, sale, and distribution are matters of constant public concern. Currently about 16 per cent of Indonesia's national income of about US$9 billion consists of rice.[6] This figure both underestimates and overestimates the importance of rice in the economy.

It is an underestimate in the sense that the price of rice is regarded as the barometer for all other prices; that wages are measured in terms

[3] The national income data cited in Table 8.3 suggest that the share of manufacturing has been about 12 per cent in most post-war years. These figures conflict with data from the Department of Labour which show a decline in the index of industrial production from 100 in 1958 to 41 in 1962, whereas it is believed that (total) national income rose slightly in the same period. See appendix to this chapter for further comments on statistical sources.

[4] Dewey (1962) shows that a high proportion of all labour engaged in marketing in East Java consists of men and women who have been forced out of farming.

[5] For further details see Thomas and Panglaykim (1966 and 1967).

[6] In 1966 approximately 10 million tons of rice were produced which, at the 1966 cost of rice imported from Burma (US$140 per ton approx.), gives a rough figure for total value of US$1·4 billion at world prices.

TABLE 8.5

RICE IMPORTS AND GOVERNMENT PROCUREMENT, 1961–7

| Year | Procurement | | | Total as proportion of gross domestic supply (%) |
	Imports '000 met. tons	Domestic '000 met. tons	Total '000 met. tons	
1961	1,013	270	1,283	13·0
1962	1,011	551	1,562	14·5
1963	1,072	462	1,534	16·3
1964	1,016	342	1,358	12·9
1965	140	318	458	4·6
1966	235	603	838	8·1
1967	265[a]	412[a]	677[a]	7·3

[a] Estimated.

Source: (domestic procurement) Food Board Reports; (imports) Biro Pusat Statistik.

of rice; that subsistence farmers use rice as their measure of value; that almost all irrigation is for rice; and that virtually no Indonesian feels he has attained a satisfactory minimum living standard if he cannot produce or buy enough rice to meet his family's consumption needs. The extremely heavy emphasis given to rice—as against all food, or all agricultural production—by the government, will be discussed at length below.

On the other hand, the figure overestimates in two significant ways the role that rice plays in the economy, particularly in the market sector. First, only a relatively small proportion, about 30 per cent, of domestic rice production enters the market, and much of this is consumed locally. Second, prices, and market forces generally, are weak determinants of the area sown to rice. This arises in part because some two-thirds of the rice is grown to meet the farmers' own needs (for consumption and for wage and other payments in kind). Farmers do not readily reduce their own demand, whatever the price of rice is relative to those of other commodities, unless they have some assurance that rice will be available in the village markets when they want it and at a price they can afford.[7] This second point also arises because, when the land is irrigated, it is almost always used for rice. Wharton (1962b: 3) has for this reason classified rice as a perennial crop. This low price

[7] Farmers in the Karo area of North Sumatra are more economic-minded than almost all others in Indonesia, yet they remain unwilling to depend wholly on the market for their basic food even though they know they can obtain much higher returns from alternative uses of their resources. But they have been prepared to reduce the area sown to rice, and by 1962 were buying some 15 per cent of their needs (Penny 1964).

responsiveness is confirmed by the work of Mubyarto and Fletcher (1966: 22), whose conclusions are as follows:

> The empirical analyses give some evidence of positive but low price responses [in Java]. There is a further suggestion that yields are more strongly influenced by prices than is the area planted. This would be consistent with production conditions under a strong monoculture system. Acreage responses to price changes were more apparent in an area where substitution of corn had become profitable. A supply elasticity of *not more than* 0·15 is indicated.

The agricultural sector lacks economic flexibility, as does the economy as a whole. It might be argued, as Bauer (1948) has, that farmers who produce rubber, coffee, or coconuts for export are responsive to changes in economic conditions, because they will shift quite readily to rice and back again as relative prices change (Thomas 1965), and, in the case of rubber, will change the intensity with which they tap their trees. They nonetheless do not act very much like modern commercial farmers, for they buy virtually no inputs, not even the seed they use. If they were truly commercial farmers they would be responsive to changes in market conditions for inputs as well. Moreover, in no part of Indonesia has the regional specialisation characteristic of modern commercial agriculture appeared in more than embryonic form.

There is indeed only one large group of economic decision-makers in Indonesia that possesses the ability to respond quickly to changes in economic conditions—traders. They are able and willing to do so because most of their capital is kept in liquid form. By contrast, the government seems to show the least ability of any group to respond to changes in economic conditions. This would not matter much if the government played a minor role in the economy, but it does not. Most of the many ills of the post-independence economy—inflation, the great instability of input/output price ratios, heavy, and, until recently, arbitrary taxes on exports, and monumental bureaucratic inefficiency—have been due to the inability of economic policy-makers and government officials generally to understand the ways in which government actions affect the overall pattern of economic activity. Certain improvements in economic policy have resulted from the change in government in 1966, but many more changes must yet be made before the government will be in a position to play a positive role in development.

To sum up: Indonesia is a fragmented island economy. Much of its agricultural production is for subsistence, and a large part of the surplus that enters market channels is produced in response to overseas rather than domestic demand. Domestic manufacturing is little-

developed, and a large proportion of the domestic market for manufactured goods is supplied from abroad. The growth of domestic trade,
which would permit profitable exchange between local manufacturers
and farmers, and also between regions, is hampered by the continued
low capacity of the inter-island shipping system. The growth of trade
in agricultural products, and industrialisation as well, have been
hindered, in turn, by the restrictionist policies that have been the
norm in the archipelago from the early colonial era to the present day.

The farmers' economic response to this situation will be described
in detail in the next section, but it should be clear by now that the
farmers have as yet been given little incentive to increase output or
to modernise their production methods. This lack of incentive mani-
fests itself in a number of ways, among which the most important are
that Indonesian peasant farmers have the most labour-intensive technology in Asia and do almost nothing to protect their plants and
animals from the ravages of pests and diseases.

Sustained agricultural development requires that farmers become
constantly willing to adopt new technology: if they fail to do so their
only possible response to increased population is to expand their
production in a 'static' way by opening up new land. The land available is fast running out.

The final section of the chapter will consider the question of
whether the incentive framework is likely to change in such a way that
development will be encouraged in the future.

THE FARM ECONOMY

Indonesian peasants work hard, with little result. Most of them are
skilled agriculturalists and fully understand the basic principles underlying the maintenance of soil fertility and successful crop production,
but what they know is best suited to conditions of unlimited supplies
of land and little or no production for the market. Like farmers everywhere, they make careful calculations before deciding which crops to
grow, how much seed they should use, or how often they should weed
their fields, but they rarely perceive, and even more rarely decide to
use, the development opportunities open to them.[8]

It is often believed that the subsistence orientation of peasants
operating small farms is due to a need to concentrate first on growing
food to meet family requirements. Their unwillingness to innovate is

[8] Development opportunities are more profitable ways of using given resources.
There are many such opportunities open to farmers in all parts of Indonesia,
except in those places (e.g. Gunung Kidul) where population pressure has led to
severe erosion. For a fuller discussion of the concept, and for some examples, see
Penny (1964, 1967).

ascribed to a lack of resources and/or to the risks they are presumed to face if they grow new crops or adopt new methods of growing old ones. These explanations of continued subsistence-minded behaviour[9] by peasant farmers in Indonesia are inadequate, and it is not difficult to show that neither lack of resources nor the size of the objective risk premiums is the major determinant of peasants' unwillingness to innovate.

If these particular hypotheses were valid one would expect farmers to become more willing to adopt modern methods once they had access to more resources. However, the behaviour of farmers in new settlements, both voluntary and government-sponsored, as in the trans-migration areas of South Sumatra, indicates that most of them are still bound by ideas and attitudes appropriate to a subsistence economy. For example, even though land may be freely available, farmers will not lay claim to more than they can handle with their traditional labour-intensive technology, rarely as much as two hectares. As they develop these small farms, their sole aim appears to be to re-create their version of an ideal peasant economy. Rice receives top priority, even though there are other crops that are more profitable, and when the physical conditions are favourable large rice surpluses are grown for sale, even though market prices for it are low.

The new settlers in Pematang Djohar, North Sumatra, were able, after ten years, to produce a marketable surplus one-and-a-half times as large as their family consumption needs. The technology they used was identical with that previously used in their densely populated villages in Java. They saw no reason to change it, or their cropping pattern, as they were now very much more prosperous.

At first the new settlements lacked the social structure that charac-terised the villages of origin: there were no landlords, no money-lenders, and no landless forced to seek a livelihood from labouring. Over time, however, these institutions reappeared: some of the settlers worked harder than the others, were more thrifty, and had smaller

9 'Subsistence-mindedness' and 'economic-mindedness' (or 'development-minded-ness') are terms that were defined and developed in Penny (1964) to describe the economic behaviour of peasant farmers.

Subsistence-minded farmers are those who think first of meeting their subsist-ence needs when they make their resource-use decisions, even when their incomes may be well above the subsistence minimum. Farmers are described as being economic-minded if they have already become active participants in the develop-ment process (i.e. they are the farmers who have decided that they want to increase their incomes, who are confident that it is possible for them to do so, and who have learned how to do so).

When the term 'subsistence-minded' is used to describe the economic behaviour of people other than peasants one aspect of subsistence-mindedness is being empha-sised, namely, the unwillingness of that person or group to use resources for investment even though he or they could afford to do so.

families, and after a while these men became landlords and money-lenders. This degeneration of the ideal society was accepted because both the fortunate and the less fortunate recognised that those who worked hard and accumulated some savings should charge a price for the land they rented or the rice they lent to those who had been lazy or unthrifty.

At present rates of increase the population of the new settlements will double every thirty years or so, thus making it inevitable that average farm size will decrease there just as it has in Java. After thirty years in one settlement area in South Sumatra (Kampto 1967: 282), approximately 73 per cent of the landholders owned less than 0·7 hectares, and 32 per cent of the heads of families owned no land at all. The original allotments were between 1·0 and 1·5 hectares per family.

The same non-developmental behaviour pattern can be observed in the densely populated villages from which the new settlers came. In field studies made by the author in North Sumatra (1962) and West Java (1966),[10] farmers from a total of sixteen villages were asked a series of questions to determine their willingness to undertake productive investment. The questions were: if you were given a certain sum of money (to the value of 60 kg, 300 kg, and 3,000 kg of paddy respectively) what would you do with it? How would you spend the money if the 'gift' could be used only in agriculture? In all but two villages (but including the new settlements where land is not yet a scarce resource) the majority of the respondents said they would buy land with the largest of the three sums of money. In the other two villages most farmers were already economic-minded, and said they would use additional capital productively, to purchase fertiliser, small tractors, irrigation pumps, etc. The economic-minded knew that farmers as a group do not benefit if each individual farmer saves in order to buy land. But the subjective economic value of land still greatly exceeds its objective value in most parts of the country, and as long as it does, the hard-working and the thrifty among the subsistence-minded peasants will prefer to strive in traditional ways to save to buy land, and will overlook the other development opportunities available to them.

The continued unwillingness of the majority of farmers (including those who have so much land that they rent some to others) to buy modern inputs and equipment or to grow the crops most profitable for their area is yet more evidence of their subsistence-mindedness—and of the irrelevance of theories of development which assert that farmers will not innovate because the (objective) costs and risks are too high.

[10] In conjunction with the Faculties of Agriculture of the University of North Sumatra and the Institute of Agriculture, Bogor.

Even in regions where land is not a scarce resource, and where many farmers have taken to growing commercial crops like rubber, coffee, or pepper, it is nonetheless rare to find farmers who are willing to grow these commercial crops in a modern way. Most of them could easily mobilise the capital needed to buy the improved planting materials, or fertiliser, or equipment, that would make their farms many times more profitable than they are now, but they are usually reluctant to do so.

The commercialisation of Indonesian agriculture is fairly well developed on the commodity-marketing side, but the reluctance of farmers to buy fertiliser, modern tools, etc., is still so great that it is unlikely that any substantial modernisation of Indonesian peasant agriculture will take place in the next decade or two. In many Sumatran villages 70 per cent and more of total output may be sold, but the same farmers' expenditure on modern factors of production is usually less than 2 per cent of the value of production.

The large marketable surplus of rice (150 per cent of consumption) from the 'new settlers' is produced wholly by traditional labour-intensive methods. In Pematang Djohar expenditure on all modern inputs is 0·13 per cent of the annual value of production (Penny 1964). For Indonesia as a whole the marketable surplus of rice is only some 30 per cent of total production.

It is what they see as the great productivity of traditional methods that deceives farmers into believing that their basic need is for more land rather than for a change in their methods of production.

There is no other country in Asia where farmers have such an ingrained preference for labour-intensive technology. Rice is still harvested head by head, with the *ani-ani*, a tool that was discarded in Europe some 2,500 years ago. The extremely low capacity of this tool makes it mandatory for farmers operating more than 0·6 hectares of rice to use hired labour at harvest time. In regions where population densities are high social pressure ensures that harvest labour is used when areas sown to rice are well below 0·6 hectares.

Continued use of the *ani-ani* encourages the maintenance of a labour pool. It also gives rise to a vicious circle: labour is needed to harvest the rice, and the rice is needed to provide the subsistence requirements of the poor people, mostly landless labourers, who do the harvesting. In most parts of Indonesia, including the new settlements, harvesters are paid a (traditional) 20 per cent share. In areas where land is cheap the harvest income earned by a poor man serves as the finance he needs to open his own farm; in the more densely populated areas the harvest income provides a large proportion of the annual food requirements of landless labourers. This reduces their incentive to seek other employment or to migrate to places where land may be available. Harvest payments also reduce the income of the landowners, and reduce their capacity to purchase the fertiliser, etc., that would make their farm operations more profitable. Geertz (1963) has described this process as 'shared poverty', an apt term indeed.

Farmers in new areas could use the sickle to harvest their rice, thereby reducing the need for such rapid population increases, but they do not.

The *ani-ani* is not the only culprit, for the tools farmers commonly use to till the soil severely limit the area they can plant. A husband and wife working hard together can plant no more than 0·8 hectares to rice when they use hand tools—and most of them do. In a study of eight representative villages in North Sumatra in 1962 it was found that buffalo and cattle were kept in seven, but that the animals were used as work-stock in only one village (Penny 1964). Ironically, this one village had the smallest farms of the eight. But even in areas where the use of work-stock is traditional (e.g. among the Javanese), farmers set their sights low: they believe that a pair of buffalo can handle at most two hectares, whereas in Burma and Thailand a figure of four hectares is accepted.

Most Indonesian farmers still think 'small'. When deciding on the use they will make of resources they also think 'narrow'. For example, it is a rare farmer who will voluntarily look beyond his own farm for the best rice seed; not one of 177 North Sumatran rice-farmers studied in 1962 (Penny 1964: 132) used other than their own seed, saved from the previous harvest. The same reluctance to seek the best planting material is also shown by rubber-growers and others. Their reluctance is explained in part by tradition—a farmer should save his own seed —and in part by the fact that farmers trust only their own judgment on a matter of such importance.

Their experience and agricultural knowledge is also narrower than it might be. There are many traditional agricultures, not just one, in Indonesia, and many farmers could improve their farming practices if they knew of, and adopted, some of the better practices of other groups. Such development opportunities include planting by the solar rather than the lunar calendar (South Tapanuli), transplanting rice once rather than twice (Bandjarese farmers in North Sumatra), and the adoption of the sickle and the flail (used in a few places) to improve labour efficiency in harvesting and threshing (Rangkuty 1966).[11] A further illustration is the willingness of Indonesian farmers to reject the use of rat-bait on the grounds that killing rats will make the gods angry, and lead to a subsequent plague of rats.

Considerable emphasis has been given to the aspects of farmer

11 Many more examples of these sorts of development opportunities will be found in two recent articles by the author (Penny 1966, 1967).

behaviour that could be described as non-developmental, but this is justified by the lack of development in the Indonesian agricultural economy as a whole. But some Indonesian farmers are much more economic-minded than the majority. There are a few areas where all, or nearly all, farmers grow commercial crops, use fertiliser and pesticides, and make their planting decisions on forecasts of the market situation at harvest time. In most other areas there are a few farmers, sometimes very few, who are already development-minded. Often, however, these men get little or no encouragement from their neighbours: they may be laughed at, or, in extreme cases, actually prevented from using a more productive technique; for example, when a farmer tried to replace the *ani-ani* with the sickle, his neighbours, many of them landless, burned his fields because they feared their livelihood was threatened.

No consideration has been given so far to the part played by such factors as poverty (level of income) or land tenure as determinants of the way farmers behave. Little need be said about poverty as a factor, for in many parts of the country there are numerous farmers who could afford to buy fertiliser or shares in a small tractor if they wished.[12] It is a factor only in the very densely populated farming areas where farmers are so poor that they cannot afford to buy fertiliser and other inputs, even if they want to.

There are many land tenure systems in Indonesia. Most of them permit strong individual rights, and land is for the most part freely transferable. The social dynamics of most Indonesian peasant communities are such, however, that land ownership tends to become concentrated in a relatively few hands, particularly as population density increases. The process by which this occurs has been described above, but the final result of the process is a landlord-dominated society, which is nonetheless quite unlike that found in India or in Latin America. Few Indonesian landlords own more than five hectares, most reside in the village, and they are not a hereditary class. They have some (local) political power and some social status but they hardly dominate in the way that landlords do in some other countries. On the other hand, few of them are agricultural innovators, as they are satisfied with their superior incomes and status.

Within their communities farmers honour those who work hard, are thrifty, and assume their share of social obligations; but when

[12] Both small tractors (5 h.p.) and portable irrigation pumps have been shown to be quite profitable in the fairly densely populated (430 persons per sq. km) village of Tjibuaja in West Java. The farmers, however, are very reluctant to buy either, and to date only three tractors have been purchased, while the only pumps used have been rented. The farmers give lack of capital as the reason, but the cost of a tractor or a pump is the same as the cost of a motor-scooter, and there are already 60 of these in the village (Penny, field work, 1967).

comparisons are made between occupations it is found that, except for labouring, farming has the lowest status of all, and farmers will, given the chance, save in order to become petty landlords or traders. The fact that the prophet, Mohammed, was himself a trader encourages the latter orientation among the Moslem majority.

Merchants in the rural areas rarely assist agricultural development by supplying fertiliser or other modern agricultural requisites, choosing to avoid the trouble of developing the market. Most of them prefer to trade in goods with a ready market and a quick turnover.[13]

What farmers undertake in groups likewise affects the level of economic activity. For the most part farmers acting together are as subsistence-minded as they are individually. They find no difficulty in organising labour, tools, and materials necessary for traditional activities like building a mosque or a village hall, or building and maintaining local roads, tracks, and irrigation systems. They are usually quite willing, too, to co-operate in building a village clinic or school. But they will seldom undertake joint investments that require the mobilisation of cash capital. It is rare to find a rice-mill or rice-huller, a tractor, a truck, a pump, or a village storehouse for fertiliser that has been financed on a co-operative basis.

The farmers and the village merchants are still confined for the most part to the subsistence orbit. The farmers, for example, work hard in traditional ways, but they rarely travel, and even when they do, are unlikely to meet anyone, from their own ethnic group or another, from whom they feel they can learn anything worthwhile. Their self-reliance and their assurance that their farming methods are about the best possible are exemplified in their unwillingness to go at all far, usually no further than their own farms, for the planting materials they will use in the next season.

The conclusion should not be drawn that they never adopt innovations. All the important crops they now grow—except rice—have been introduced from abroad since Europeans first came to Indonesia. These 'new' crops include maize, cassava, coffee, tobacco, and rubber. However, with the exception of tobacco—which is given fertiliser—these crops have either been wholly absorbed into the subsistence economy (e.g. maize and cassava), or are grown in the inefficient ways preferred by subsistence-minded farmers (e.g. rubber and coffee). Even though farmers have shown a willingness to grow new crops, they are still very reluctant to make changes that matter—to adopt new farming methods, new inputs, new tools, or new ways to meet group needs. A modern commercial agriculture will not evolve until the farmers are

---

[13] The inflationary conditions that have prevailed since independence also encourage trade in goods with a quick turnover.

prepared to recognise that good farming is more than just producing a surplus above subsistence needs by customary methods. The inter- dependence of farmers, consumers, suppliers, manufacturers, and others, which is characteristic of a progressive modern economy, has barely begun to emerge in Indonesia.

The majority of Indonesian peasant farmers, then, are not yet behaving in a way conducive to development. But the lack of develop- ment cannot be ascribed wholly to the continued subsistence-minded- ness of the majority of the peasantry. The farmers are part of a wider economy and society, and what happens away from their farms and villages has a profound impact on their willingness to assume the costs and risks of economic change.

THE GOVERNMENT AND THE AGRICULTURAL ECONOMY

Agricultural produce has for centuries been sold in the cities and on export markets, but Indonesian farmers have never felt that these markets have offered sufficient economic incentive to warrant their making a sustained effort to increase output and adopt innovations. Until very recently much of the economic value that has gone from the farms has been captured by the ruling groups: initially local aristo- cracies and later colonial overlords.

In the era of the great kingdoms in Java, most of the marketable surplus was taxed away to meet the cost of wars, monuments, and the other expenses of a feudal type of government. The *wong tjilik* ('little people') received few material goods in return, but they were provided with the security and leadership required for the building of their irrigation systems, while drama, art, sculpture, and literature flourished at the courts.

The Dutch ensured profits from the East India trade by gaining effective control over the purchase and collection of important com- modities—spices, sugar, coffee, and rice, in particular. During the period of the Culture System, which was in its heyday from 1830 to 1870, farmers were also forced to plant certain crops for delivery to the colonial government.[14]

The 'marketable surpluses' of the colonial era (to about 1900) consisted mainly of goods produced for European markets, and prices received by farmers were usually low. The Dutch used a variety of means to keep prices below those that would have ruled on a free market. These included compulsory plantings and subsequent com- pulsory free deliveries (mentioned above), low fixed purchase prices,

[14] For a good description of Dutch agricultural policy to 1900 see Furnivall (1944).

K

heavy export taxes, and a judicious use of import and export pro-
hibitions.

Farmers received relatively low market prices for rice, for example.
Since the local market for rice was limited, and the price elasticity
of demand for rice was low, it was easy for the Dutch colonial government to depress its price through imports (de Vries 1937).

Before the Dutch, there were some trading towns on the north coast
of Java, but the indigenous traders were given no chance to participate
once the export trade with Europe was opened up. The Dutch and
other Europeans dominated the foreign trade field throughout the
whole of the colonial era, and they also controlled domestic trade,
except for actual collection, which was dominated by Chinese and
other non-Indonesians.

It is often believed that Chinese merchants had and still have a
stranglehold on domestic and foreign trade, but data on per capita
income by groups (for 1930 and 1939) show that it was the Europeans,
rather than the Chinese or Indonesians, who profited most from
colonialism (Table 8.6). Since independence Europeans have, of
course, lost their economic power. Reliable data on the relative
importance in trade of Indonesians and Chinese (citizens or foreign)
at the present time are lacking.

Great profits can be and are earned from trade in Indonesia today.
It is doubtful, however, whether the 'low' prices the farmers are
always complaining about are due in any important degree to monopoly or monopsony powers of merchants, whatever their origin.
Regular markets (daily, weekly, etc.) have long been a feature of rural
Indonesia, and there are so many buyers and sellers in these markets
that neither party to a transaction is likely to receive any monopoly
profits. A 1962 study of marketing margins in North Sumatra showed

TABLE 8.6

INDEX OF INCOME PER HEAD OF WORKING POPULATION
BY RACIAL GROUPS, 1930 AND 1939

(Indonesian income = 1)

Year	Europeans	Foreign Asiatics[a]	Indonesians[b]
1930	47	5	1
1939	61	8	1

[a] Mainly Chinese.

[b] Per capita income of working population was 171 guilders in 1930 and 87 guilders
in 1939. Per capita incomes of Indonesians on Java were 76 and 74 per cent of those
earned by Indonesians in the Outer Islands in 1930 and 1939 respectively.

Source: Polak 1943.

that, for four commodities widely traded within the province, market-ing margins were much lower than they are in a developed, high-income economy, and that so many marketing channels were open to producers that no single trading organisation could possibly obtain, let alone maintain, a dominant position (Zulkifli 1962).

Marketing costs are, however, higher than they might otherwise be because of various inefficiencies: techniques of storage and packing are often primitive and cause losses, while the delays in inter-island shipping raise storage costs and lead to further losses.[15]

On-farm prices would certainly rise if transport costs fell, if traders were encouraged to adopt more modern methods of handling goods, and if there were more competition in some marketing channels, but the main cause of 'low' prices at the farm level must be sought elsewhere.

The greatest single factor leading to low on-farm prices, however, has been government price policy. Ever since independence it has been government policy to keep agricultural prices below those that would have ruled on a free market. The government has found it relatively easy, for example, to depress the prices of agricultural export commodi-ties by overvaluation of the rupiah at the official rate of exchange and through export taxes. The price of rice has long been kept below the world market price by prohibiting exports and through open-market sales of imported rice by the government. The professed aim of the rice 'injection' policy is to stabilise prices, but the price has always been stabilised at a low level. The government has been even more successful in reducing the rupiah price of rubber and other export crops (Table 8.7).

The government has justified these measures in various ways—'if we change the exchange rate, inflation will be encouraged'; 'if the price of rice increases, this will mean wages will rise, and inflation will be encouraged'; and 'in a nation that wants to develop economically, resources must flow from the rural sector to help finance development projects in other sectors'. These low-price policies have failed to achieve their stated aims, for inflation is still not under control (about 112 per cent in 1967), real wages fell by more than 50 per cent between 1958 and 1964, and there has been very little development in other sectors. Changes in the domestic terms of trade within Java show a trend against the farmers in recent years (Table 8.8).

A substantial proportion of the farm income has been taken by the government to finance the costs of the greatly expanded bureau-cracy (up sevenfold since 1940), the armed services, and the prestige projects (to 1965). The farmers have received little in return: the roads and railways have deteriorated (they are being rehabilitated

---

[15] As far as is known there are no studies of agricultural marketing efficiency other than for rice (Mears 1961) and maize (Survey Agro-Ekonomi in press).

TABLE 8.7

RUBBER-RICE PRICE RATIOS[a] ON MAJOR WHOLESALE MARKETS IN
INDONESIA AND MALAYSIA, 1955–66

Year	Indonesia[b]	Malaysia[c]
1955	4·1	5·9
1956	2·5	4·9
1957	2·4	4·5
1958	2·5	3·9
1959	4·4	4·9
1960	5·3	5·4
1961	2·6	4·1
1962	2·0	3·5
1963	2·5	3·4
1964	1·9	3·3
1965	2·2	3·5
1966	2·1	3·1

[a] Price per kg of rubber divided by price per kg of rice.

[b] Indonesian data are for domestic wholesale prices in the Djakarta market. The wholesale price of rubber includes export duties. The rupiah price of rubber is usually higher in Djakarta than at other ports. The rubber prices are for RSS I, and the rice prices are for milled rice of average quality.

[c] Malaysian data are for domestic wholesale prices in the Singapore market. The rubber prices are for RSS I, and the rice prices are for local white No. 1.

Source: (Indonesia) Nugroho 1967; (Malaysia) Department of Statistics 1967.

TABLE 8.8

DOMESTIC TERMS OF TRADE[a] FOR RURAL AREAS IN JAVA, 1960–6

Year	Index (1953 = 100)
1960	33
1962	94
1964	64
1966[b]	57

[a] Index of food prices divided by index of clothing prices for rural areas in Java, as a percentage of 1953 base.

[b] August 1966 only. An unweighted average for the seven years, 1960–66 inclusive, shows that the buying power of the farmers was an average of 35 per cent lower than in 1953.

Source: Biro Pusat Statistik 1967: 5.

slowly now), and the delivery capacity of the irrigation systems has declined. Often fertiliser and other agricultural requisites have been unavailable to farmers or have been available late and only at black-market prices.

Government services to farmers have not been commensurate with the (total) taxes they have paid. It has often been argued (see Selosoemardjan (1962), for example) that farmers have not paid their fair share of taxes since independence, and it is certainly true that the visible taxes paid by farmers have declined substantially since before the war. However, the government's price and exchange rate policies have proved effective substitutes.

In any case these measures have by no means been the only ways used to transfer income from farmers to other groups. Others have been the government rice-buying program, the enhanced prices farmers must pay for fertiliser and other inputs whenever there are dislocations in the official distribution system, import duties and sales taxes levied on many consumption items, and the many unofficial levies.[16]

It is widely recognised that bribing is inevitable. Salaries are very low and officials must live! But there are so many officials and so many regulations that business people, farmers, and the public-at-large have come to feel that regulations have been made for the benefit of the officials.[17]

The government of independent Indonesia has severely hampered agricultural development through its rice-buying program. Part of the salaries of officials, civil and military, and of labourers in essential industries is paid in kind, and the largest single component of these payments-in-kind is rice. The government has often had to buy, at home or abroad, more than one million tons of rice per annum to meet its commitments (Table 8.5). The government's domestic buying price has been consistently below the open market price in most provinces, while the open market price in turn has been consistently below the world market price.[18] Farmers have sold their rice to the

[16] In 1966 a truck carrying rice from Krawang to Djakarta (60 miles) was subject to 13 levies which represented about 14 per cent of the retail price. In North Sumatra a skilled mechanic who sought permission to build a rice-mill had to pay more to acquire the 9 licences necessary than it cost him to erect the building and install the equipment. Indeed, any person conducting a business for which licences were required would have to pay 'extra' to get the licences and to make sure that the licences were not revoked after an inspector had paid a visit.

[17] Since the Suharto government came to power in 1966 regulations have been simplified or cancelled in some fields, particularly foreign trade, but in many others, e.g. rice-milling and manufacturing generally, little has so far been done. See also Wertheim (1964: 125-7) for an extended discussion.

[18] Except in the period December 1967 to March 1968. The April 1968 retail price of rice was once again well below the world market price.

government reluctantly, and buying targets have rarely been met, but even so it has often been difficult for them to resist the pressure exerted by officials.

Almost the whole burden of the rice-buying program has been borne by the farmers of Java, the one major island with a severe population problem. Before 1963 government rice-buying was rarely attempted in the Outer Islands because it was well known that the farmers there would stoutly resist efforts to get them to sell rice at prices below those ruling on the market. The government domination of the rice trade on Java reflects the weak economic position of the peasants and the extent to which they are conditioned to accede to authority.[19]

Government activities also reduce farmer incentives in other ways. Farmers are discouraged from growing crops other than rice on their irrigated fields even when the other crops are more profitable. Many regulations have been issued to protect farmers from 'rapacious middlemen', but the consequence of replacing competition with monopsony in this way has been that farmers have received lower, not higher, prices. Farmers are also placed at a disadvantage by the penchant of provincial authorities to forbid the movement of rice and other agricultural commodities to other provinces, particularly when local prices are low.

This list is by no means exhaustive, but it is adequate to indicate that the continued subsistence-mindedness of the majority of the peasants is in part due to factors beyond their control.

The crux of the problem is that policy-makers are as subsistence-minded as the peasants.

Not all policy-makers are subsistence-minded, of course, but then neither are all peasants. Most policy-makers and most peasants are, however, reluctant to exploit the economic opportunities available; both policy-makers and peasants are pessimistic about the likely returns from long-term investments; and both groups show an excessive concern over the pattern of income distribution, because neither feels there is any likelihood that the 'size of the cake' will increase in the foreseeable future.

The Indonesian economy is static, or nearly so, and it is not surprising therefore, that the economic behaviour of the policy-makers is analogous to that of the majority of the peasants. The subsistence-mindedness of the political élite, which differs only in the forms it takes, is illustrated with the following examples:

19 In July 1967 the rupiah price of imported rice at Tandjung Priok (port for Djakarta) was Rp22 per kg. In the Krawang area of West Java the government buying price was less than Rp11 per kg. For details on rice prices for earlier years see Mears (1961), and Mubyarto and Fletcher (1966).

1. Rice is the major subsistence crop and has always been given the highest priority in agricultural programs, even though many other crops are more profitable.

2. The limited supplies of fertiliser have been allocated to rice-growers, even when there is an unsatisfied demand from growers for other crops.

3. Extension workers have been required to act as agents in the government's rice-buying program.

4. Rice imports have always been given a much higher priority than imports of fertiliser and other production requisites.[20]

5. Imported rice has almost always been sold at less than the world market price, which has benefited the government establishment and urban consumers generally, but not the farmers.

6. The government has rented land for sugar-growing and facilities for rice-milling at well below the market prices for each: it feels that it is the people's duty to serve the government in this way, and that, in any case, the government will make better use of the resources than the farmers or rice-millers themselves.

7. The government has long given high formal priority to export production, export drives, and so on, but, as has been mentioned, the rupiah has been grossly overvalued at the official rate since independence.[21] Exports have been further discouraged by the operations of government trading monopolies (e.g. copra, kapok), and by regulations, illegal taxes, and the like.

8. The first claim on government receipts—from all sources—has always been the consumption needs of government personnel. Even in the period of retrenchment, from 1966 to the present (1968), it has been the development budget, not the routine budget, on which pressure has been put.

9. Until the fall of the Soekarno government, highest priority was given to government expenditure of the sorts that the kings of Mataram would have approved: magnificent stadia, monuments, and great public buildings, one large irrigation project, and several foreign adventures. While the new government has ended the extravagance, it continues to give highest priority to maintaining the official class.

20 The most ever spent for imported fertiliser was US$23 million, in 1962. In the same year over $100 million was spent to import rice. The average annual value of fertiliser imports in the three years 1964-6 was $6 million. Less fertiliser was available for sale in Indonesia in 1966 than in any year in the 1927-31 period.

21 Before 1966 the rupiah was subject to deliberate overvaluation: since 3 October 1966, however, there has been a relatively free market in foreign exchange, but the price of the dollar has been lower than it might otherwise have been because the foreign credits (some US$200 million over and above the US$470 million received from non-oil exports) were sold on the market.

There is little the Indonesian government has done since independence, in either the Soekarno or the post-Soekarno period, that indicates that it knows what it should do to create an environment conducive to agricultural development. As has been pointed out, the government élite as yet has no real commitment to economic development. It nonetheless continues to assume that only the government can lead the people towards a better social and economic future. It persists in this belief despite its continued failure to do anything substantial to improve the overall climate for entrepreneurial activity in the agricultural sector.

In sum, modern Indonesian society retains many characteristics of what Polanyi (1957) has described as an 'agrarian empire'. Agrarian empires, such as the Chinese empires and the Khmer kingdom in Cambodia, were large and imposing civilisations built on a surplus derived from 'unproductive' peasant agriculture. They usually had a static economy, for the agricultural surplus needed to sustain the non-farming groups can be maintained for a very long time—without development—if new land can be opened up as population increases.

Most of the empires established in the Indonesian archipelago prior to the Dutch era were also of this type. The only major exception was Sriwidjaja, an empire the power of which was based on control over trade routes. The centres of the other empires, Mataram, Modjapahit, etc., were all inland, and the economic basis of their power was the control of the rulers over the production of the peasants. The economic surpluses they captured from the peasants were not used for development. One of the reasons for this was, of course, that new technology was not available, but another and more important reason was that the political leaders attached no importance to general prosperity, in part because they did not see it as a possibility, and in part because they felt it was the duty of the people to serve and honour them.

Relationships between the Indonesian government and the people are little different now. Most officials are paternalistic, and feel that their duty is to do everything they can for the common people. People who accumulate wealth from production in any field, be it trade, industry, or agriculture, still tend to be looked down upon. Their attempts to accumulate wealth are discouraged, and many means have been devised for transferring any accumulation to the government. The state still considers it is the only agency in society that has the right or ability to tell the common people what to do; and the people who have government jobs feel that it is they alone, by virtue of their position, who should live well—and high military and civil officials do live well.

It should be stressed, however, that the government establishment is not monolithic and that the generalisations made here do not apply to each and every government official. There are exceptions, many of them, but by no means enough to invalidate the general argument.

The continued pre-eminence of the official class in Indonesian society explains the general form of economic policy since independence—the officials consider that economic activities should be guided and controlled, and they have built up a mammoth complex of regulations. But for the most part they lack the knowledge and the experience needed to frame the sort of regulations that would encourage, rather than hinder, productive economic activity.

In the field of agricultural policy, programs are drawn up on the assumption that farmers are so poor, and so exploited by landlords, merchants, and creditors, that no improvement in agricultural productivity is possible unless the government assumes full responsibility. Little is left to the initiative of the farmers themselves. But the fact is that most farmers can still provide for their own needs through their own efforts, whereas the government has shown that it can fulfil no more than a fraction of its promises. It is the government, not the peasants, that has failed economically: the government has failed in its promises to provide better seed on a regular basis; it has failed to develop a reliable system for distributing fertiliser, tools, and other production requisites, to employ as agricultural extension agents men who can win the respect and confidence of farmers, and the practical research programs essential for modern agriculture. In the few cases where a government program has been successful—the Bimas (Mass Guidance) Program[22] is perhaps the best example—the positive results obtained have been swamped by the effects of other problems that beset the farmers, such as population increase, inflation, and price policies.[23]

Many farmers would welcome the opportunity to use more productive methods; but at present they have few opportunities to make their wishes and their capabilities known to those who exercise political power. At the moment they can only resist passively, by failing to deliver the whole of their rice quota or to pay their taxes in full, for example, or negatively, through smuggling or banditry.

[22] For details of the Bimas Program see Roekasah and Penny (1967).

[23] This chapter was completed in May 1968. By November 1968, however, there were encouraging signs that some, at least, of the general points made in this section were not as applicable as they were in May: two substantial changes were the appointment of a new Minister for Agriculture (a man with a distinguished agricultural background) and the decision, since carried out, to import much more fertiliser than in 1966 or 1967. These two changes are hopeful portents for further improvement in agricultural policy, and in government policy generally.

It will be a long time before the government, or any other source, will provide the peasants with all the information they need on market conditions and new techniques to modernise their farming operations. It will take even longer for commodity markets to become sufficiently safe and 'steady' for farmers to rely on them when making their resource allocation decisions. It will take perhaps the longest time before it becomes possible for farmers to make their own needs and wishes known to the authorities, and before authorities feel that there is a need to consult farmers before formulating agricultural policies or programs.

PROSPECTS FOR THE FUTURE

The short-term outlook is gloomy and, if present trends in production and population continue unchanged, the long-term outlook is no better.

There is little likelihood that Indonesia's economy would have developed if Indonesia had remained a Dutch colony. Independence did lead to an upsurge in agricultural development activity in a few areas, but the momentum provided by independence has long since been lost in the country as a whole. Sustained development can get under way only after the trends of centuries have been reversed. This will require a fundamental change in the relationships between the government and the people; modern methods of birth control will need to come into widespread use; and the whole community will need to make a commitment to development.

Indonesia is not a country rich in untapped resources, and if the attitudes of the rulers—and the ruled—towards development remain as they are now, Indonesia could well become the first country to fulfil Malthus's fateful prediction.

It is not as if Indonesia has yet used up all its resources, or that the supply of development opportunities has become exhausted, but rather that neither farmers, nor policy-makers, nor any other group, are sufficiently aware of the severity of the population problem.

Modern agriculture is more productive than subsistence agriculture, but most Indonesian farmers nonetheless continue to prefer their old methods, because subsistence farming appears to offer them more certain returns than they would obtain if they adopted better varieties, artificial fertiliser, labour-saving equipment, pesticides, or other elements of modern agriculture. Moreover, the peasants will continue to follow this self-defeating course until they can see that modern agriculture is indeed more productive and more certain than their present system. The vast majority of the peasant farmers will likewise continue to be self-reliant for food and for the means of production as long as markets remain uncertain. The farmers will be able to gain confidence in markets, (i.e., in the wider economic world), only after there has been a fundamental change in the economic, social, and political relationship between farmers and non-farmers.

APPENDIX

A Note on Statistical Sources

The best source of data on the Indonesian agricultural economy is *Indonesia: Facts and Figures* by Nugroho (1967). In its 608 pages there is not only a wealth of statistical information, but also a list of the publications of the Biro Pusat Statistik (Central Bureau of Statistics), bibliographies of other publications, and brief notes on the methods of collection and the probable accuracy of the data published in the volume.

Other useful publications of the Biro Pusat Statistik include those on the Population Census (1961), the Agricultural Census (1966b), the Census of Estates (1966c), and the National Sample Surveys (1966d).

Some data on the economy as a whole may be found in Nugroho (1967). Other useful sources are the *Economic Survey of Asia and the Far East* (ECAFE, annual); the *Bulletin of Indonesian Economic Studies* (ANU, tri-annual since June 1965); and the writings of Paauw (1959, 1963a, 1963b).

The main original sources for published data on agriculture are the Ministry of the Interior (including the Land Tax Office) and the Agricultural Extension Service. Data from these two sources are not consistent, and it is not known which of the two is the more accurate. The Land Tax Office makes its estimate of agricultural production (for Java and Madura only) from the results of 'crop cutting' from a sample of 18,000 fields—but the sample was drawn thirty years ago. The Extension Service relies on estimates made by local agricultural officials, and the estimates of production, etc., from this source are mostly higher than those of the Land Tax Office.

Table 8.9 gives an indication of the discrepancies that exist in agricultural data.

Data from different sources on other important aspects of the agricultural economy are likewise inconsistent. For example, data from the Department of Agriculture show that, in 1961, 472,000 tons of fertiliser were imported. The Biro Pusat Statistik figure for the same year was 598,000 tons, a discrepancy of 126,000 tons. In 1964 the discrepancy was in the opposite direction but it was still large—75,000

TABLE 8.9

AGRICULTURAL LAND IN JAVA

('000 hectares)

Province	Registration (1963)	Ag. Census (1963)	N.S.S.I.[a] (1963)	N.S.S.II.[b] (1964)
West Java	2,155	1,491	1,570	1,998
Central Java[c]	2,763	1,813	2,813	2,839
East Java	2,942	2,138	2,347	2,741
Total	7,860	5,633	6,730	7,578

[a] National Sample Survey, Round 1.
[b] National Sample Survey, Round 2.
[c] Includes Jogjakarta Special Area.
Source: Kartono and Susilo 1967.

tons. In that year the Department of Agriculture reported that 92,000 tons of fertiliser were imported.

The figures for rice imports are similarly uncertain (Table 8.10).

A portion of the revealed discrepancies is simply explained: the Department of Agriculture reports relate to the ordering of the fertiliser, etc., while the Biro Pusat Statistik reports show the actual quantities received some months later. Part of the big difference in the 1965 rice import figures can be explained in this way also.

Data on some other variables (e.g. prices) are much more reliable.

Many published data have, of course, been cited in the text, and inferences have been drawn from them, but care has been taken not to undertake any sophisticated analysis because for the most part the data are not reliable enough to support it. Furthermore, it is not possible to say whether the data from a given source are sufficiently consistent from year to year to permit more than the most general

TABLE 8.10

RICE IMPORTS FOR THREE SELECTED YEARS

('000 tons)

Year	Source of data	
	Dept. Agric.	C.B.S.[a]
1961	1,013	694
1963	1,072	487
1965	186	796

[a] Biro Pusat Statistik.

inferences about trends. It is well known, for example, that smuggling has not occurred at the same rate each year.[24]

Other data of interest to students of the agricultural economy are scarce to non-existent: very few farm management studies have been published (but quite a few unpublished studies have been made by the students and staff of the Institute of Agriculture, Bogor); there is little information indeed on marketing costs and margins except from Mears (1961); and data on interest rates, prices, rents, and wages at the village level in the various regions are almost non-existent. Quite an amount of information on the operations of the estate sector exists, but little is published.

Since 1966 the Indonesian government has taken steps to improve data coverage and accuracy: the main burden of this work, the Agro-Economic Survey of Indonesia, has been carried out by an *ad hoc* body under the Minister for Agriculture. The Survey is generating new data on a number of important problems and is also checking the accuracy of data collection methods currently in use. The reports of the Survey will be published.

Yet if we compare the situation in Indonesia with that in many other tropical countries it may be said that Indonesia's social statistics are probably more complete and more accurate than most. Paauw (1961: 189) has commented:

> With the exodus of the Dutch the flow of data continued, but their aggregation, analysis and presentation virtually ceased during the first several years of independence . . . Much of the reporting system set up by the Dutch . . . whose statistical coverage of economic phenomena was unusually good . . . has continued to function.

The Biro Pusat Statistik has plenty of expertise and could make a dramatic contribution towards making more and better data available if only it had the money: Nugroho's great volume had to be privately printed.

---

[24] See Richter (1966: 43-4) for some estimate of the amount of rubber smuggled into Malayan ports in recent years.

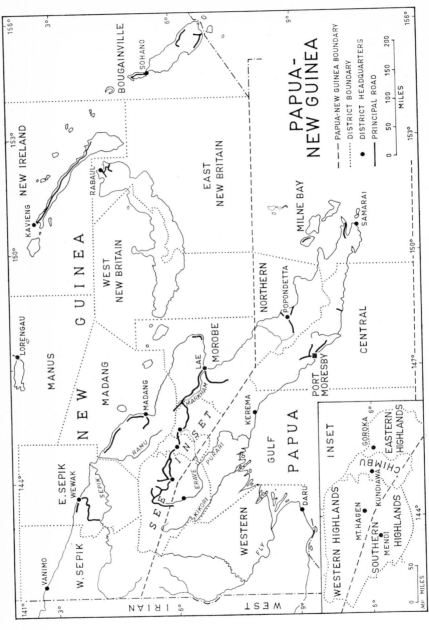

Map 10 Papua-New Guinea

DEPARTMENT OF HUMAN GEOGRAPHY, A.N.U.

9

Papua–New Guinea

R. T. Shand

PAPUA–NEW GUINEA began substantial contact with the outside world in 1884, when Papua was annexed as a British protectorate, preparatory to becoming an Australian territory in 1906, and New Guinea was annexed as a German colony.

From 1884 until World War II, the economy[1] comprised a large subsistence sector and a small slow-growing monetary sector. Economic activity of the indigenous population in the subsistence sector was primarily agricultural, supplemented by hunting, fishing, and foraging. The staple foods were sweet potatoes in the highlands, and taro, yams, and bananas in the lowlands. Agricultural producers followed a cultivation/bush fallow system for cropland use; production technology was simple, but effectively adapted to local conditions.

Fisk (1962: 467-8) has aptly described the traditional standard of living in the subsistence sector as one of 'primitive affluence'. Subsistence needs have been, and in most areas still are, met with inputs of land and labour well within supplies potentially available, and the level of output per economic unit has been governed by a ceiling on demand rather than on supply.

Economic units within the subsistence sector have typically been based on the nuclear family. Each unit has been largely self-sufficient, producing most of its own basic requirements of food, clothing, and shelter. Trade within the sector has been of quite marginal significance in most areas, though ceremonial exchanges of substantial quantities of agricultural and other produce were common. Each unit supplied most of the resources it needed for subsistence production, though reciprocal arrangements were often made for assistance from relatives and friends to meet peak labour demands.

1 Papua and New Guinea remained administratively and economically separate entities until World War II, since when they have been jointly administered by Australia. For convenience, however, the pre-war economies will be discussed here as one Territory economy. In this economy, two sectors will be distinguished: one is a monetary or advanced sector in which all transactions are monetised and economic action is guided by the profit motive, and the second is a subsistence sector, which consists of initially self-sufficient economic units.

Growth of the monetary sector during the fifty-five years from annexation to World War II was slow and spasmodic, and was based almost exclusively on the development of coconut estates. In New Guinea, German planters had established some 53,000 hectares by the outbreak of World War I. During the war, while New Guinea was under an Australian military administration, there was a further expansion of planting.[2] These estates were expropriated after the war and were mostly taken over by Australians. The new owners continued to expand in a minor way until World War II, and by 1940 there were some 517 estates with about 111,000 hectares under cultivation. About 106,000 hectares were under coconuts and another 800 were under each of three other crops—cocoa, coffee, and rubber. At that time there was also a small timber industry, and a number of other crops were being tried experimentally, such as tea, oil palm, and cinchona (Commonwealth of Australia 1941a: 73-5).

Development in Papua was on a smaller scale but was also dominated by the coconut industry. Estate planting took place mainly from 1907 until 1923, when there were 25,000 hectares under cultivation, of which 19,000 were under coconuts, 3,000 under rubber, and 2,500 under sisal. After 1923 there was no significant change in coconut area, rubber showed an increase to about 7,000 hectares in 1940, and sisal production disappeared. In total, there were some 26,000 hectares under estate cultivation in Papua in 1941 (Commonwealth of Australia 1941b: 26).

Since the development of the monetary sector was concentrated in export crops, the performance of export income gives a measure of the progress achieved. From negligible levels in the 1900s, total export income of Papua–New Guinea reached a rather unspectacular peak of a little more than A$8 million in 1939-40. Until the 1930s it was almost exclusively derived from coconut products, though rubber from Papua did contribute up to 10 per cent in the late 1920s. During the 1930s export income from the coconut industry declined drastically as a result of the price collapse in international markets, and it was in this period that the only other significant industry, goldmining, made its chief contribution. In 1939-40 gold exports earned 75 per cent of export income, and, over the decade, served to maintain a slow trend of expansion in total export income in the face of declining earnings from agricultural exports.

The participation of the indigenous population in the monetary sector of the economy was at best marginal until after World War II, apart from employment on estates. In the early years, as plantations were being established, 'trade' copra formed an important proportion

2 German planters were unable to remit profits and chose to re-invest them, possibly in the hope of higher compensation if Germany lost the war, or of placing them in a sound post-war position if Germany was victorious.

of total exports from the Territory. This generally represented a surplus of production over subsistence needs, bought, or more generally bartered, by Europeans from producers in coastal villages. Despite efforts of the administrations in both Papua and New Guinea to encourage village plantings and production specifically for the market, the smallholder contribution in later years failed to expand at the same pace as estate production (Shand 1963: 42-54). One source estimated the contribution at only 3-4 per cent of total copra exports in 1934-5 (Klein 1937: 571-2). The only other involvement of the indigenous population was in labour employment on European estates. As many as 51,000 were employed in this way before the war, but since wages were low and were paid partly in kind, their involvement produced only a small impact on the pre-war growth of the monetary sector.

The metropolitan governments of Germany and Australia contributed little to the development of their charges before World War II. The local administrations were almost wholly dependent upon internal revenue,[3] which severely limited the scope of their activities. They were concerned largely with the extension of control and law and order (West 1966: 18), and while their policies encouraged the inflow of foreign capital and indigenous development in agriculture, the scarcity of government resources prevented any significant state investment in economic development.

The limited inflow of private capital invested in agriculture before World War II arrived mainly in the early part of the century. In New Guinea, World War I interrupted progress and effectively eliminated Germany as a source of capital. During the inter-war period no marked inflow took place. Even the transfer of ownership of expropriated estates in New Guinea involved little injection of new capital, for the new owners purchased them on twenty-year terms without substantial down-payments. But whatever the reasons for the sluggish investment performance during the 1920s, there is no doubt that the collapse of agricultural prices during the 1930s discouraged further investment, and market recovery was hardly in sight before World War II overtook Papua and New Guinea.

The monetary sector of the economy received a severe set-back during World War II. In Papua production was more or less maintained, but New Guinea, where economic activity was most developed, suffered considerable damage in the fighting during the early 1940s. In particular, many coconut estates were badly damaged and pre-war levels of output were not regained until 1951. The war did, however, have the incidental effect of providing a widespread demonstration of material goods, which stimulated a new or increased receptivity

3 New Guinea was wholly dependent, while Papua did receive a small grant, which, however, never exceeded A$85,000 per annum.

amongst the indigenous population to economic opportunities offered after the war.

Overall, the first sixty years after contact produced very limited progress towards the establishment of a modern market economy. The dominant subsistence sector remained separate from, and largely unaffected by, the growth of the small monetary sector. Growth in the monetary sector was generated by a small number of expatriates, and was highly localised—mainly on the islands of New Guinea, such as New Britain and New Ireland. It depended heavily on the population of the subsistence sector for labour but stimulated relatively little production amongst them for the market.

Yet despite the modest growth performance, there were useful legacies of this period. The post-war administration was able to draw upon a small but experienced agricultural staff with accumulated knowledge of agricultural production and extension techniques under Territory conditions. Research had been carried out on a number of crops, such as coffee and cocoa, which became commercially attractive after the war. Furthermore, although little development had taken place in the countryside, at least peaceful conditions had been established over a wide area so that attention and resources could be concentrated on the tasks of economic development. This extension of administrative control had, in fact, expanded the human and natural resources available for development. In particular, the exploration and pacification of the highlands in the 1930s opened up extensive new development possibilities. Owing to limitations of staff and finance, however, this task had by no means been completed before the war, and consequently a proportion of the resources of the post-war administration had still to be applied to this end.

TWENTY YEARS OF GROWTH (1946-66)

Under the stimulus of a large inflow of private, and particularly public capital, Papua–New Guinea has recorded a substantial rate of economic growth since World War II. Gross national production (Territorial), in constant 1966 prices, expanded from A\$172 million in 1949-50 to A\$373 million (US\$418 million) in 1965-6, at an average compound growth rate of 4·9 per cent per annum (Zmudski 1969-70). Expansion was particularly fast during the five years to June 1966, when the growth rate averaged 6·1 per cent per annum.[4]

4 These growth rates can only be regarded as rough approximations, since gross national product figures in the early post-war period were backward projections from actual estimates prepared for a few later years. Several of the assumptions underlying these projections, particularly with regard to the estimation of subsistence sector output, are open to question.

This performance was achieved despite the influence of the large subsistence sector component, which expanded only slowly, at a rate of $2 \cdot 3$ per cent per annum between 1949-50 and 1965-6, probably more or less in step with the growth of the indigenous population. The main impetus came from a fast expansion of the monetary sector, where gross product grew at $10 \cdot 5$ per cent per annum for the whole period and at $12 \cdot 8$ per cent per annum between 1961 and 1966. Two points should be noted here. First, these high growth rates were partly due to the small starting base of the monetary sector after the war. Second, they reflect the rapid growth of financial assistance from Australia, rather than the expansion of productive capacity within the Territory.

The Australian grant has grown from A$0·5 million in 1945-6 to A$62 million in 1965-6. In addition there has been a considerable and expanding volume of direct expenditures by Australian government departments and instrumentalities in the Territory during the post-war period, and especially in recent years. In 1965-6, for example, direct expenditures amounted to A$33·7 million, and, combined with the grant, gave a total financial contribution of almost A$96 million. It was estimated that between 1961 and 1966 Australia financed just over two-thirds of all public expenditure in the Territory (Anon. 1967: 52). Furthermore, a large proportion of the commercial and industrial activity in the private sector has been stimulated by the expansion of the Australian financial contribution, so that the total economic effects of aid are far greater than the direct effects of the contribution.

Expansion of agricultural output has made a major contribution to overall growth in the economy. In the 1961-5 period (Table 9.1), the increase in total agricultural production accounted for 44 per cent of the increase in national product. However, three-quarters of this came from an expansion of non-marketed agricultural output in the subsistence sector, and only one-quarter from marketed agricultural production. Indeed, one-third of the increase in national product came from the subsistence sector alone.

The continuing importance of the subsistence sector can be further appreciated from the fact that in 1964-5 it still accounted for 84 per cent of total agricultural production (Table 9.1). Despite the comparatively fast rate of expansion in marketed agricultural production in recent years, the decline in the relative importance of the subsistence sector has been slow.

In the monetary sector, the rate of growth in non-agricultural production greatly exceeded that of agricultural production from 1960-1 to 1961-5 (Table 9.1). This caused a decline in the relative importance of the latter within the monetary sector and also of total agricultural

TABLE 9.1

AGRICULTURAL PRODUCTION IN NATIONAL OUTPUT[a]

	1960–1		1964–5		Annual growth rate (%)[b]
	A$ million	%	A$ million	%	
Marketed agricultural production	27·5	33	37·2	26	8·0
Non-agricultural production	56·6	67	103·7	74	16·3
Monetary sector product (at factor cost)	84·1	100	140·9	100	13·8
Marketed agricultural production	27·5	14	37·2	16	8·0
Non-marketed agricultural production[c]	174·1	86	202·8	84	3·9
Total agricultural production	201·6	100	240·0	100	4·4
Total agricultural production	201·6	78	240·0	70	4·4
National product (at factor cost)	258·2	100	343·7	100	7·4

[a] Agricultural production here includes crop, livestock, forestry, and fishery output. All output and production data are in gross terms.

[b] Compound, average.

[c] Non-marketed agricultural production includes output in the subsistence sector and non-marketed output which may be consumed by those partially or wholly engaged in the monetary sector.

Source: Estimates of gross agricultural production in monetary sector from Department of Territories, private communication, 1968. The remainder are drawn from Zmudski 1969–70.

production within national product. This again is largely explained by the expansion of Australian assistance. Aid expenditures have benefited principally the service and construction industries. Relatively little was spent directly on agriculture. Indirect expenditures, such as those on the economy's infrastructure, which have been a rising proportion of total administration expenditures in recent years, take effect in the long term and have not as yet greatly influenced the rate of growth of agricultural production.

Growth in non-marketed agricultural output has been almost exclusively in those commodities traditionally produced and consumed within the subsistence sector, and has been a simple response to rising population within the sector.

Growth in marketed agricultural production has taken place partly for the domestic but mainly for the export markets. The increase in domestic sales has occurred principally in horticultural crops and timber, but it is an interesting feature of this economy that although the subsistence sector is self-sufficient in food, the market sector is not. Imports of foodstuffs increased from a level of A$8 million in 1953-4 to A$22·6 million in 1965-6, and in the latter year were little short of 50 per cent of total export income. Most of the horticultural output marketed has been traditional foodstuffs, purchased by indigenes employed in the monetary sector. There has as yet been only minor progress in local production of the types of foodstuffs demanded by the expatriate population and increasingly by the indigenous population entering the monetary sector.

TABLE 9.2

RELATIVE IMPORTANCE OF COMMODITIES IN
AGRICULTURAL EXPORTS, 1950–1 TO 1966–7

	Coconut products (%)	Cocoa beans (%)	Coffee beans (%)	Timber (%)	Rubber (%)	Peanuts (%)	Other (%)
1950–1 to 1954–5	81	3	—	3	11	—	3
1955–6 to 1959–60	65	7	2	11	10	1	3
1960–1 to 1964–5	49	16	13	9	8	2	3
1965–6	50	11	21	9	6	1	2
1966–7	35	21	23	10	6	1	4

Source: Based on Department of Territories 1967: 21A, 21B.

The growth of agricultural exports has been almost wholly responsible for the expansion of total export income from A$16·7 million in 1950-1 to A$49·8 million in 1965-6,[5] almost a threefold increase in fifteen years. The expansion has been due partly to additions to output of pre-war export commodities, coconut products, and rubber, and partly to diversification into new commodities such as cocoa, coffee, and timber (Table 9.2). By 1966-7 the traditional dominance of the coconut industry in exports had been considerably modified by the emergence of these new industries. Other minor contributors have been peanuts, crocodile skins, passionfruit products, marine shell, and recently pyrethrum and tea on a small scale.

In the coconut industry, pre-war production levels were regained by the early 1950s, and since then marketed output has expanded at a modest rate (Table 9.3). Between 1951-2 and 1965-6 the average annual rate of growth of output was 4 per cent, accelerating somewhat

[5] During the post-war period the only non-agricultural commodity of any note exported was gold. This industry has declined, however, as known deposits have dwindled. Export income from it diminished from A$4·3 million in 1952-3 to A$0·9 million in 1965-6.

TABLE 9.3

MARKETED OUTPUT OF MAJOR AGRICULTURAL PRODUCTS, 1946–7 TO 1966–7

(metric tons)

Years	Coconut products[a] [b]	Rubber[a] (raw)	Cocoa beans[a]	Coffee beans[a]	Timber[c]	Peanuts[a]	Tea
1946–7 to 1950–1	44,552	1,434	180	25	n.a.	—	—
1951–2 to 1955–6	91,320	3,310	864	90	160·4	131	—
1956–7 to 1960–1	100,733	4,345	4,497	1,099	245·1	1,403	—
1961–2	107,672	4,755	10,175	3,499	301·7	2,247	—
1962–3	110,636	4,836	14,297	4,923	345·6	2,097	19
1963–4	103,748	5,085	15,891	6,941	421·3	1,732	30
1964–5	114,571	5,405	20,469	8,826	485·2	1,631	18
1965–6	123,616	5,474	17,050	10,977	611·6	1,558	11
1966–7	112,222	5,705	21,938	13,118	702·2	1,670	n.a.

[a] Quantities exported.
[b] Excludes minor quantities of coconuts exported in whole form.
[c] Logs harvested locally and for export, in '000 cubic metres.

n.a. = not available.

— = nil or insignificant.

Source: Department of Territories 1967: 10, 23A, 23B; Territory of Papua-New Guinea 1965–6.

in recent years. Despite this moderate performance, the absolute size of this industry has made it an important factor in the expansion of total marketed primary production in the post-war period. Expansion in rubber, the other established export industry, has been slow and output has not risen far beyond the pre-war maximum of around 1,422 metric tons in 1939-40, though there are areas of immature high-yielding rubber.

Performances have been more impressive in the new industries—cocoa, coffee, and timber. Plantings of coffee and cocoa were particularly heavy from the mid-1950s and a fast expansion of output followed from 1960 onwards. The timber industry, which has supplied domestic and export markets, has also recorded a fast rate of growth in the 1960s. By international standards, however, these three industries are still small.

In recent years progress has been made in the development of a number of other potentially important rural industries. Beef cattle numbers have risen from around 12,000 head in 1959 to 34,000 in 1966, with the primary objective of replacing imports. So far, however, efforts have been directed towards increasing the size of quality breeding herds rather than expanding the marketed supply of beef. The tea industry is also in an early phase and most of the planted area is as yet immature. The fishing industry has been relatively inactive, and increases in local demand since World War II have been met mainly from imports.

The expansion in marketed agricultural production has come from two sources—expatriate-owned estates and indigenous smallholdings, with the former making the largest contributions (Table 9.4). In 1965-6 estates produced about 64 per cent of the gross value of marketed primary production. They have been the chief source of growth in the coconut, cocoa, and rubber industries, as they were in the coffee industry until 1964, since which time smallholders have produced the greater share of output.

Coconut products provided about half the gross value of estate production in 1965-6 despite the more rapid growth in newer export commodities. Smallholder income was much less concentrated, with 36 per cent derived from copra,[6] almost as much from coffee, and a further 22 per cent from horticultural crops. Horticultural production has expanded in response to the demands of increasing numbers of indigenes partly or wholly employed in the monetary sector.

The expansion of marketed agricultural production has taken place in a geographically uneven fashion in Papua-New Guinea, both before

[6] Marketed production of coconuts is considerably below total output on smallholdings, since a proportion is consumed within the village. Total smallholder area under coconuts is actually greater than the estate area.

and since World War II. Location of the various industries has been decided primarily on the availability of areas with suitable physical conditions. The location of areas of smallholder export crop production, however, has differed markedly from the geographical distribution of the indigenous population (Table 9.5). For example, the islands of New Britain, New Ireland, Bougainville, and Manus, with 14 per cent of the population in 1966, accounted for 61 per cent of the total planted area of major export crops, while the highlands (Eastern, Western, Southern, and Chimbu) contained 39 per cent of

TABLE 9.4

ESTATE AND SMALLHOLDER CONTRIBUTIONS TO GROSS OUTPUT VALUE OF MAJOR AGRICULTURAL COMMODITIES, 1965–6

Crop	Estate production (A$ '000)	%	Smallholder production (A$ '000)	%	Total (A$ '000)
Coconut products	14,686	71	5,936	29	20,622
Cocoa beans	4,022	78	1,152	22	5,174
Coffee beans	3,211	36	5,628	64	8,839
Rubber	3,002	100	15	—	3,017
Peanuts	337	64	190	36	527
Pyrethrum	—	—	84	100	84
Horticultural crops for local sale	—	—	3,636	100	3,636
Sub-total, crops[a]	25,258	60	16,865	40	42,123
Timber	4,497[b]	100	—	—	4,497
Total marketed agricultural production[c]	29,755	64	16,865	36	46,620
Total non-marketed agricultural production for consumption[d]	1,903	1	150,122	99	152,025
Total agricultural production	31,658	16	166,987	84	198,645

[a] These crops are valued at average export prices for 1965–66. Each contains value added from processing operations, which somewhat inflates the contributions. The smallholder total includes a minor contribution from passionfruit products.

[b] The figure is an estimate of the value of all timber production for 1965–66 in log form (i.e. before processing), using the average export price (A$7·35 per cubic metre) of export logs for 1965–66.

[c] Excludes fisheries and marine products and livestock products.

[d] If private and community investment within the subsistence sector is added, a figure for total non-marketed agricultural production is obtained which is comparable to those given in Table 9.1 for 1960–61 and 1964–65. The value for smallholders represents a projection from 1962–63 based on population increase.

Source: Territory of Papua and New Guinea (1965–6); Department of Territories (1964, 1967); private communication from Department of Territories, 1968.

the population but only 3 per cent of the total planted area. It can be appreciated from this pattern that *in situ* opportunities for participation in smallholder production for the market were quite uneven amongst the indigenous population. It is also clear from Table 9.5 that, as in the pre-war period, export crop production has been largely concentrated in New Guinea.

TABLE 9.5

GEOGRAPHICAL LOCATION OF INDIGENOUS POPULATION AND AREAS OF MAJOR EXPORT CROPS[a]

District	Indigenous population (June 1966) ('000)	(%)	Smallholder area of export crops (1965) (%)	Total area of export crops (1965) (%)
Western	60·9	3	2	1
Gulf	55·4	3	8	4
Central[b]	130·5	6	3	7
Milne Bay	99·1	5	8	6
Northern	56·5	3	3	3
Southern Highlands	183·9	9	—	—
Total Papua	586·1	27	24	21
Eastern Highlands	201·8	9 }	3	2
Chimbu	167·0	8 }		
Western Highlands	291·6	14	1	1
West Sepik	99·1	5 }	9	4
East Sepik	157·5	7 }		
Madang[c]	150·4	7	6	9
Morobe[d]	204·9	10	4	3
West New Britain	43·9	2 }	27	30
East New Britain[e]	104·9	5 }		
New Ireland	49·2	2	14	16
Bougainville	71·8	3	11	12
Manus Island	20·2	1	2	3
Total New Guinea	1,562·1	73	77	80
Total Papua-New Guinea	2,148·3	100	100	100

[a] Includes coconuts, cocoa, coffee, and rubber.

[b] Includes Port Moresby, the capital and centre of government (total population 32,222).

[c] Includes Madang township (7,422).

[d] Includes Lae township (13,321).

[e] Includes Rabaul township (6,947).

Source: (population data) Territory of Papua and New Guinea 1968; (export crop areas) Department of Agriculture, Stock and Fisheries, private communications.

DETERMINANTS OF POST-WAR AGRICULTURAL DEVELOPMENT

The Resource Base

The availability of plentiful land and labour resources has been an important factor in agricultural development since World War II. The average population density in 1966 was low, at around 4·5 per sq. km (12 per sq. mile) though wide variations were apparent, ranging from only 0·6 per sq. km in the Western District of Papua to around 200 per sq. km in a few locations in the highlands of New Guinea. Total land area amounts to some 47 million hectares (nearly 117 million acres), of which 85 per cent is accounted for by the eastern part of the New Guinea mainland and the rest by some 600 islands, among which New Britain, New Ireland, and Bougainville are the most important. Mountainous terrain, swamplands, and infertility resulting from the widespread leaching process of a tropical climate render large areas unusable. Nevertheless it was estimated in 1963-4 that 5 per cent or more (upwards of 2·4 million hectares) had a potential for crop production, that 4 million of an estimated 5·5 million hectares of grasslands could support animal production, and that 8 to 12 million of around 31 million hectares of forest land had a commercial potential (IBRD 1965: 66-7). In 1961-2 only a little over 400,000 hectares of cropland were under cultivation, the areas of grasslands were virtually untouched, and only about 400,000 hectares of forest lands were under administration control with a view to exploitation. No radical increase in the extent of land use has taken place since 1961-2, although there are plans for an expanded utilisation of forest, grass, and croplands.

In this perspective it can be readily appreciated that most indigenous communities not only had sufficient land for current and future subsistence needs, but also had reserves which could be tapped for cash crop production. The administration, moreover, was able to secure land for lease to expatriates, without prejudicing indigenous interests.[7]

The position with respect to labour was no less favourable for development. Economic units in the subsistence sector were able to satisfy their traditional production needs with only a partial commitment of their available labour resources. When an adequate incentive offered for the production of a cash crop, a labour reserve was avail-

[7] This is not to say that conflicts of interest over land have not arisen in the Territory. As Crocombe (1967: 208-9) has pointed out, the percentage of high-quality land close to roads and harbours which has been alienated is high, and has generated a political problem in some districts. This is a result of pre-war policies, however. Since World War II the administration has been careful to avoid this problem.

able which could be devoted to market production without disturbance to the customary level of subsistence production. The stimulus for participation in the monetary sector also maintained a reasonably adequate supply of indigenous labour for estates during the post-war period.

While a general assessment of resources shows Papua-New Guinea was well placed for agricultural development, its physical features have greatly complicated resource exploitation. Areas of land with a high potential for development were often relatively inaccessible. Until air transport was introduced, for example, mountain barriers on the New Guinea mainland cut off the highlands from coastal areas and isolated one highland valley from another. At low altitudes there was a similar problem arising from the wide scatter of islands and the long mainland coastline. The high costs of providing an adequate infrastructure for agricultural development, whether in the form of coastal and inter-island shipping facilities, of an airport system in the highlands, or of feeder road systems into the hinterlands beyond coast seaports or inland airports, have greatly influenced the location of, and have limited the pace of, post-war agricultural expansion.

The Role of Government

The avowed post-war policy of the Australian government was to generate economic development of the Territory through the participation both of the indigenous population and of expatriate private investors. It was clear from the nature of the problems involved that this participation could only be achieved through strong government intervention with substantial resources of finance and manpower.

In fact, the growth of financial assistance from Australia, coupled with modest increases in internal Territory revenue, has enabled a rapid expansion of administration expenditures (from a mere A$0·6 million in 1945-6 to an estimated A$120 million in 1966-7),[8] of administration staff, and of the range and scale of their activities.

The nature of this assistance created certain favourable conditions for economic development which are largely denied to countries having to depend more heavily on internal resources. The expanding annual grant permitted a wide gap in the balance of trade, with imports continually and substantially in excess of exports. Since there were no restrictions on imports, there was no choice to make, in particular, between capital and consumer goods. This allowed imports of capital goods in volume required for public and private investment and thus avoided inflationary pressures. It also gave full rein to the development of consumer choice, which was of particular significance

[8] Budget estimate for the Territory.

for the indigenous population, for whom consumption goods have contained an important incentive component. Finally, skilled manpower made available by Australia also prevented any serious bottlenecks in the implementation of development programs arising from an initial shortage of locally-trained personnel.

A large proportion of administrative expenditures has been devoted, directly or indirectly, to the promotion of economic development, particularly in recent years. Details of allocations, while lacking for the early post-war period, are available for the 1960s (Table 9.6). They

TABLE 9.6

RELATIVE IMPORTANCE OF ADMINISTRATION EXPENDITURES IN
PAPUA-NEW GUINEA, 1962–3 TO 1966–7

Category	1962–63	1963–64	1964–65	1965–66	1966–67
Commodity-producing sectors:					
Agriculture, stock, and fisheries	5·3	4·8	5·4	5·2	4·8
Forestry	1·7	1·5	1·9	1·7	1·6
Lands, surveys, and mines	2·6	2·8	2·4	2·7	2·9
Trade and industry	1·9	1·9	1·4	1·4	1·5
Development finance	—	—	—	—	0·9
Other	0·2	0·2	0·2	0·2	0·1
Sub-total	11·6	11·2	11·2	11·1	11·9
Economic overheads:					
Roads and bridges	6·1	6·9	9·0	9·5	8·9
Ports	0·4	0·7	0·8	1·0	0·7
Coastal shipping	0·7	0·6	1·3	1·6	1·1
Aerodromes	0·8	0·5	0·7	0·6	0·4
Electric power	4·5	2·7	4·5	4·6	4·4
Post and telecommunications	4·1	3·8	3·8	4·0	3·8
Other	0·2	1·0	1·3	1·2	0·8
Sub-total	16·7	16·1	21·5	22·5	20·2
Total economic	28·2	27·3	32·7	33·7	32·1
Social services					
Education	14·9	16·7	15·6	15·6	16·9
Other	24·7	21·9	18·5	18·5	19·3
General Administration	25·4	26·3	23·7	24·6	24·3
Other	6·8	7·9	9·6	7·6	7·5
Total expenditure	100·0	100·0	100·0	100·0	100·0

Source: Department of Territories, unpublished data, 1968.

show that the proportion devoted to economic development has risen from a little more than one-quarter in 1962-3 to around one-third more recently. The inclusion of education raises the proportions to 43 per cent in 1962-3 and to almost 50 per cent since 1964-5. Expenditures of Australian Commonwealth departments in the Territory (which were additional to the grant, and which rose from around A$8 million in 1961 to roughly A$28 million in 1966), while heavily weighted towards defence purposes, also contained a significant economic component, especially where construction of roads, bridges, ports, and aerodromes was involved (Anon. 1967: 54).

The rural sector has been heavily favoured in recent economic expenditures. The bulk of those on commodity-producing sectors was allocated to the Departments of Agriculture, Stock and Fisheries, and Forestry. Portions of the allocations to Land, Surveys and Mines, and to Trade and Industry were also to some extent concerned with rural production, and large parts of the provisions for economic overheads were linked with the rural sector, particularly the provisions for roads, bridges, and aerodromes. The extension of administrative influence in the post-war period (under the heading of general administration expenditures in Table 9.6), as pointed out above, considerably enlarged the resource base available for economic development. Furthermore, the mobilisation of village labour resources by administrative officers for community projects such as feeder road and airstrip construction materially promoted cash crop production by smallholders.

The commodity pattern evident in marketed primary production (Table 9.3) reflects the broad policy objectives of promoting expansion in established agricultural industries, such as coconuts, and of diversifying into new or previously minor industries, such as cocoa, coffee, timber, and more recently tea, pyrethrum, beef cattle, and palm oil. The pattern of contributions to this expansion (Table 9.4) reflects a policy like the pre-war one, of encouraging participation by the indigenous population on a smallholder basis, and of attracting local expatriate and overseas investment into the more capital-intensive and larger-scale estate form of production organisation. We now move to a more detailed review of the means by which the administration has implemented these policies since World War II.

Research and extension. One of the major problems of agricultural development has been to find commodities which could be produced commercially under Territory conditions, for, with the exception of coconuts, none of those traditionally grown has had a ready market internationally. Much of the credit for the post-war diversification of agriculture must be given to research, particularly applied research, carried out both prior to and since World War II. This work led to the development of the cocoa industry based on the use of a *trinitario* variety, accepted as a distinctive 'flavour' cocoa on world markets; to the recent rising trend in rubber yields with imported, high-yielding Malayan strains; to the development of *arabica* coffee production in the highlands of New Guinea; to the testing and selection of high-quality varieties of tea as a basis for another highland industry, and of beef cattle breeds suitable for Territory conditions. Research designed to raise productivity in subsistence food production was also undertaken, but so far this has not registered much progress.

Having found suitable crops, a second major problem has been to stimulate a fast rate of production. A large proportion of the Department of Agriculture's resources has been devoted to extension work. The Department provided expatriate estates with technical advice, but most notably it functioned as the principal agent for promoting indigenous smallholder production of cash crops.

For the promotion of indigenous smallholdings the administration adopted an approach, called here a 'village garden' strategy, whereby villagers were encouraged to establish an area of cash crops on land located within traditional clan or tribal boundaries, in addition to their normal activities associated with subsistence production. Provided that land suitable for that crop was chosen, the decision as to the locality of cash crop gardens was made by the villagers themselves, according to customary systems of land allocation and land rights, and there was no pressure to establish consolidated holdings. Initially the administration favoured planting and ownership on a communal basis, but soon switched to encouragement of individual (family) ownership.

This strategy was implemented through the establishment of a network of district agricultural extension stations, from which periodic extension patrols were sent to areas suitable for cash crop production. Each patrol typically visited a number of villages in turn, first to persuade the inhabitants of the merits of growing a particular cash crop, then to provide them with the planting material and technical advice necessary for crop establishment, and subsequently to supervise activities on the smallholdings.

Reliance on this strategy meant that the growth of indigenous participation in cash cropping was strongly influenced by the extent of field coverage by the extension service. This in turn depended upon the number of staff and other resources available. In 1949, for example, there were only eighty-nine on the staff of the whole Department of Agriculture, Stock and Fisheries. Such limitations necessitated a concentration of effort on few crops and in few locations for each crop. But even though effort was in practice concentrated in the areas selected, patrolling officers still had large numbers of villages to visit, and since these were often widely scattered over difficult terrain generally accessible only on foot, many villages could only be visited once or twice a year. This was often insufficient to ensure that recommended techniques for crop establishment and maintenance were followed, with consequent effects on output.

The growth of available administration finance led to a fast expansion in staff numbers over the period. By 1967, for example, there were more than 1,600 within the Department of Agriculture, Stock and Fisheries, of which over 1,000 were in the Extension and Marketing Division. This, of course, has permitted a greatly expanded extension effort both in crops encouraged and in the areas of concentration.

Within the areas selected, the promotional efforts of extension officers generally met with considerable initial success, in that a high proportion of villagers actually planted an area of the recommended cash crops.[9] This was in part because widespread interest in money-earning activities already existed, and in part because, as mentioned above, there was a favourable population/resource balance in the villages.

Whether the initial level of participation was subsequently maintained by smallholders depended essentially on their assessment of whether the monetary returns warranted the (almost exclusively labour) inputs required. This calculation has been interpreted by Fisk (1964: 160) as:

> . . . a comparison by the subsistence producer of the disutility of additional labour (or negative leisure) necessary to earn money with the utility of the goods and services that money will enable him to buy.

In other words, the calculation depended on one hand upon the cash return per unit of labour, and on the other, upon what the money earnings could buy. These were the elements of the terms of trade for the village smallholder. In the early stages of contact, both elements were heavily influenced by the high per unit costs of processing, transporting, and marketing goods flowing to and from the village.

Processing, transporting, and marketing. In areas where smallholdings were located in close proximity to estates, costs of processing, transportation, and marketing were often substantially reduced. The relatively high level of estate output attracted processing, transport, and marketing services (often provided by the estates themselves) which enjoyed substantial economies of scale. Where these were offered to smallholders, they too were able to benefit.

[9] For a more detailed analysis of the factors which determined the response of subsistence producers see Fisk (1964: esp. 157-60).

Agricultural extension patrols operated in many areas which were beyond the limits of the existing infrastructure and beyond estate influence. The first problem in such areas was to establish an effective linkage between village and market. The indigenous population itself made a significant contribution to solving this problem with a self-help program. Under the guidance and exhortation of administration officers, villages contributed up to one day's free labour per week on public works, such as road building and airstrip construction. Most of the access roads to villages evolved in this fashion from bush tracks first used to establish administrative control and to provide medical assistance. The monetary expenditures on economic overheads given in Table 9.6 exclude this contribution and therefore substantially understate total infrastructure investment. One estimate put the value of work (including both new and maintenance investment) on district roads, rest houses, airfields, etc., at A$15 million for 1962-3 (Department of Territories 1964: 53), which is not far short of the A$21 million of administration expenditure in that year for comparable types of work (purchase of capital assets, capital works and services, and maintenance of works and services).

The development of transport linkages, however, did not necessarily entice needed services to the village. In some areas low population densities and/or limited production of cash crops per grower resulted in a level of total output and cash income which was insufficient to encourage private enterprise to provide either processing, transport, and marketing services for the crops produced, or trade store services to meet consumer needs. The administration was then faced with the choice either of accepting a decline in interest among growers, disappointed by the low cash returns on labour and the lack of opportunities to buy consumer goods locally, or of providing the needed services itself.

In most instances the administration chose the latter course, either by direct action, or by supporting the establishment of producer and consumer co-operatives. While its policy was to operate such ventures at cost, some element of subsidy usually intruded. Often this subsidy proved economically justified. Cash returns not only were sufficient to maintain the initial output level, but they generated further expansion by stimulating less venturesome villagers to commence production or encouraging others to extend their cash crop holdings. The subsequent expansion of total output then attracted private interests into transportation, processing, and marketing, or it placed co-operatives on a sounder financial footing, thus allowing the administration to withdraw from these activities.

Administration involvement did not always succeed in expanding output levels or perpetuating initial levels, however, even where services were subsidised. In some cases producers were too scattered and isolated, and transport linkages could only be established at prohibitive expense; in other cases physical conditions for production proved to be marginally economic. Producers then either reverted to leisure activities or switched to other relatively more remunerative occupations (e.g. as labourers on estates).

Incentives for private investment. The administration also offered a wide range of incentives for private investment in agricultural industries, primarily as an enticement for estate development. These included low direct rates of taxation, at about half the Australian rates or less; tax allowances on capital expenditures for primary production, at rates higher than in Australia; and the opportunity to carry taxable losses forward. Tax holidays were also granted to industries granted 'pioneer' status, a measure which has benefited at least one agricultural processing firm. Duty-free imports or low rates of duty also assisted industry by damping costs of living and of production.

Subsidies were granted to the rubber industry to cover Malayan export duties and air-freight costs on high-yielding clones, and to the cattle industry to reduce freight and other costs on stock imported from Australia; and tea seed has been made available to planters at concessional prices.

The administration assisted the development and operation of the wage-labour market, particularly to meet estate needs, and was responsible for locating and securing suitable land for estate development and for leasing these areas to expatriates at low rentals. Recently the administration announced its willingness to participate financially, on a partnership basis, with private enterprise. One agreement has already been made to contribute 50 per cent of the capital for a large oil palm estate (to become a nucleus estate, around which indigenous smallholdings of oil palm will be developed), and in so doing the administration has been able to secure the participation of a large overseas firm.

Market Factors

The expansion in marketed agricultural output in the Territory has been primarily in export crops, and trends in international markets for these commodities have greatly influenced the pattern and pace of development.

High world prices after World War II were largely responsible for generating expatriate investment in coffee in the highlands and cocoa in coastal areas, and for administration emphasis on these two as cash crops for smallholders.

L

The outlook for copra was also favourable in the early post-war years as prices rose steadily to a peak in 1951. Territory producers, however, did not derive much benefit from the buoyant market since (from 1949) output was marketed under a nine-year contract between the United Kingdom and Australian governments. While this contract provided security against adverse price trends, it largely removed the windfall gains. This factor, together with the relatively more attractive prospects of profits in cocoa and coffee, probably account for the disappointing lack of investment in the coconut industry. The World Bank Report of 1963 (IBRD 1965: 85) noted that:

> About 13 per cent [of palms] planted before 1900 are now senescent or senile; 21 per cent dating from 1900 to 1914 have passed their prime; only 53 per cent established between 1915 and 1950 are in their prime; while 13 per cent planted since 1950 are juvenile or just beginning to bear.

A 1951 survey (Bureau of Agricultural Economics 1951: 20) had previously noted that new planting since the war had barely been sufficient to maintain output in the long term and had not made up the losses suffered during the war.

After 1951 prices declined as market competition from other fats and oils increased, making the industry even less attractive to expatriates as an investment prospect. But the administration has continued to regard coconut-growing favourably for smallholders, particularly when interplanted with cocoa, as it is a familiar crop for coastal people, requiring low inputs of labour and a minimum of extension supervision.

Rubber prices in the early post-war period were low and unattractive for investors. They rose briefly in the early 1950s, with a peak in 1951, but declined thereafter as a result of increased competition from synthetic substitutes. A survey carried out in 1953 concluded that the Territory had favourable physical conditions for rubber production and had areas suitable for expansion of the industry, but that to become a low-cost producer it needed to adopt a more rational system of employment of plantation labour and to introduce high quality planting material (Mann 1953). Under the existing terms of employment, indigenous labourers recruited for planting were repatriated to their villages after only eighteen months' service. For rubber much of this time was required for training labour in tapping techniques, and repatriation meant a continual loss of skilled manpower. Yields in the industry were low because most of the existing area under rubber had been planted with common unselected seedlings of low-yielding capacity, and because a large proportion of trees were nearing or had reached the end of their economic life. Efforts to introduce higher-yielding clones have been partially successful and will soon begin to influence yields in the industry, but the declining prices and gloomy market projections for this crop have effectively discouraged any large-scale expansion in estate holdings in the post-war period. Furthermore, until recently the administration has taken the view that rubber would be a difficult crop for smallholders, and has promoted other crops, such as coffee, cocoa, and coconuts, in preference to it.

The coffee and cocoa markets remained attractive to investors throughout most of the 1950s, and encouraged a fast annual growth of planting. For coffee, prices began to fall from 1957 as world market supplies commenced to outstrip demand, declining until 1963, their lowest level since 1949. The obvious surplus of productive capacity amongst world exporters led to the International Coffee Agreement, designed in the short run to restrict the quantity marketed internationally through export quotas, and to re-establish long-term market balance by exerting pressure on producers to adopt programs of tree eradication and agricultural diversification. Australia and Papua-New Guinea signed the agreement jointly and, as a single member, was classified as a net importer, free of export quota restrictions. However, even at that time it was clear that, as planted areas in the Territory came more fully into bearing in the late 1960s, Australia-Papua-New Guinea would become a net exporter, requiring an application to the International Coffee Council for a quota. The administration accordingly has had to discourage further expansion of coffee planting, both on estates and smallholdings.

The continued growth of this promising but still infant industry has thus been curtailed in the Territory. Since coffee was the only major market-oriented rural industry in the highlands until tea was recently introduced, this has been a particularly serious check to development.

The international cocoa market remained buoyant until 1959 when signs of excess supply became evident. Prices declined sharply from 1959 to 1962. Attempts have subsequently been made to conclude an international agreement, but evidence of permanent imbalance has not been as clear as for coffee. Since 1962 the market has fluctuated widely, with a partial price recovery over the 1962-4 period, a collapse in 1965, and a further recovery since then. Despite these fluctuations, estate investment in cocoa, and planting by smallholders, have continued at a substantial pace during the 1960s, although somewhat slower than that recorded in the previous decade.

Attractive domestic and export markets were instrumental in stimulating growth in the timber industry, particularly after 1959. Expansion of domestic demand was mainly the result of the growth in administration expenditures, while the expansion in Japanese demand created attractive conditions in the export market. Until recently the mixed nature of Territory timber stands discouraged their exploitation, but the rate of market expansion in Japan, combined with technical improvements in milling and greater knowledge of the usefulness of Territory timber, have largely overcome this disadvantage.

Political Factors

The question of political independence for Papua-New Guinea has undoubtedly had a bearing on the extent and pattern of expatriate investment. Until the late 1950s the issue remained fairly remote and the security of private investment was not seriously questioned. There was a net inflow of private capital from Australia and a considerable reinvestment of profits within the Territory. From that time, however, independence became an increasingly real prospect and business confidence among expatriate investors was seriously affected. Estimates of private capital movements[10] indicated net outflows annually from 1958-9, reaching a peak in 1962-3. In 1963-4 and the following year small net inflows were recorded, and more recently there appears to have been a general resurgence of confidence and a net inflow of more substantial proportions, though undoubtedly much of this has been connected with the heavy increases in administration expenditure.

The extent to which the political climate has specifically affected investment in agricultural industries cannot be determined. A proportion of the outflow, particularly in the early 1960s, probably did come from estate owners who transferred a portion of their investible funds to Australia as a precautionary measure, while uncertainty as to investment security may well have discouraged some intending investors from entering the Territory.

Nevertheless, new estate plantings were made in the coffee and cocoa industries even in 1962-3, which seems to indicate that business pessimism was by no means universal. The fact that it was concentrated in these rather than in the coconut and rubber industries probably indicates a general preference for crops with shorter pay-off periods and high returns to compensate for the risks thought to be involved.

More recently the investment picture has improved, with substantial investments being made in tea and palm oil industries, and the recent decision of the administration to assume the role of partner in estate development could well help to increase confidence further among private investors.

10 See White (1964); Department of Territories (1964: 131-5).

A GENERAL COMMENT

In reviewing the encouraging but modest achievements of the past twenty years, the question arises as to whether there were alternative policies which could have generated a faster rate of growth in agricultural output. It was shown above, for example (Table 9.1), that the non-monetised subsistence sector continued to make a sizeable contribution to the economy and to its growth over the period. The administration fostered this trend by advocating continued self-sufficiency amongst agricultural producers beginning to grow crops for the market, and thus may have contributed to a slower monetisation of the subsistence sector. Under the circumstances this policy seems to have been justified, disappointing though it may have been for those who would have wished to see a higher proportion of output marketed. While the growth of small urban centres encouraged limited localised production of foodstuffs, the lack of development of transport facilities in the hinterlands imposed extremely high costs on the transport of commodities (both inward- and outward-flowing) beyond a short radius of these townships and effectively eliminated economic gains from trade and specialisation in locally produced foodstuffs within the rural sector itself (Shand 1965).

A second question is whether the concurrent encouragement of estates and smallholders was the most productive policy for the period —that is, whether total marketed output could have been increased by placing relatively greater emphasis on one or the other. Here again this author's view is that the policy followed was broadly appropriate for the circumstances, principally because there was in fact little serious competition between estates and smallholders for development resources. Indeed, it can be argued that their simultaneous development led to a fuller utilisation of available resources than would have occurred, for example, if a reduced participation by estates had been preferred.

This case can be argued on various grounds. There was no major shortage of, and therefore no competition for, land for agricultural development. Areas made available to estates on leasehold were sold to the administration by indigenous groups who considered they were well able to spare it, taking into account their own future needs for development. Estate labourers chose their employment freely, in the light of opportunities available (or not) for growing cash crops in their own villages. The retention of labour in the villages would not necessarily have led to any substantial increase in smallholder output, whereas it would undoubtedly have reduced estate output.

The major capital expenditures which directly benefited estate development (airports, sea ports, roads, bridges, etc.) in most instances assisted smallholders as well. Furthermore, as has been argued above, the presence of estates and smallholdings in the same area often led to the availability of efficient large-scale processing, transport, and marketing facilities which were of benefit to indigenous producers. In such cases the presence of estates probably involved some savings for the administration, which otherwise might have had to provide these services itself.

Given that the balance between estate and smallholder development was roughly appropriate from an economic viewpoint, a third question remains, as to whether the specific policies of the administration directed towards estates and smallholders separately were most conducive to their development.

Estate development has required the provision of basic infrastructure facilities and of research and extension services. These have in fact been made available by the administration as the demand arose. Progress with regard to the former lagged, but it is doubtful whether this factor alone discouraged much investment.

Beyond the provision of adequate facilities and services, and of other financial incentives which were provided in generous array by the administration, the rate of estate expansion was influenced chiefly by three factors. The first was the prospective rates of return on capital, which have depended primarily on price trends in world markets. These were, of course, beyond the control of the Territory. The second was the awareness in the business community at large of the investment prospects. Until recently the Territory has relied almost exclusively upon Australia as a source of external private capital. It is possible that further private investment might have been stimulated, had a more global and aggressive policy of overseas promotion been adopted. Third, estate investment was influenced by security factors. It is possible that the provision of investment guarantees by the Australian government might have led to a higher and more sustained level of investment, particularly during the early 1960s, but a discussion of this question would introduce issues beyond the scope of this chapter.

Finally, the choice of the 'village garden' strategy for smallholders appears to have been broadly appropriate for the circumstances, though, as indicated below, these circumstances have recently undergone some changes, and now call for a modified approach in the future.

The strategy suited a situation where land and labour resources were plentiful, allowing a simple addition of cash cropping activities to traditional subsistence activities. Part of the available labour reserve could be mobilised for public works to effect linkages with the market economy. At a time when the financial grant from Australia was still small, this labour proved to be a valuable substitute for scarce capital.

The most feasible alternative strategy would have been an enlarged program of land settlement schemes. This might have generated a higher average level of cash crop output per grower than was achieved in the villages, but the high capital requirements generally characteristic of these schemes, and the restricted availability of capital in this period, would have meant that only a limited number could have been undertaken. The likely net result would have been a lower total contribution from smallholders over the period.

FUTURE PROSPECTS

Prospects for short-run growth in marketed rural output are largely decided already, since the most important contributions will be from perennial tree crops which take a number of years to come into bearing. Additions to output of coffee, cocoa, rubber, coconuts, and to some extent of tea will be determined principally by the age structure of trees already planted. Since a considerable proportion of trees of all crops (coconuts to a lesser extent) are immature or have yet to reach full production, a continued expansion can be expected at least until the early 1970s. This should ensure a substantial increase in the volume of exports, and in income as well, if prices do not decline, at least until 1971-2. One projection suggests that the expansion in rural exports will be the chief factor in raising export income from A$45 million in 1966-7 to A$71 million or more in 1971-2, an overall increase of 58 per cent, or an average compound rate of increase of 9·5 per cent per annum (Table 9.7).[11]

The main increases to 1971-2 are projected for timber (over A$7 million), coffee (almost A$5 million), and cocoa (A$3 million), with a smaller but fast-growing contribution from tea. In addition, some expansion in production for the domestic market is expected in timber, meat, and horticultural crops.

Longer-term trends in agricultural production will be greatly affected by development activity over the next few years. This final section is primarily concerned with a broad consideration of the factors affecting prospects for agricultural development over this period, and with the development strategies appropriate for these conditions.

From the viewpoint of physical resources, the Territory is still well placed for further agricultural development. An accelerated program of field assessments and surveys in recent years has greatly expanded

---

[11] The only prospect of a sizeable contribution from non-rural industries lies with a large-scale copper sulphide deposit on Bougainville, at present being assessed for its commercial potential. The development of this deposit, however, would probably not affect export income until the early 1970s.

TABLE 9.7

ACTUAL AND PROJECTED EXPORT INCOME,[a] 1964–5 TO 1974–5

(A$ million)

Item	Actual				Projected							Average growth rate[b] (%)
	1964–65	1965–66	1966–67	1967–68	1968–69	1969–70	1970–71	1971–72	1972–73	1973–74	1974–75	
Coconut products	19·8	20·9	15·8	22·5	22·6	22·7	22·6	22·5	22·4	22·2	22·0	1·1
Cocoa beans	7·0	4·4	9·5	8·4	9·5	10·7	11·2	12·1	12·9	13·8	14·8	7·7
Rubber	2·6	2·6	2·5	3·2	3·5	3·7	3·8	4·1	4·6	4·9	5·4	7·6
Coffee	7·3	8·8	10·2	12·2	14·0	14·4	14·8	15·2	15·6	16·0	16·4	8·4
Tea	—	—	—	—	0·1	0·4	1·2	2·4	3·7	5·0	6·1	
Oil palm products	—	—	—	—	—	—	0·1	0·3	0·9	1·5	2·6	
Pyrethrum extract	0·1	0·1	0·4	0·4	0·4	0·4	0·4	0·4	0·4	0·4	0·4	14·9
Peanuts	0·5	0·5	0·5	0·4	0·4	0·4	0·3	0·3	0·3	0·3	0·3	−5·0
Passionfruit pulp and juice	0·2	0·2	0·1	0·3	0·3	0·3	0·3	0·3	0·3	0·3	0·3	4·1
Crocodile skins	0·9	1·0	0·7	0·5	0·5	0·5	0·5	0·5	0·5	0·5	0·5	−5·7
Marine shell	0·1	—	0·1	—	—	—	—	—	—	—	—	—
Timber	3·5	3·7	4·3	6·7	7·8	9·7	10·9	11·8	13·3	13·3	13·3	14·3
Gold bullion	1·1	0·9	0·9	1·0	0·9	0·9	0·8	0·8	0·8	0·7	0·7	−4·4
Other	0·3	0·2	0·2	0·2	0·2	0·3	0·3	0·3	0·4	0·4	0·4	2·9
Total	43·3	43·5	45·7	55·8	60·2	64·4	67·2	71·0	76·1	79·3	83·2	6·7

a Excludes re-exports.

b Average annual rate of increase, 1964–5 to 1974–5.

— = Less than A$0·05 million.

Source: Based on Territory of Papua and New Guinea 1967: T. 17.

estimates of the area of forest with known commercial potential, has located promising marine fishing grounds, and has also revealed good prospects for crop and livestock production on large tracts of previously unused or little-used land. Two factors, however, which will assume increasing significance in the coming years should be noted. First, some areas which recorded substantial progress over the post-war period, such as the Gazelle Peninsula of New Britain, are now experiencing some degree of land shortage. In Territory terms this means that in such areas villagers are finding that further expansion of cash crop areas affects the long-term output levels of subsistence food crops, since fallow time has to be shortened. At this point cash crop production no longer supplements subsistence production, but becomes competitive with it for land. Second, although much land of high potential remains, it is less accessible, and its exploitation will be costly. The added cost does not necessarily mean that the returns to such exploitation will be lower than in the past—indeed, there is a real possibility they will be as high or higher—but it does mean a high scale of development finance will be needed.

Market prospects for Territory products vary greatly from commodity to commodity, and are different from those which influenced the pattern of output over the past twenty years (Shand 1966). The surplus of production capacity on the world market continues to preclude any possibility of large-scale expansion of coffee output. The recent steep decline in rubber prices will almost certainly discourage estates from expanding their areas or undertaking considerable replanting, and will seriously hinder administration plans to launch smallholders into the industry through land settlement schemes. Copra prices have revived somewhat from depressed levels in the early 1960s, and may be high enough to encourage further planting by smallholders, but it is questionable whether longer-run prospects are bright enough to warrant much replanting or new planting by estates except where the crop is interplanted with cocoa. World cocoa prices suffered a serious decline in 1965, but have since recovered and would still appear high enough to justify further expansion both by estates and smallholders. Prospects in the palm oil and quality tea markets are sufficiently encouraging to warrant current plans for large-scale development, particularly in view of the excellent production potentialities of these crops under Territory conditions. The local and domestic market outlook for timber and the domestic market prospects for beef are also favourable.

The political climate for private investment is difficult to predict and is notably changeable. On present indications, however, it will be stable for some years, and favourable for a continued inflow of private capital from overseas and for reinvestment within the Territory, at least in those industries which offer good prospects of a high rate of profit. Subjective interpretations by investors of the risk involved will vary, so that while further inflows are likely, there will probably be some outward movement as well. Overall, however, there seem to be good possibilities for a substantial net inflow over the next few years, particularly in the direction of the newer industries such as oil palm and tea.

Providing that staffing problems are overcome, prospects for an expansion of administrative programs to support agricultural development are sound. The Territory will continue to possess a limited though growing capacity to finance the program now envisaged and will lean heavily on external funds. However, there is reason to expect that the Australian government will increase future annual grants to the Territory, though possibly not at the high rates of the past. These grants will be supplemented by loan funds obtained from international sources,[12] and these latter funds will be of particular significance for agriculture.

The availability of technical staff will be a crucial factor in program implementation. In the absence of trained indigenes during the past twenty years, the administration has relied on expatriates. In recent years career uncertainty has led to serious losses, particularly among more experienced officers. Efforts are being made to stem this outflow with offers of more attractive conditions for expatriates, while at the same time trained indigenous officers are now beginning to enter the service. In the short run, however, the successful implementation of programs will continue to hinge largely upon the availability of expatriates.

In the light of these projected conditions, the broad policy of concurrently encouraging estate and smallholder development of Territory resources is still appropriate, though some modifications of past strategies may be needed, designed particularly to maintain a fast rate of indigenous development.

There still are areas in which the 'village garden' strategy can be expected to generate a substantial growth of smallholder output, but it is becoming evident that diminishing returns are setting in for many areas. In some instances this is because smallholders are reaching the limits of land available or suitable for cash crop production. In other instances the provision of transport facilities in new and more remote areas, with perhaps low population densities, would involve costs relatively high in relation to the output response which could be expected.

[12] An application is currently being made to the International Development Association for a portion of the finance required for the agricultural development program.

These factors require a more selective application of the 'village garden' strategy, restricting its use to those areas where an extension effort, coupled with the provision of transport facilities, will generate a concentrated and substantial output response, and which will therefore bring benefits from higher returns to growers through lower processing, transport, and marketing costs. They also suggest the need to explore alternative strategies for agricultural development.

One way of achieving concentration of output and its benefits within the village context is to encourage the establishment of relatively large blocks of cash crops, located contiguously, by individual farmers who are provided with credit to permit a fast rate of planting, and with technical assistance and regular supervision by extension officers. Villagers utilise their own tribal land, on which some rationalisation of land tenure is introduced, and, depending on the location of villages, varying investments in infrastructure formation are made to ensure market linkages. This so-called 'village concentration' strategy was first advocated by the 1963-4 World Bank Mission (IBRD 1965: 20) and has been incorporated into the administration's development programs (Territory of Papua and New Guinea 1967: 16-22). The problem with this strategy is that it requires a fair-sized population in an area of high agricultural potential (preferably with developed communications), and such areas are not common in Papua-New Guinea.

A further path for agricultural development, opened up by the prospect of more funds from Australia and from international sources, is the possibility of large-scale development of hitherto unexploited land through resettlement schemes. These hold particular promise for providing opportunities for enterprising individuals currently located in areas with limited development potential.

Administration programs now include a number of formal resettlement schemes. Some are to be organised in conjunction with estate development, others are to be organised independently. Some will be located in isolated areas, as growth points for further development; others will be located in more settled areas, where they can create 'spill-over' effects by stimulating peripheral development amongst local villagers, bringing to them the same benefits from external economies as those provided by estates.

The policy of integrating smallholdings and estates in the same area has economic merits, and these are appreciated in current administration programs. For example, expansion in the tea and palm oil industries is being organised on a 'nucleus-estate' basis, whereby an expatriate estate is established and is encouraged to install factory processing capacity which will be able to absorb not only its own output but also that of smallholders located around its periphery (either in resettlement schemes or on local village blocks). This is a particularly suitable arrangement for tea and palm oil, which need to be processed immediately after harvest and which, for efficient processing, require capital outlays in factory facilities beyond the means of smallholders. Provided equitable price agreements are reached by the parties concerned, this arrangement enables the administration to avoid an expensive commitment itself, and at the same time brings the benefits of external economies of scale in processing to the smallholder.

A close integration of large- and small-scale enterprise is also being encouraged in the developing beef cattle industry. At present breeding herds are being expanded on expatriate ranches, and to a lesser extent on administration livestock stations. These will later provide nucleus stock for indigenous cattle projects. By 1974 the program envisages ownership of 40 per cent of all stock by indigenous producers, mainly as an outgrowth of this distribution system.

The achievement of targets for smallholder development will be influenced by institutional factors which hitherto have not been of major significance. In particular, land and labour markets and a system of agricultural credit have not been important influences until recent years. While there was ample land for cash cropping and subsistence production within tribal boundaries, there was no necessity for a land market. While smallholdings were of a size which could be operated by a family unit, there was no need for hired labour, and so long as production inputs were those available within the subsistence sector, there was no demand for agricultural credit. Furthermore, the plentiful supply of resources meant that cash cropping could be accommodated within the traditional socio-economic organisation with a minimum of friction (Epstein 1968: 101, 171).

This situation is now changing significantly. An increasing number of smallholdings are reaching a size which necessitates extra casual or even permanent labour, and this has become apparent in a strengthening demand on the wage-labour market in more advanced areas. The continued expansion of cash crop production has also led to localised land shortages, to the emergence of a market for land, and to an accompanying demand for clearly established ownership rights to it. There is also an expanding demand for credit among smallholders, due partly to the growing market transactions in land and labour and to the new strategies being followed by the administration. Resettlement and village concentration schemes envisage substantial plantings at a fast rate by participants, most of whom have no capital reserves of their own. Demand is also growing as a consequence of the gradually expanding use of purchased inputs for agriculture, such as weedicides, pesticides, and fertilisers.

These trends are already requiring modifications in the structure of traditional institutions, particularly with respect to land tenure, and the creation or adaptation of external institutions, as witnessed by the demand for credit from the newly established Development Bank. As yet no major bottlenecks have become apparent, but since the new needs are in their infancy, these institutions, new and old, have not yet been fully tested.

Timber is one industry in which substantial indigenous participation is not envisaged within the development program. Until recently production was in the hands of a number of expatriate firms, whose small scale of operations has greatly limited the rate of exploitation of the forest potential. Attempts are now being made to attract large-scale overseas operators to the Territory, with the offer of large concession areas—one recently advertised, for example, covers more than 300,000 hectares. It is possible that a degree of local participation in some of these ventures will be achieved through partnerships between private interests and the administration, as in the oil palm industry.

Attention is also being given to the encouragement of indigenous fishermen in various ways, and to attracting overseas capital for larger-scale fishing operations with a view to import replacement and to export marketing.

The major impact of the agricultural development program will not be felt for about ten years (roughly 1977), owing to the delayed effects of infrastructure investment and to the types of crops concerned. The full impact will not be evident until the 1980s.

The successful implementation of the program will make a sizeable contribution to the growth of the economy, but major problems will remain. Much of the population will still be operating largely within the subsistence sector, and regional unevenness in development will still be apparent, though somewhat ameliorated by the program. Agricultural settlement schemes are a way out of the latter difficulty, but it is unlikely that they can be undertaken on a sufficient scale to meet the problem fully, and additional solutions will have to be sought. Expansion of employment opportunities in other sectors of the economy will help to absorb the growing labour force, and in the agricultural sector the continuing search for new marketable commodities may reveal some which can be introduced to backward areas. It is doubtful, again, however, whether these two will be sufficient to provide opportunities for all.

Because of these limitations, particular attention should be paid to the possibilities of intensifying production on land currently in use. Research is in progress on genetic, agronomic, and other aspects of export crop production. Some success has already been achieved in raising yields, though tree crops are not notably responsive to modern inputs. Greater returns may be forthcoming from a vigorous program of research into increasing productivity of food crops. In most areas the subsistence food crop cycle is highly land-demanding. Technical advance, which could reduce the fallow period and could permit more intensive, continuous cropping, could ease local pressures on land, permit the expansion of other land uses, and, in combination with investment to develop the rural infrastructure, might encourage increased specialisation and trade in these food crops and thus channel an increasing share of subsistence production through the market.

10

Perspectives on Asia

R. T. Shand

THIS final chapter attempts to interpret the agricultural performances of the nine countries under review here in terms of the interactions of four factors: the availability of resources (land, labour, capital, and production inputs associated with capital), the productivity of these resources, the production incentive, and the institutional factor. The first two need no special introduction, but some preliminary comment is required for the latter two.

In broad terms, the production incentive is determined by the farmer's comparison of the prospective satisfactions and costs of a particular resource-use decision. Where labour is involved, it is necessary to compare the utilities (derived from real money income earned or income in kind) and disutilities (either from leisure or alternative employment forgone) of additional labour inputs. Where capital is involved, the comparison lies with the monetary benefits and costs, expressed as a prospective rate of return on investment.

In both situations the prospective satisfactions stem from the utility of additional real money income. They therefore depend on two things: how much is earned (i.e. physical product and prices received); and the purchasing power of this income (i.e. the range and prices of available goods and services).

Physical product from additional inputs depends on environmental factors and on the level and combination of inputs already committed. Prices received by farmers depend on supply/demand factors in the market and on costs and profit margins in intermediate processing, transport, and marketing industries. In turn, costs in these industries are affected by scale of operation, whilst competition between firms will influence profit margins.

The range and prices of goods available to the farmer will be determined partly by transport and marketing facilities in rural areas and partly by the size of the rural market (i.e. the number of rural consumers and the level of disposable cash income).

One further aspect of the incentive factor should also be noted. Since farmers must make decisions on the basis of forecast outcomes, risk and uncertainty influence their expectations and therefore affect the margin of expected profit required to induce an investment. Risk and uncertainty arise from variability in physical conditions of production, from commodity price variability, and from price and supply variability in resource markets. The degree of risk and uncertainty shows itself in a rate of discount placed by the farmer on anticipated returns.

The institutional factor is defined here as the socio-economic framework within which agricultural production takes place. It encompasses such important considerations as the pattern of resource ownership in agriculture, the social structure and cultural attitudes of the agricultural community, and the organisation of factor and commodity markets and of the economic infrastructure serving agricultural producers.

This determinant affects output through its influence on each of the three foregoing factors (i.e. on the availability and productivity of resources and by modification of the production incentive). In relation to the production incentive, it can operate through commodity and input prices, through marketing, transport and processing costs and margins, and through retail prices for consumer goods. It can also modify risk and uncertainty both in the market (price control and stabilisation) and in the field (flood and drought control).

The interpretation of past growth performances in the nine countries discussed here, in terms of the interactions of these four determinants, would be an imposing task were it not for the fact that the situations with regard to the availability and productivity of resources fall into a narrow range. In all cases, with the exception of Japan, the agricultural labour force expanded in the periods under review. Capital was a scarce resource in all nine countries, while the availability of land needs to be considered for only two situations: where it was plentiful (unlimited land) and where it was scarce (limited land) in relation to demand.[1] Resource productivity can similarly be considered in terms of two broad situations: one where it was, overall, either constant or declining (static technology), and the other where it was increasing (dynamic technology).[2] There are therefore four basic resource situations for consideration: unlimited land—

[1] Limitations on land use are seldom absolute. They may reflect a given technology or economic circumstances, e.g. where new land is of poor quality and/or is remote from population concentrations. They may also be imposed for political reasons, e.g. Burma, or for economic reasons, e.g. in the interests of national conservation.

[2] This distinction need not be interpreted too rigidly. It is possible, for example, for a technical innovation to be introduced which raises the productivity of one resource, such as labour, but also for its effect to be counterbalanced by declining productivity of another, such as land, so that the net effect gives the appearance of a static technology.

static technology; limited land—static technology; unlimited land—
dynamic technology; and limited land—dynamic technology. The influence of the production incentive and the institutional factor can be
examined within this framework.

UNLIMITED LAND — STATIC TECHNOLOGY

The first situation is one where a growing labour force operates on
an expanding area under cultivation, with limited capital and a static
set of production techniques. This characterised colonial Burma,
Indonesia excluding Java, and, until recent years, Thailand and the
Philippines.

Under these circumstances, the growth of agricultural output
depends heavily upon the quality of land brought under cultivation.
In colonial Burma, Richter's evidence that rice production grew
roughly in proportion to the shift in population into the delta and
river valleys of Lower Burma suggests that resource productivity was
probably at least maintained. In Thailand the trend in output per
labour unit prior to World War II may have been slightly downward.
Silcock suggests that the only significant technological change was the
introduction of the steel ploughshare, which would have increased
the working capacity of farm labour, but this was counterbalanced by
a decline in rice yields. Hooley and Ruttan suggest that, in the
Philippines, the increase in area under cultivation took place on land
of declining quality, and that although labour productivity was maintained, overall resource productivity showed some decline.

In these three countries, the increase in area cultivated was accompanied by the production of a considerable agricultural surplus which
was primarily marketed internationally. In each case the size of the
surplus was strongly influenced by export prices. Favourable rice
prices, for example, stimulated settlement and export production in
Lower Burma from 1850 until 1927, after which the steep decline of
the depression greatly discouraged any expansion of market production. Coconut and rice production in the Philippines and rice and
rubber production in Thailand reacted to the varying price incentives
offering on world and domestic markets.

In each case the increase in area under cultivation and the level of
output per farm in these new areas depended heavily on the facilities
and services made available to the settlers. These in turn depended
largely on the extent and nature of government intervention. In
Burma, the British government invested in market and transport
facilities (railways and canals) and succeeded in attracting private
enterprise to serve the needs of farmers, especially for the developing
rice export industry in Lower Burma, where measures taken for flood

protection also reduced production risks for rice farmers. In Thailand, the railway system was developed prior to World War II, and a road network was built, somewhat belatedly according to Silcock, after the war. Tax inducements were also offered in Thailand for settlers opening up new areas. In the Philippines there were some schemes of government-assisted migration and settlement during and after the period of American colonisation. More effective, however, was the provision of roads, for example in Northern Luzon and on Mindanao, which encouraged substantial voluntary migration and resettlement.

Action to raise productivity in Philippine agriculture was notably lacking, at least until recently. Some investment did take place in irrigation facilities but this enabled a very limited expansion of double-cropping and was probably more effective in reducing the risk of crop failures induced by floods and droughts. In this country the only notable productivity gains were recorded in the sugar industry and were the result of a substantial research program which developed higher-yielding crop varieties and improved inputs and farm practices.

The production incentive in both new and existing areas under cultivation was substantially influenced by agricultural price policies, at least in Thailand and the Philippines. Silcock argues that the heavy export tax on rice kept farm prices low, and largely denied farmers the benefits of higher international prices until recently. This reduced the incentive for rice production and induced the cultivation of alternative crops in some areas. In the Philippines, rice price policy since World War II also appeared to favour consumers with announced support prices generally set below ruling market levels, which were themselves held down with the injection of imported rice. Incentives for production of agricultural exports were also reduced between World War II and the late 1950s by overvaluation of the domestic currency. This resulted from the emphasis placed on industrialisation at the time, requiring cheap imports of raw materials and capital goods. The protection of domestic industry also raised prices of consumer goods which, in turning terms of trade against agriculture, indirectly acted as a disincentive.

These cases show, however, that when market conditions are favourable, and when the institutional commitment of government is adequate (involving, as it generally does, considerable public investment), substantial economic benefits can be realised. Growth of a marketed surplus raises aggregate farm income and therefore total consumer purchasing power in the sector and can, in addition, supply the food needs of increasing numbers outside agriculture. Where the surplus includes exportable commodities, rising export income can substantially increase a country's import capacity. Furthermore an expanding surplus can, if exploited, provide an increasing source of government revenue which, when directed as public investment towards the exploitation of new areas, can serve to maintain the growth of this surplus.

These benefits are subject to serious limitations, however. Since resource productivity is at best only maintained, average per capita farm incomes (and rural living standards) cannot be raised for long, though they may, of course, rise from time to time with windfall market gains. Although total purchasing power in the sector expands, it does so on a geographically extensive basis, as new areas are opened up for cultivation, with a repetition of expenditure patterns in these new areas. Farm incomes also vary through fluctuations in production and market conditions. These factors restrict the range and scale of industries supplying the rural market and the efficiency of firms using agricultural commodities as inputs. Furthermore, most countries have only limited opportunities for maintaining this kind of growth in agricultural output. Either the quality of new land declines, as it has in the Philippines, or it is located in progressively more remote areas, as in Indonesia. Either factor may deter a voluntary flow of population to the frontier, and deter governments from encouraging such settlement; the declining benefit/cost ratios may result in a growing concentration of population on a fixed area under cultivation.

LIMITED LAND — STATIC TECHNOLOGY

The second situation is one in which the agricultural labour force continues to increase, the extension of cultivated area at the margin ceases, while resource productivity in agriculture remains constant or declines. This describes Burma since independence and Indonesia as dominated by Java. It is also relevant for Thailand and the Philippines where increases in total cultivated area are not expected to continue for much longer.

In Burma, the limitations on area cultivated have had a political origin, arising from problems of insurgency. In Indonesia limitations have arisen from difficulties in finding new areas with promising productive potential, combined with the prohibitive costs of shifting large numbers from Java to other islands and of providing basic facilities for the settlers.

The consequences of an increasing concentration of population on a limited land area have been somewhat similar in both cases. Marketed surpluses have declined, leading to a reduction in the volume of agricultural exports. In Indonesia, shortages of domestic food supplies have appeared, particularly in climatically unfavourable years, which have led to price inflation and to the need for imports to make up the deficiencies. The consequences within the rural sectors have also been severe. While subsistence supplies have not as yet become inadequate in Burma, large numbers in Java have been reduced to a state of chronic malnutrition in average climatic years and have faced starvation in unfavourable seasons. In both countries, declining and fluctuating purchasing power within the rural sectors has severely limited commercial and industrial expansion, while diminishing farm incomes have yielded declining government revenues.

Burma's difficulties derive from five institutional influences. The first has been the lack of law and order, which has limited not only the area cultivated, but also the incentive to produce in unsecured areas (e.g. tribute taxes to rebels, etc.). Second, the pricing system for rice has consistently taxed surpluses away from farmers to finance capital formation in public utilities and in industry and has discouraged farm investment. Third, the agrarian reforms, although transferring income to the majority of farmers, reduced the scale and profitability of the larger farms and failed to provide smallholders with adequate credit and managerial services to replace those formerly provided by private enterprise. Fourth, where farm investment opportunities still existed, exploitation was impeded by shortages of key inputs (e.g. irrigation pumps, etc.). Finally, the government failed to build production incentives into its credit system or to provide an adequate institutional basis for achieving increases in resource productivity in the sector.

Penny has shown the negative influence of the institutional factor in Indonesia's agricultural record, principally through its effects on production incentive. This has been so from colonial times, until recently some limited indications of policy change towards agriculture have become evident. Prices of export commodities have been depressed by an overvalued currency or heavy export taxes, while internal food prices (of rice in particular) have been held well below international levels by export prohibition, injection of imports, and enforced government purchases at low prices. Failure to develop, or even maintain, transport and marketing facilities has depressed farm prices, particularly outside Java, and has discouraged migration to these areas. Since independence, many of the existing facilities (e.g. transport and irrigation) have actually deteriorated. Again, government efforts to raise agricultural productivity have been inadequate. Research and extension services have been poor, and, as Penny has pointed out, the extension service has been used for other purposes. It is only recently with the Bimas Program and the formulation of a five-year economic development plan that some signs of positive approach to agricultural development have been forthcoming. Broadly, government has been willing to siphon off much of the surplus from agriculture, but has shown little interest in maintaining the surplus, let alone increasing it, in the face of growing population pressure.

The important lessons from these cases are that inadequate or inappropriate government development policies towards productivity in agriculture, especially smallholder agriculture, can progressively reduce the capacity of an economy to initiate such development. Declining export income and a rising food component in imports reduce the capacity to import needed capital goods and to meet repayment schedules on loans raised internationally. The decline in government revenues, on the other hand, reduces the capacity for public investment from internal resources, and correspondingly increases the need for external finance. Declining farm incomes reduce the capacity of the agricultural sector to take advantage of the investment opportunities available and increase the need for government to provide credit, thus placing a further strain on the supply of scarce public capital. It is clearly in a country's own interest to commence its program of agricultural development before this stage is reached.

UNLIMITED LAND — DYNAMIC TECHNOLOGY

Taiwan and Malaysia have been able to increase resource productivity while the extension of area under cultivation was still in progress. This is also true for Papua-New Guinea, though the increase has benefited crops newly introduced to the economy and has not yet been achieved for crops within the traditional subsistence sector.

These countries illustrate the potential benefits of the institutional factor. In Taiwan and Malaysia, government intervention has played a vital part in setting the stage for the research effort which generated higher resource productivity in the agricultural sectors. The operative factors were higher-yielding crop varieties, more efficient use of water, better quality fertilisers, and improved practices. Intervention was crucial in the supply of inputs required for the successful adoption of these innovations. Public investment provided the enlarged water storage capacity and reticulation systems which enabled double-cropping to be expanded. The governments promoted the supply of fertilisers, distributed new seeds, and augmented the supply of agricultural credit to help meet expanded capital needs of producers. Most important too, government extension services, utilising existing channels of communication in the agricultural sector, effectively brought information about innovations to the farm.

Aside from inflation control measures instituted in the years immediately following World War II, there was little direct government intervention in agricultural prices in Taiwan, probably because, for most of the period reviewed, these already favoured agricultural producers. However, there were ways in which the production incentive was affected by government activity. Expansion of transport and marketing facilities, which stimulated the extension of cultivated area and the production of marketed output, and the provision of law and order had all been accomplished largely before productivity gains were achieved. There was also substantial intervention in the pattern of resource ownership through land reforms. The gradual, three-stage series of reforms, as Myers has shown, effectively reduced the incidence of taxes and rentals on the farmers themselves, and whilst landlord-tenant relations were retained until after World War II, tenure security of tenants was improved. Both factors have raised the production incentive.

Fisk has shown the institutional factor to have been no less significant for Malaysia. Here the government relied as much as possible upon the private sector to stimulate the growth of agricultural output, and the development of high-yielding rubber clones through the research programs carried out by estates and on an industry basis was a notable achievement in this respect. The government's role was confined to activities which supplemented the institutional framework of the private sector (e.g. through provision of transport and market facilities). Intervention was more substantial where the benefits of large-scale private operations were not available. A far wider range of assistance was provided, for example, to smallholders in the rice industry, comprising not only basic transport and market facilities but water supplies, improved seeds, fertilisers on occasions, capital, and skills. Even here, however, the government encouraged self-help amongst farmers wherever possible, through co-operatives and farmers' associations, in such activities as bulk buying and distribution of farm inputs and consumer goods.

The historical isolation of Papua-New Guinea from the rest of the world until recent times, combined with the economic independence of the small, subsistence-oriented agricultural communities within the country, have given rise to development problems which differ from those in most parts of the region.

With a plentiful supply of land, a continued growth of non-marketed subsistence production has been possible as population has expanded without the necessity for government intervention. However, the encouragement of marketed output has required a heavy commitment of public resources, in order to find crops in demand on world markets, to disseminate the skills and knowledge required for their production, and to provide an incentive which would encourage participation amongst the indigenous population. As in Malaysia, there has been a heavy reliance upon large-scale estate operations run by expatriates, for which intervention mainly centred around the provision of transport and marketing facilities. For smallholders, however, it was necessary to create a research service for the introduction and adoption of new crops, an extension service to spread information on these crops to the village and to develop an economic infrastructure to facilitate the transport and marketing of smallholder output. It was necessary to supply these services themselves where private enterprise could not, at least initially, be encouraged to provide them; and finally it was necessary to stimulate, directly and indirectly, the sale of consumption goods within the agricultural sector.

These case studies show that agricultural development through increasing resource productivity places heavy demands on available capital, particularly where a major effort is directed towards foodgrain production, in which substantial investments are required to ensure adequate water supplies. Government participation may be needed in the development of seed and fertiliser industries, while in newly developing areas, extensive investment may be required in transport and marketing facilities. The provision of other services, such as research, extension, agricultural planning, and administration, also requires a large-scale and continuing financial commitment. Furthermore, the credit needs of farmers, although not individually large, are substantial in the aggregate, where the attempt is made to spread technological advances widely in the sector. Taiwan and Malaysia have been all the more able to meet these financial requirements internally because development programs have been instituted while marketed output has been expanding as the area under cultivation increased. In Malaysia, for example, this has meant increasing government revenue from agriculture and also increasing export income, which has placed the country in a better position to borrow internationally.

The studies show that rising resource productivity has crucial implications for the development of agriculture itself, and for the development of the rest of the economy. Where agricultural development is intensive (i.e. where the level of marketed output is raised per unit of land), there is a geographically more concentrated growth in marketed output. This makes available scale economies in transport, marketing, and processing operations, and possibly greater competition in these industries, which can result in higher farm prices and a further stimulus to production. Higher farm incomes induce a greater and more concentrated rural purchasing power which can lead to an expanded and more diverse consumer demand and therefore to the availability of a wider range of consumer goods at lower prices. Activity is accordingly also stimulated in industries supplying agriculture with inputs.

Rising farm incomes lead to a greater capacity for financing farm investments internally. As the farm becomes more capitalised, farmers become more credit-worthy, and this may in the long run increase the supply of commercial credit and reduce its cost. With a higher capital intensity, farmers may place a lower rate of discount on new investment, since the consequences of investment failures become less critical, and this may therefore effectively raise the volume of farm investment. As farmers gain in their capacity to expand farm operations, there may be correspondingly less need in the long run for government intervention in the supply of credit. Finally, as Fisk has shown for Malaysia, efficient agricultural industries are in a better competitive position *vis-à-vis* international competitors and substitute industries. Improvements in efficiency enable agricultural industries (and the economy indirectly) to absorb temporary market adversity, and can offer more profitable returns on investments for the extension of the area under cultivation.

LIMITED LAND — DYNAMIC TECHNOLOGY

The fourth situation is one in which the area under cultivation shows no increase, and where marketed output can only expand through more intensive application of other inputs. This characterises Japan (1868-1920), Taiwan since World War II, and India since the Third Plan.

It is clear from the foregoing studies that whilst the rate of gain in agricultural productivity has been an adequate basis for generating economic prosperity in Japan and Taiwan, it has been inadequate thus far for economic development in India. Crawcour suggests in his analysis that an average growth rate of around 2 per cent per annum was sustained in Japan over the period; a modest achievement by comparison with Taiwan since World War II. It was adequate for Japan principally because at that time the agricultural labour force was stationary, so that output increases were manifest in increased per capita farm incomes. In India, on the other hand, the slow growth in resource productivity was only acceptable so long as there was a sizeable addition to total output from an increasing area under cultivation. Once the latter ceased to be a factor, as it did after the Second Five-Year Plan, the contribution from productivity gains proved to be inadequate in a situation of rapid population growth.

In Japan, agricultural development followed from a combination of favourable market conditions for agricultural commodities combined with positive government intervention during the period. Once again, the basis for sustained growth of output was a stream of technical innovations. Crawcour argues that these began to make a substantial contribution around and after the turn of the century. Until then, progress was based on a more widespread adoption of known high-yielding crop varieties and better existing agricultural practices. This was achieved primarily through the development of a national extension system, based on co-operation with leading influential members of farming communities and farmers' associations. This extension system, together with the network of agricultural experiment stations and the system of agricultural education, was responsible for the development and adoption of more numerous technical innovations which produced the somewhat faster rate of growth after the turn of the century.

Crawford argues that a lack of strong government support for agriculture explains the inadequate Third Plan performance. Increases in resource productivity were technically possible, but the institutional basis for their exploitation was lacking. There was apparently a considerable potential for raising foodgrain yields using high-quality traditional seed varieties, combined with adequate supplies of water and fertilisers, and appropriate agricultural practices, but the inputs necessary were in short supply and the production incentive was inadequate. The record shows a poorly developed seed industry, inadequate domestic production and imports of fertilisers, a weak impact of investment in irrigation, a shortage of credit, and a relatively ineffective extension system. In addition, foodgrain price policy consistently favoured consumers, at least until 1964, and consequently weakened the incentive for farm investment needed to achieve output targets.

Japan was relatively well placed to mount a program of agricultural development in the Meiji era, for, as Crawcour pointed out, farmers were producing a sizeable marketed surplus in most areas at the time of the Restoration. This provided funds for on-farm investment and yielded substantial revenues which could be reinvested by the government within the sector. By contrast, India is now in a more difficult position with a fast rate of growth in its agricultural population. A rate of output growth must be attained which first meets the growing internal subsistence needs of the sector and then provides a large market surplus which will meet the needs of the non-farm population, raise per capita farm incomes, and stimulate activity in other sectors of the economy as well. This must be achieved at a time when income for large numbers within the sector has already been reduced to a bare subsistence minimum, or less in climatically unfavourable years.

If countries within the region assign agriculture a high priority in overall programs of economic development, as some already have, the sector's performance will depend primarily on the progress of individual smallholders who comprise the vast majority of agricultural producers in the region. It has been argued above that the most effective contribution which smallholders can make towards economic development comes from opportunities for intensification of production through increasing resource productivity. It has also been argued that these opportunities depend very largely on the effectiveness of the institutional framework within which agricultural production takes place, and that this in turn depends crucially on the willingness and capacity of governments to commit the resources required, within a comprehensive program for the sector.

Such programs should give adequate attention to technical research programs designed to create a flow of innovations adapted to local production and market conditions. Fortunately, much of the technical progress in agriculture achieved in recent years is relevant to the Asian scene. High-yielding varieties are now available for a wide range of important food crops (rice, wheat, maize, sorghum, and millet). Provided adequate adaptive research is carried out on these varieties, they should provide the basis for a fast increase in output.

These programs must also ensure the adequate and timely availability of supplies of inputs for farmers who are being encouraged to adopt these innovations. For improved varieties of foodgrains these include the planting material itself, fertilisers, water, plant protection materials, skills, and credit.

Adequate incentives are required, as well, for a fast rate of adoption of technical innovations. Government policy should aim at striking and maintaining a balance in agricultural price and tax measures between the interests of agricultural producers, consumers, and other sectors of the economy, which will retain for farmers the motivation for the desired investments. There is similarly a need for balance in product and factor markets to protect farmers against monopsonistic and monopolistic pressures which may unduly depress commodity prices and raise input costs at the farm level. Finally, government intervention may be necessary to reduce the element of risk and uncertainty in farmers' decisions, for example, by stabilising commodity prices, by public water and flood control, and by appropriately modifying conditions of tenancy which inhibit farm investment.

These activities require a comprehensive and co-ordinated government program for the sector, in which priorities and implementation procedures for the various components have been established, and which is backed by a large and well trained administrative staff. Such a public program also clearly requires a heavy investment of public capital.

The capacity of those economies with lagging agricultural production to mount a substantial program from internal resources varies greatly within the region. In particular, the extent of past neglect of the sector has conditioned the availability of public finance for government investment in agriculture, the capacity of farmers to meet on-farm investments from their own resources, and, through the external trade position, the ease with which overseas borrowing can be undertaken and repayment commitments can be met. However, even where a serious institutional effort has already been made towards agricultural development, these case studies show there are still substantial areas in which local resources are in short supply and in which international assistance, in capital and skills, can relieve short-term bottlenecks and can assist in raising the long-term capacity of these countries to meet their own needs for agricultural development.

If this analysis ends without a confident prediction of a successful solution to the agricultural problems of the region, it is because so few governments have as yet made the sustained institutional commitment required for the development of the agricultural sector.

References

Chapter 1: Japan, 1868-1920

Andō, S. (1958). *Kinsei Zaikata Shōgyō no Kenkyū* [A Study of Rural Commerce in the Tokugawa Period]. Yoshikawa Kōbunkan, Tokyo.

Chihōshi Kenkyū Kyōgikai [Council for Local History Research] (ed.) (1960). *Nihon Sangyōshi Taikei* [History of Japanese Industry Series]. 7 vols. Tōkyō Daigaku Shuppan Kai, Tokyo.

Crawcour, E. S. (1965). 'The Tokugawa Heritage', in W. W. Lockwood (ed.), *The State and Economic Enterprise in Japan*. Princeton U.P., Princeton.

Dore, R. P. (1960). 'Agricultural Improvement in Japan: 1870-1900', *Economic Development and Cultural Change*, **IX** (1) Pt II: 69-91.

Furushima, T. (ed.) (1963). *Kinsei Nihon Nōgyō no Tenkai* [The Development of Pre-modern Japanese Agriculture]. Tōkyō Daigaku Shuppan Kai, Tokyo.

Hayami, Y. and Yamada, S. (1966). 'Technological Progress in Agriculture'. Paper presented at the International Conference on Economic Growth—Case Study of Japan's Experience. Tokyo (mimeo.).

Horie, E. (ed.) (1963). *Bakumatsu Ishin no Nōgyō Kōzō* [The Structure of Agriculture in the late Tokugawa and Restoration Periods]. Iwanami Shoten, Tokyo.

Jorgensen, D. W. (1966). 'Testing Alternative Theories of the Development of a Dual Economy', in I. Adelman and E. Thorbecke (eds.), *The Theory and Design of Economic Development*. Johns Hopkins Press, Baltimore.

Kaneda, N. (1967). 'Long-term Changes in Food Consumption Patterns in Japan 1878-1964'. Discussion Paper No. 21, Yale University Economic Growth Center, New Haven (mimeo).

Nakamura, A. (1963). 'Mensaku Chitai no Nōgyō Kōzō' [The Structure of Agriculture in the Cotton-growing Region], in E. Horie (ed.), *Bakumatsu Ishin no Nōgyō Kōzō* [The Structure of Agriculture in the Late Tokugawa and Restoration Periods]. Iwanami Shoten, Tokyo.

Nakamura, J. I. (1966). *Agricultural Production and the Economic Development of Japan 1873-1922*. Princeton U.P., Princeton.

328   REFERENCES

Noda, T. (1956). 'Nōsambutsu Juyō no Chōki Henka to Shotoku Danryokusei' [Long-term Changes in Demand for Agricultural Products and Income Elasticities], in S. Tōbata and K. Ohkawa (eds.), *Nihon no Keizai to Nōgyō* [Japan's Economy and Agriculture]. 2 vols. Iwanami Shoten, Tokyo.

—— (1963). 'Juyō to Bōeki no Chōki Henka' [Long-term Changes in Demand and International Trade], in K. Ohkawa (ed.), *Nihon Nōgyō no Seichō Bunseki* [Growth Analysis of Japanese Agriculture]. Daimeidō, Tokyo.

Nōrinshō [Ministry of Agriculture and Forestry] (1955). *Nōrinshō Ruinen Tōkei Hyō 1868-1953* [Cumulative Annual Statistics of the Ministry of Agriculture and Forestry, 1868-1953]. Nōrin Tōkei Kyōkai, Tokyo.

Ogura, T. (ed.) (1963). *Agricultural Development in Modern Japan.* Fuji Publishing Co., Tokyo.

Ohkawa, K. (1964). 'Concurrent Growth of Agriculture with Industry: A Study of the Japanese Case', in R. N. Dixey (ed.), *International Explorations in Agricultural Economics.* Iowa State U.P., Ames.

—— *et al.* (1957). *The Growth Rate of the Japanese Economy since 1878.* Kinokuniya, Tokyo.

—— and Rosovsky, H. (1960). 'The Role of Agriculture in Modern Japanese Economic Development', *Economic Development and Cultural Change*, **IX** (1) Pt II: 43-67.

—— and —— (1965). 'A Century of Japanese Economic Growth', in W. W. Lockwood (ed.), *The State and Economic Enterprise in Japan.* Princeton U.P., Princeton.

——, Shinohara, M., and Umemura, M. (eds.) (1966). *Chōki Keizaı Tōkei* [Estimates of Long-term Economic Statistics of Japan since 1868] (referred to as LTES). 13 vols. Tōyō Keizai Shimpō Sha, Tokyo.

Rosovsky, H. (1961). *Capital Formation in Japan.* Free Press of Glencoe, New York.

Sakai, H. (1963). 'Kōshin Chitai no Nōgyō Kōzō' [The Structure of Agriculture in a Backward Region], in E. Horie (ed.), *Bakumatsu Ishin no Nōgyō Kōzō* [The Structure of Agriculture in the Late Tokugawa and Restoration Periods]. Iwanami Shoten, Tokyo.

Sheldon, C. D. (1958). *The Rise of the Merchant Class in Tokugawa Japan.* J. J. Augustin, New York.

Smith, T. C. (1955). *Political Change and Industrial Development in Japan.* Stanford U.P., Stanford.

—— (1959). *The Agrarian Origins of Modern Japan.* Stanford U.P., Stanford.

Takebe, Y. (1960). 'Sekkasen no Mengyō' [The Cotton Industry of Settsu, Kawachi, and Izumi], in Chihōshi Kenkyū Kyōgikai [Council for Local History Research] (ed.), *Nihon Sangyōshi Taikei* [History of Japanese Industry Series]. 7 vols. Tōkyō Daigaku Shuppan Kai, Tokyo.

Toya, T. (1949). *Kinsei Nōgyō Keieishi Ron* [Agricultural Management in Pre-modern Japan]. Nihon Hyōron Sha, Tokyo.

Tōyō Keizai Shimpō Sha [Oriental Economist] (ed.) (1927). *Meiji Taishō Kokusei Sōran* [Survey of National Economic Statistics in the Meiji and Taisho Periods]. Tōyō Keizai Shimpō Sha, Tokyo.

Umemura, M. and Yamada, S. (1962). 'Nōgyō Kotei Shihon no Suikei [The Estimates of Agricultural Capital Stock], *Nōgyō Sōgō Kenkyū* [Quarterly Journal of Agricultural Economics], XVI (4).

Yamada, S. (1963). 'Nōgyō ni okeru Tōnyū Sanshutsu no Chōki Hendō [Long-term Changes in Agricultural Inputs and Outputs], in K. Ohkawa (ed.), *Nihon Nōgyō no Seichō Bunseki* [Growth Analysis of Japanese Agriculture]. Daimeidō, Tokyo.

Chapter 2: Taiwan

Andō, Yasuo (1925). 'Horai Kome ni Tsuite' [P'eng-lai Rice], *Taiwan Nōjihō* 19 (1): 24-41.

Bank of Taiwan (1958). *Jrh-chu Shih-tai T'ai-wan Ching-chi Shih* [Taiwan's Economic History under Japanese Rule]. Taipei.

—— (1962). *T'ai-wan Chih Nung-yeh Ching-chi* [Taiwan's Farm Economy]. Taipei.

—— (1966). *Export and Import Exchange Settlements for the Year 1965*. Taipei.

Barclay, G. W. (1954). *Colonial Development and Population in Taiwan*, Princeton U.P., Princeton.

Bureau of Accounting and Statistics, Provincial Government of Taiwan (1962). *Taiwan Statistical Abstract* No. 22.

Chang, Han-yu and Myers, R. H. (1963). 'Japanese Colonial Development Policy in Taiwan, 1895-1906: A Case Study of Bureaucratic Entrepreneurship', *Journal of Asian Studies*, 22 (4): 443-9.

Ch'en, Ch'eng (1951). *Ju-ho Shih-hsien Keng-che yu Ch'i-t'ien* [How to see that Farmers own their own Land]. Cheng-chung shu-chu, Taipei.

Cheng, Chia-ying (1963). *T'aiwan T'u-ti Chih-tu K'ao-ch'a Pao-kao shu* [A Report on the Survey of Taiwan's Land System]. Bank of Taiwan, Taipei.

China Publishing Co. (1952). *China Handbook: 1952-53.* Taipei.

Department of Agriculture and Forestry (1967). *Taiwan Agricultural Yearbook, 1966 Edition.* Taiwan.

Directorate-General of Budgets, Accounts, and Statistics (1964). *National Income of the Republic of China.* Taipei.

Gallin, B. (1966). *Hsin Hsing, Taiwan: A Chinese Village in Change.* University of California Press, Berkeley.

Ho, Yhi-min (1966). *Agricultural Development of Taiwan 1903-60.* Vanderbilt U.P., Tennessee.

Horng, Wei-huai (1959). *A Report on the Taiwan Field Rat Control Campaign.* Joint Commission on Rural Reconstruction, Taipei.

Hsieh, S. C. (1963a). 'The Role of Local Government in Rural Development', *Industry of Free China,* **20** (1): 2-8.

—— (1963b). 'Farmers' Organisations in Taiwan and their Trends of Development', *Industry of Free China,* **20** (6): 23-38.

—— and Lee, T. H. (1966). *Agricultural Development and its Contributions to Economic Growth in Taiwan — Input-Output and Productivity Analysis of Taiwan Agricultural Development.* Joint Commission on Rural Reconstruction, Taipei.

Industry of Free China (1967). *Taiwan Economic Statistics,* **27** (2): 45-159.

Ino, Kanori (1965). *Taiwan Bunkashi* [A Cultural History of Taiwan]. Vol. II. Tōkōshoin, Tokyo.

Johnston, B. F. (1966). 'Agriculture and Economic Development: the Relevance of the Japanese Experience', *Food Research Institute Studies,* **6** (3): 251-312.

Joint Commission on Rural Reconstruction (1956). *Taiwan Agricultural Statistics: 1901-55.* Taipei.

—— (1966). *T'aiwan Sheng Keng-yun chi Li-yung Tiao-ch'a Pao-kao* [A Report of a Survey on the Use of the Rotary Tiller Plough in Taiwan Province]. Department of Agriculture and Forestry of the Taiwan Provincial Government, Taichung.

Kajimoto, Michiyoi (1932). *Taiwan Nōgyō Ron* [Essays on Formosan Agriculture]. Shinkodo Shoten, Taipei.

Kao, C. H. (1967). 'An Analysis of Agricultural Output Increase on Taiwan, 1953-64', *Journal of Asian Studies,* **26** (4): 611-27.

Kuznets, S. (1966). *Economic Growth and Structure: Selected Essays.* Heinemann Educational Books Ltd, London.

Lewis, A. B. (1967). 'The Rice-fertiliser Barter Price and the Production of Rice in Taiwan, Republic of China: 1949-65', *Journal of Agricultural Economics*. Research Institute of Agricultural Economics. Chung-hsing University, Taichung.

Li, Teng-hui (1962). 'T'ai-wan chih Nung-ch'an Chia-ko' [Agricultural Commodity Prices in Taiwan], in Bank of Taiwan, *T'ai-wan chih Nung-yeh Ching-chi* [Taiwan's Farm Economy]. Taipei.

Liu, Fan-cheng (1965). *T'ai-wan Lü-hsing chi* [A Record of Travels in Taiwan]. Bank of Taiwan, Taipei.

Lo, Tsong-tseat (1961). *A Brief Report on Plant Diseases and their Control in Taiwan*. Joint Commission on Rural Reconstruction, Taipei.

Mochiji, Rokusaburō (1912). *Taiwan Shokumin Seisaku* [Taiwan Colonial Policy]. Fuzambo, Tokyo.

Moyoshi, Shirō (1933). 'Taiwan Chisōshi Kōron' [A Short Essay on the History of the Formosan Land Tax], *Taiwan Nōjihō*, (315): 81-113.

Myers, R. H. and Ching, A. (1964). 'Agricultural Development in Taiwan under Japanese Rule', *Journal of Asian Studies*, 23 (4): 555-70.

Nakamura, Zekō (1905). *Taiwan Tochi Chōsa Jigyō Gaiyō* [A Summary Account of the Land Survey Enterprise in Taiwan]. Taipei.

Rada, E. L. and Lee, T. H. (1963). 'Irrigation Investment in Taiwan —An Economic Analysis of Feasibility, Priority and Repayability Criteria'. Economic Digest Series No. 15. Joint Commission on Rural Reconstruction, Taipei.

Rinji Taiwan Kyūkan Chōsakai (1901). *Taiwan Shi-hō* [Private Law in Taiwan] II. Toyo Insatsu Kabushiki Kaisha, Tokyo.

Rinji Taiwan Tochi Chōsa Kyoku [Temporary Commission to Survey Taiwan's Land] (1900). *Shin-Fu Ichi-Han* [A Survey of Chi'ng Taxation]. Taiwan Nichi-nichi Shimpōsha, Taipei.

Shen, T. H. (1964). *Agricultural Development on Taiwan since World War II*. Comstock Publishing Associates, Ithaca.

Shigeno, Shin'ichi (1925). 'Hontō Nōson Seisaku to Hokō Seido no Zenyō [Village Policy in Taiwan and the Effective Use of the *Pao-chia* System], *Taiwan Nōjihō*, 19 (4): 292-308.

Society of Soil Scientists and Fertiliser Technologists of Taiwan (1963). *Soils and Fertiliser in Taiwan*. Taipei.

Suzuki, Shin'ichiro (1927). 'Hontō no Tochi Seido ni Tsuite' [The Formosan Land Tenure System], *Taiwan Nōjihō*, 21 (3): 2-13.

Taiwan Keizai Nempō Kankokai (1941). *Taiwan Keizai Nempō* [Taiwan Economic Yearbook]. Kokosai Nihon Kyōkai, Tokyo.

M

Taiwan Nōjihō (1924). *Hontō Kakushuchō ni okeru Suitō Naichishu Saibai ni tai suru Shisetsu* [Measures taken to Encourage the Cultivation of Japanese Rice in Various Prefectures of Formosa], **19** (5).

Taiwan Provincial Civil Affairs Bureau (1946). *T'ai-wan Sheng Wu-shih-i nien lai T'ung-chi T'i-yao* [A Statistical Summary of the Past 51 Years of Taiwan Province]. Taipei.

Taiwan Provincial Department of Agriculture and Forestry (1966). *Chung-pu Ch'en-hsing Nung-chia Shih-nien-lai Nung-ch'ang Ching-ying Chi-chang Tzu-liao chih Fen-hsi* [An Analysis of the Managerial Accounting Records of the Ch'en Farmstead in Central Taiwan for a Ten-Year Period]. Taichung.

Taiwan Provincial Food Bureau (1967). *Food Production and Activities of the Taiwan Provincial Food Bureau.* Taiwan.

Taiwan Sōtokufu Minseibu Zaimu Kyoku Zeimuka (1918). *Taiwan Zeimushi* [A History of Taiwan's Fiscal Affairs]. Vol. I. Nichi-nichi Shimpōsha, Taipei.

Taiwan Sōtokufu Shokusan Kyoku (1926). *Kakushu Kosaku Kankō Chōsa* [A Survey of Tenant Customs in Each Prefecture of Formosa]. Seishin Shōkō Insatsubu, Taipei.

—— (1930). *Taiwan ni okeru Kosaku Mondai ni Kansuru Shiryō* [Materials Concerning the Tenant Problem in Formosa]. Taihoku Insatsu Kabushiki Kaisha, Taipei.

—— (1931a). *Taiwan ni okeru Kosaku Kankō:Taichū* [Tenant Customs in Taiwan: Taichung Prefecture]. Shōchō Honten Insatsu Kōjō, Taipei.

—— (1931b). *Taiwan ni okeru Kosaku Kankō: Ako* [Tenant Customs in Taiwan: Kaohsiung Prefecture]. Kyōdō Insatsusho, Taipei.

—— (1931c). *Taiwan ni okeru Kosaku Kankō: Taitō* [Tenant Customs in Taiwan: Taitung Prefecture]. Taihoku Insatsu Kabushiki Kaisha, Taipei.

—— (1936). *Hontō Kosaku Kaizen Jigyō Seiseki Gaiyō* [A Summary of the Results Achieved in Improving the Conditions of the Formosan Tenant]. Taihoku Sanka Shoten, Taipei.

Takegoshi, Kōjirō (1905). *Tiwan Tōchishi* [A History of Japan's Rule of Taiwan]. Hakubun Insatsusho, Tokyo.

Tsui, Y. C. (1959). *A Summary Report on Farm Income of Taiwan in Comparison with 1952.* Joint Commission on Rural Reconstruction, Taipei.

—— and Lin, T. L. (1964). *A Study of Rural Labour Mobility in Relation to Industrialisation and Urbanisation in Taiwan.* Joint Commission on Reconstruction and Development, Taipei.

Wang, I-t'ao (1966). *Kuang-fu T'aiwan chih T'u-ti Chih-tu yü T'u-ti Cheng-ts'e* [The Land System and Land Policy in Taiwan Before the Return of Taiwan to China]. Bank of Taiwan, Taipei.

Yuan, M. H. (1964). 'The Fertiliser Industry of Taiwan', *Industry of Free China*, 21 (1): 2-21.

Chapter 3: India

Central Statistical Organisation (1964-). *Estimate of National Income*. Department of Statistics, New Delhi.

Chandrasekhar, S. (ed.) (1967). *Asia's Population Problems*. George Allen and Unwin, London.

Crawford, J. G. (1966). 'Planning Under Difficulties', *Australian Journal of Politics and History*, 12 (2): 155-76.

—— (1967). 'India: Planning and Administration', *Public Administration*, 26 (3): 227-44.

—— (1968). 'The Malthusian Spectre in India', *Australian Journal of Science*, 30 (10): 383-90.

Fertiliser Association of India (1965-). *Fertiliser Statistics*. New Delhi.

Hopper, W. David (1965). 'The Mainsprings of Agricultural Growth', *Dr Rajendra Prasad Memorial Lecture*, (to the Indian Society of Agricultural Statistics).

Indian Council of Agricultural Research (1966). *Annual Report 1964/5*. New Delhi.

Ministry of Finance (1967a). *Economic Survey 1966-67*. Government of India Press, New Delhi.

—— (1967b). *Indian Economic Statistics*. Government of India Press, New Delhi.

Ministry of Food and Agriculture (1963). *Indian Agriculture in Brief*. 6th ed. Government of India Press, New Delhi.

—— (1965a). *Agricultural Production in the Fourth Five Year Plan — Strategy and Programme*. Government of India Press, New Delhi.

—— (1965b). *Report of the Committee on Fertilizers*. Government of India Press, New Delhi.

—— (1966). *Growth Rates in Agriculture 1949/50 to 1964/5*. Government of India Press, New Delhi.

Ministry of Food and Agriculture, Community Development, and Cooperation (1967). 'Report on High Yielding Varieties Programme (Studies in Eight Districts, Kharif, 1966-67)'. New Delhi (mimeo.).

Panse, V. G., Amble, V. N., and Abraham, T. P. (1964). 'A Plan for Improvement of Nutrition of India's Population', *Indian Journal of Agricultural Economics*, **19** (2): 13-40.

Planning Commission (1953). *First Five Year Plan*. Government of India Press, New Delhi.

—— (1956). *Second Five Year Plan*. Government of India Press, New Delhi.

—— (1959). *Reports of the Committees of the Panel on Land Reform*. Government of India Press, New Delhi.

—— (1961). *Third Five Year Plan*. Government of India Press, New Delhi.

—— (1964). *Memorandum on the Fourth Plan*. Government of India Press, New Delhi.

—— (1966a). *Fourth Five Year Plan: A Draft Outline*. Government of India Press, New Delhi.

—— (1966b). *Annual Plan 1966/7*. Government of India Press, New Delhi.

—— (1967). *Annual Plan, 1967/8*. Government of India Press, New Delhi.

Reserve Bank of India (1955). *1951 Rural Credit Survey*. Bombay.

—— (1964). *Performance of Co-operative Credit*. Bombay. Rockefeller Foundation (1967). *The President's Review and Annual Report*. New York.

Sharma, P. (1965). 'A Study of the Structural and Tenurial Aspects of Rural Economy in the Light of the 1961 Census', *Indian Journal of Agricultural Economics*, **20** (4): 46-82.

Subramaniam, C. (1964). 'Speech by the Minister for Food and Agriculture to the Indian Council of Agricultural Research'. New Delhi (mimeo).

Sukhatme, P. V. (1965). *Feeding Asia's Growing Millions*. Asia Publishing House, Bombay.

Chapter 4: Thailand

Caldwell, J. C. (1967). 'The Demographic Structure', in T. H. Silcock (ed.), *Thailand: Social and Economic Studies in Development*. A.N.U.P., Canberra.

Chapman, E. C. and Allen, A. B. (1965). 'Internal Migration in Thailand'. Paper presented to ANZAAS Conference, Hobart, Tasmania, 1963.

Chuchart, Chaiyong, *et al.* (1959a). *Economic Survey of Land Co-operatives in the North East of Thailand*. Kasetsart University Economic Report No. 4. Bangkok.

—— (1959b). *Costs and Returns on Korat Farm Enterprises in 1957*. Kasetsart University Economic Report No. 5. Bangkok.

—— (1961). *Production and Marketing Problems Affecting the Expansion of Kenaf and Jute in Thailand, 1961*. Kasetsart University Economic Report No. 7. Bangkok.

—— (1962). *Production and Marketing Problems Affecting the Expansion of Corn Growing in Thailand*. Kasetsart University Economic Report No. 8. Bangkok.

Dept of Customs (1962). *Annual Statement of Foreign Trade of Thailand*. Bangkok.

—— (1963). *Annual Statement of Foreign Trade of Thailand*. Vol. II. *Exports and Re-exports*. Bangkok.

Embree, J. F. (1950). 'Thailand—A Loosely Structured Social System', *American Anthropologist*, 52.

Evers, H. D., and Silcock, T. H. (1967). 'Elites and Selection', in T. H. Silcock (ed.), *Thailand: Social and Economic Studies in Development*. A.N.U.P., Canberra.

Ingram, J. C. (1955). *Economic Change in Thailand since 1850*. Stanford U.P., Stanford.

IBRD (International Bank for Reconstruction and Development) (1959). *A Public Development Program for Thailand*. Johns Hopkins Press, Baltimore.

Mahapol, S. (1954). *Teak in Thailand*. Ministry of Agriculture, Bangkok.

McFadyean, A. (1944). *The History of Rubber Regulation*. George Allen and Unwin, London.

McFarland, G. B. (1944). *Thai-English Dictionary*. Stanford U.P., Stanford.

Ministry of Agriculture (1955-). *Agricultural Statistics of Thailand*. Bangkok.

—— (1959-). *Crop Reports*. Product Series, Bangkok.

—— (1962). *Annual Report on Rice Production in Thailand*. Rice Department, Bangkok.

Nakhorn Sawan Provincial Government (1964). *Annual Report*. Government Office (in Thai).

Narkswasdi, Udhis (1963). *Agricultural Credit System in Certain Countries*. Kasetsart University, in co-operation with USOM/Thailand, Bangkok.

NEDB (National Economic Development Board) (1964a). *The National Economic Development Plan, 1961-1966, Second Phase: 1964-66*. Bangkok.

—— (1964b). *National Income Statistics of Thailand*. 1964 ed., Bangkok.

—— (1966). *Performance Evaluation of Development in Thailand for 1965*. Bangkok.

National Statistical Office (1960). *Census of Population*. Bangkok.

—— (1963a). *Report of the Labor Force Survey (Round 1)*. Bangkok.

—— (1963b). *Report of the Labor Force Survey (Round 4)*. Bangkok.

—— (1965a). *Census of Agriculture, Nakhorn Sawan*. Bangkok.

—— (1965b). *Statistical Year Book of Thailand*. Bangkok.

Office of the Prime Minister (1962-). *Budget in Brief*. Budget Bureau, Bangkok.

—— (1965). *Budget Documents*. Budget Bureau, Bangkok (in Thai).

Phanupongse, Chira (1964). *Rai-ngan Kansüksa Saphab Ang-kep-nam Cholaprathan nai Phak Tawan-awk-chiang-nüa* [Report of the Study on Irrigation Tanks in the Northeast]. Ministry of National Development, Bangkok.

Samapuddhi, Krit (1957). *The Forests of Thailand and Forestry Programs*. Ministry of Agriculture, Bangkok.

Sharp, L. *et al.* (1953). *Siamese Rice Village: A Preliminary Study of Bang Chan, 1948-1949*. Cornell Research Center, Bangkok.

Silcock, T. H. (ed.) (1967a). *Thailand: Social and Economic Studies in Development*. A.N.U.P., Canberra.

—— (1967b). 'The Rice Premium and Agricultural Diversification', in Silcock (ed.), *Thailand*. A.N.U.P., Canberra.

—— (1967c). 'Promotion of Industry and the Planning Process', in Silcock (ed.), *Thailand*. A.N.U.P., Canberra.

—— (1968). 'Economic Effects of Thai Policy at the End of World War II'. Paper presented at ANZAAS, Christchurch, N.Z., 1968.

—— (in press). *The Economic Development of Thai Agriculture*. A.N.U.P., Canberra.

Skinner, G. W. (1957). *Chinese Society in Thailand*. Cornell U.P., Ithaca.

United Nations/ECAFE (1964). *Committee for the Co-ordination of Investigations of the Lower Mekong Basin*. Bangkok.

Usher, D. (1966). 'Income as a Measure of Productivity: Alternative Comparisons of Agricultural and Non-Agricultural Productivity in Thailand', *Economica*, 33 (132): 430-41.

—— (1967). 'The Thai Rice Trade', in T. H. Silcock (ed.), *Thailand*. A.N.U.P., Canberra.

USOM/Thailand (1960). *Economic Survey of the Korat–Nongkai Highway Area*. Bangkok.

Wijeyewardene, J. (1967). 'Some Aspects of Rural Life in Thailand', in T. H. Silcock (ed.), *Thailand*. A.N.U.P., Canberra.

Chapter 5: The Union of Burma

Andrus, J. Russell (1947). *Burmese Economic Life.* Stanford U.P., Stanford.

Aye Hlaing, U (1958). *Agro-economic Problems in Burma* and *Some Aspects of Seasonal Agricultural Loans in Burma.* University of Rangoon, Rangoon.

Aye Maung, U (1954). *Country Review CR-7.* United Nations FAO, Rome.

Burma Provincial Banking Enquiry Committee (1930). *Report* (1929-30). Rangoon.

Cady, J. F. (1958). *A History of Modern Burma.* Cornell U.P., Ithaca.

CSED (Central Statistical and Economics Department) (1952-). *Quarterly Bulletin of Statistics.* Rangoon.

—— (1962-). *Selected Monthly Economic Indicators.* Rangoon.

—— (1967-). *Statistical Pocketbook.* Rangoon.

Christian, J. L. (1942). *Modern Burma: A Survey of Political and Economic Development.* University of California Press, Berkeley.

DI (Directorate of Information) (1957). *Premier U Nu on the 4-Year Plan.* Rangoon.

—— (1962a-). *Forward.* Rangoon.

—— (1962b). *Is Trust Vinidicated?* Rangoon.

—— (n.d., c. 1965). *Burma: National Economy, 1963-64.* Rangoon.

ESB (Economic and Social Board) (1954). *Pyidawtha: The New Burma.* Rangoon.

Fisher, C. A. (1964). *South-east Asia.* Methuen, London.

Furnivall, J. S. (1931). *An Introduction to the Political Economy of Burma.* Peoples' Literary Committee and House, Rangoon.

—— (1948). *Colonial Policy and Practice: A Comparative Study of Burma and Netherlands India.* Cambridge U.P., Cambridge.

—— (1958). *The Governance of Modern Burma.* Institute of Pacific Relations, New York.

Guardian, The. Rangoon.

Hagen, E. E. (1956). *The Economic Development of Burma.* National Planning Association, Washington.

—— (1962). *On the Theory of Social Change.* Massachusetts Institute of Technology, Cambridge, Mass.

Hall, D. G. E. (1956). *Burma.* 2nd ed., Hutchinson, London.

—— (1958). *A History of South-East Asia.* Macmillan, London.

Harvey, G. E. (1925). *History of Burma From the Earliest Times to 1824.* Longmans Green, London.

—— (1946). *British Rule in Burma, 1824-1942.* Faber and Faber, London.

KTA (Knappen Tippetts Abbett, associated with Pierce Management, Inc., and Robert R. Nathan Associates, Inc.) (1952). *Preliminary Report on Economic and Engineering Survey of Burma for Burma Economic Council*. Hazell, Watson and Viney, London.

———— (1953). *Comprehensive Report, Economic and Engineering Development of Burma*, 2 vols. Hazell, Watson and Viney, London.

Jacoby, E. H. (1949). *Agrarian Unrest in Southeast Asia*. Columbia U.P., New York.

Laurence French Publications (1946-). *Far East Trade and Development*. London.

Levin, J. (1960). *The Export Economies: Their Pattern of Development in Historical Perspective*. Harvard U.P., Cambridge, Mass.

Mali, K. S. (1962). *Fiscal Aspects of Development Planning in Burma, 1950-1960*. University of Rangoon, Rangoon.

MNP (Ministry of National Planning) (1951a-). *Economic Survey of Burma*. Rangoon.

———— (1951b-). *The National Income of Burma*. Rangoon.

———— (1961). *Second Four-Year Plan, 1961-62 to 1964-5*. Rangoon.

Nash, Manning (1965). *The Golden Road to Modernity*. Wiley, New York.

Nation, The. Rangoon.

Richter, H. V. (1968). 'State Agricultural Credit in Postwar Burma', *Malayan Economic Review*, 13 (1): 101-17.

Spate, O. H. K. (1941). 'The Beginnings of Industrialisation in Burma', *Economic Geography*, 17 (1): 75-92.

Thet Tun, U (1960). 'Burma's Experience in Economic Planning', *The Open Mind*, 1 (12): 1-47.

Tinker, H. (1957). *The Union of Burma: A Study of the First Years of Independence*. Oxford U.P., London.

Trager, F. N. (1958). *Building a Welfare State in Burma 1948-1956*. Institute of Pacific Relations, New York.

Tun Wai, U (1953). *Burma's Currency and Credit*. Orient Longmans, Calcutta.

———— (1961). *Economic Development of Burma from 1800 to 1940*. University of Rangoon, Rangoon.

Union Bank of Burma (1952-). *Monthly Review*.

UBRC (Union of Burma Revolutionary Council) (1964-). *Report to the People by the Union of Burma Revolutionary Council on the Revolutionary Government's Budget Estimates*, 1964-65, 1965-6, 1966-7. Rangoon.

United Nations FAO (June 1967). *Monthly Bulletin of Agricultural Economics and Statistics*, Rome.

—— (1962). *Progress in Land Reform*, 3rd Report. New York.

University of Rangoon, Dept of Economics, Statistics and Commerce (1957). *Village Study Series*, No. 1, Okpo; No. 2, Wanetkon; No. 3, Kyungale; No. 4, Kyaukanya. Rangoon.

—— (n.d., *c.* 1959). *The Economic Development of Burma*. Rangoon.

—— (1959-61). *Economic Papers*, No. 5 (1959); Nos. 6 and 7 (1961). Rangoon.

Walinsky, L. J. (1962). *Economic Development in Burma 1951-60*. Twentieth Century Fund, New York.

Wickizer, V. D. and Bennett, M. K. (1941). *The Rice Economy of Monsoon Asia*. Stanford U.P., Stanford.

Working People's Daily. Rangoon.

Chapter 6: Malaysia

Asian Development Bank (1968). *Asian Agricultural Survey*, Vol. 1. Manila.

Bank Negara Malaysia (Central Bank of Malaysia) (1967). *Annual Report and Statement of Accounts 1966*. Kuala Lumpur.

Barlow, Colin and Lim Sow Ching (1968). 'Natural Rubber and West Malaysia'. Paper presented at the Singapore meeting of the South-East Asia Business Committee, 12-15 May (mimeo.).

Bauer, P. T. (1948). *The Rubber Industry*. Longmans, London.

—— (1961). 'Malayan Rubber Policy', in T. H. Silcock (ed.), *Readings in Malayan Economics*. Eastern U.P., Singapore.

Ding, Eing Tan Soo Hai (1963). *The Rice Industry in Malaya*. Department of History, University of Singapore, Singapore.

Federal Agricultural Marketing Authority (1967). *Annual Report and Statement of Accounts 1965 and 1967*. Kuala Lumpur.

Federation of Malaya (1954). *Report of the Mission of Enquiry into the Rubber Industry of Malaya*. Government Printer, Kuala Lumpur.

—— (1955). *Taxation and Replanting in the Rubber Industry*. Government Printer, Kuala Lumpur.

—— (1959a). *Household Budget Survey of the Federation of Malaya, Report for Year 1957-58*. Department of Statistics, Kuala Lumpur.

—— (1959b). *Report of the Drainage and Irrigation Department for the Years 1955, 1956 and 1957*. Government Printer, Kuala Lumpur.

———— (1960a). *National Accounts of the Federation of Malaya 1955-1960*. Department of Statistics, Kuala Lumpur.

———— (1960b). *Malaya Rubber Statistics Handbook, 1960*. Department of Statistics, Kuala Lumpur.

———— (1961a). *Monthly Statistical Bulletin of the Federation of Malaya, August 1961*. Department of Statistics, Kuala Lumpur.

———— (1961b). *Second Five-Year Plan 1961-1965*. Government Printer, Kuala Lumpur.

Ferguson, C. G. (1965). 'The Story of Development in Malaya (now Malaysia)—Some Aspects', *Journal of Local Administration Overseas*, 4 (3): 149-64.

Fisk, E. K. (1962). 'Establishment Costs of Small Rice Farms: An Analytical Model of Returns to Capital', *Malayan Economic Review*, 7 (2): 45-63.

———— (1964). *Studies in the Rural Economy of South East Asia*. University of London Press, Singapore.

Freeman, J. D. (1955). *Iban Agriculture*. H.M.S.O., London.

Grist, D. H. (1955). *Rice*. 2nd ed. Longmans Green, London.

IBRD (International Bank for Reconstruction and Development) (1955). *The Economic Development of Malaya*. Government Printer, Singapore.

Jackson, James C. (1968). *Planters and Speculators*. University of Malaya Press, Singapore.

Koh Theam Hee and Liao Hsing Chia (1966). *Farm Surveys and Farmers' Association Business Service Plan*. Department of Agriculture, Kuala Lumpur.

Malaysia (1964), *National Accounts of West Malaysia 1955-1964*. Department of Statistics, Kuala Lumpur.

———— (1965a). *Rubber Statistics Handbook 1965*. Department of Statistics, Kuala Lumpur.

———— (1965b). *First Malaysia Plan 1966-1970*. Kuala Lumpur.

———— (1966a). *Rice Statistics for West Malaysia 1966*. Department of Statistics, Kuala Lumpur.

———— (1966b). *Rubber Statistics 1966 of West Malaysia (Preliminary Release)*. Department of Statistics, Kuala Lumpur.

———— (1966c). *Portions of Draft Budget Speech for Hon'able The Minister of Finance Dec. 1966*. Department of Statistics, Kuala Lumpur.

———— (1967a). *Malaysian Balance of Payments Estimates 1961 to 1966 and Forecasts for 1967 and 1968 (1st Half)*. Department of Statistics, Kuala Lumpur.

———— (1967b). *Portions of Draft Budget Speech for Hon'able The Minister of Finance 1967*. Department of Statistics, Kuala Lumpur.

—— (1967c). *Monthly Statistical Bulletin of West Malaysia, August 1967*. Department of Statistics, Kuala Lumpur.

—— (1967d). *Malaysia 1965 Official Year Book*. Government Printer, Kuala Lumpur.

Mohamed bin Jamil and Koh Theam Hee (1966). *The Development of Farmers' Association in Malaysia as a Unit for Extension Programme Planning and Implementation of Agricultural Projects*. Department of Agriculture, Kuala Lumpur.

Narkswasdi, Udhis (1968). *A Report to the Government of Malaysia of the Rice Economy of West Malaysia*. FAO, Rome.

—— and Selvadurai, S. (1967a). *Economic Survey of Padi Production in West Malaysia—Report No. 1—Selangor*. Ministry of Agriculture and Co-operatives, Kuala Lumpur.

—— and —— (1967b). *Economic Survey of Padi Production in West Malaysia—Report No. 2—Collective Padi Cultivation in Bachang, Malacca*. Ministry of Agriculture and Co-operatives, Kuala Lumpur.

—— and —— (1967c). *Economic Survey of Padi Production in West Malaysia—Report No. 3—Malacca*. Ministry of Agriculture and Co-operatives, Kuala Lumpur.

Rubber Research Institute of Malaya (1966). *Annual Report 1965*. Kuala Lumpur.

Rueff, Jacques (1963). *Report on the Economic Aspects of Malaysia*. Government Printer, Kuala Lumpur.

Silcock, T. H. and Fisk, E. K. (eds.) (1963). *The Political Economy of Independent Malaya*. A.N.U.P., Canberra.

Singh, Sumer (1965). 'Economic Aspects of Three New Land Development Schemes Organised by the Federal Land Development Authority in the Federation of Malaya', Ph.D. thesis, Australian National University, Canberra.

Tan Siew Sin (1968). *Malaysia, the 1968 Budget*. Federal Department of Information, Kuala Lumpur.

Chapter 7: The Philippines

Abelarde, P. E. (1947). *American Tariff Policy Towards the Philippines*. King's Crown Press, New York.

Barker, Randolph (1966). 'The Response of Production to a Change in Price', *The Philippine Economic Journal*, 5 (2): 260-76.

—— (1968). 'The Role of the International Rice Research Institute in the Development and Dissemination of New Rice Varieties'. Paper presented at the University of Reading International Seminar on Changes in Agriculture, September 3-14, 1968.

Barrett, O. W. (1946). *Coconuts.* Panama Pacific International Exposition, San Francisco.

Bureau of Census and Statistics (1918-). *Philippine Census of Agriculture.* Manila.

––– (1947). *Yearbook of Philippine Statistics, 1946.* Manila.

––– (1960). *Census of the Philippines, 1960, Agriculture.* Summary, Vol. 2. Manila.

Bureau of Commerce and Industry (1929). *Statistical Bulletin of the Philippine Islands.* Manila.

Caintic, C. U. *et al.* (1959). *Management Practices, Costs and Returns of Sugarcane Farms in the Victorias Milling District.* Technical Bulletin No. 10, College of Agriculture, University of the Philippines, Laguna.

Central Bank of the Philippines (1950-). *Statistical Bulletin.* Manila.

Coronas, José (1920). *The Climate and Weather of the Philippines, 1903-18.* Bureau of Printing, Manila.

Department of Agriculture and Natural Resources (1955-). *Philippine Agricultural Statistics,* Vols. 1 and 2. Bureau of Printing, Manila.

––– (1958-9). *Crop and Livestock Statistics.* Quezon City (also earlier issues).

––– (1964). 'Rice: Area, Production and Yield per Hectare by Region'. Quezon City (mimeo.) (also earlier years).

Estanislao, J. P. (1965). 'A Note on Differential Farm Productivity by Tenure', *The Philippine Economic Journal,* 4 (1): 120-4.

Fonollera, Raymundo E. (1966). 'Labour and Land Resources in Philippine Agriculture: Trends and Projections'. M.A. thesis, University of the Philippines, Laguna.

Golay, Frank H. (1961). *The Philippines: Public Policy and National Economic Development.* Cornell U.P., Ithaca.

––– and Goodstein, Marvin E. (1967). *Rice and People in 1990— Philippine Rice Needs to 1990: Output and Input Requirements.* United States Agency for International Development, Manila.

Goodstein, Marvin, E. (1963). *The Pace and Pattern of Philippine Economic Growth, 1938, 1948 and 1959.* Cornell U.P., Ithaca.

Hartendorp, A. V. H. (1958). *History of Industry and Trade in the Philippines.* American Chamber of Commerce of the Philippines, Manila.

Hayden, Joseph R. (1945). *The Philippines: A Study in National Development.* Macmillan & Co., New York.

Hicks, George L. (1967). *The Philippine Coconut Industry: Growth and Change, 1900-1965.* Field Report No. 17, National Planning Association, New York.

Hooley, R. W. (1966). 'Long Term Economic Growth in the Philippines', in Hooley and Barker (eds.), *Growth of Output in the Philippines*. International Rice Research Institute, Laguna.

—— and Barker, Randolph (eds.) (1967). *Growth of Output in the Philippines*. International Rice Research Institute, Laguna.

Hsieh, S. C. and Ruttan, V. W. (1967). 'Environmental, Technological and Institutional Factors in the Growth of Rice Production: Philippines, Thailand and Taiwan', *Food Research Institute Studies*, 7 (3): 307-41.

Ilag, L. M. (1964). 'Farm Management Analyses of Some Sugarcane Farms in the Victorias Mill District, 1961-62'. M.A. thesis, University of the Philippines, Laguna.

Krishna, Raj (1967). 'Agricultural Price Policy and Economic Development', in Southworth and Johnston (eds.), *Agricultural Development and Economic Growth*, Cornell U.P., Ithaca.

Lawas, José M. (1968). 'Output Growth, Technical Change and Employment of Resources in Philippine Agriculture: 1948-1975'. Ph.D. thesis, Purdue University, West Lafayette.

Legarda, Fernández B. (1962). 'Foreign Exchange Control and the Redistribution of Income Flows', *The Philippine Economic Journal*, 1 (1): 18-28.

Mangahas, Mahar, Recto, Aida E., and Ruttan, V. W. (1966a). 'Market Relationships for Rice and Corn in the Philippines', *The Philippine Economic Journal*, 5 (1): 1-27.

—— (1966b). 'Price and Market Relationships for Rice and Corn in the Philippines', *Journal of Farm Economics*, 48 (3): 685-703.

Mitchell, Kate (1942). *Industrialization in the Western Pacific*. Institute of Pacific Relations, New York.

Morrow, Robert B. (1966). 'Palay Production Differentials by Tenure Class and School Achievement', *The Philippine Economic Journal*, 5 (2): 380-5.

Nyberg, A. J. (1966). 'The Growth of the Philippine Coconut Industry 1901-66', in Hooley and Barker (eds.), *Growth of Output in the Philippines*. International Rice Research Institute, Laguna.

Philippine Sugar Association (1964). 'Handbook of the Sugar and Other Industries in the Philippines'. Manila.

Quintana, E. U. (1965). 'Resource Productivity Estimates for Five Types of Philippine Farm', in Sicat *et al.*, *Economics and Development*. University of the Philippines, Quezon City.

Recto, Aida E. (1965). 'Price and Market Relationships for Corn in the Philippines'. M.A. thesis, University of the Philippines, Laguna.

Ruprecht, T. K. (1966). 'Labour Absorption Problems and Economic Development in the Philippines', *The Philippine Economic Journal*, 5 (2): 289-312.

Ruttan, V. W. (1965). 'Land Reform and National Economic Development', in Sicat, G. P. (ed.), *The Philippine Economy in the 1960's*. University of the Philippines, Quezon City.

—— (1966). 'Tenure and Productivity in Philippine Rice Producing Farms', *The Philippine Economic Journal*, 5 (1): 42-63.

—— (in press). 'Agricultural Product and Factor Markets in Southeast Asia', *Economic Development and Cultural Change*.

Sicat, G. P. (ed.) (1965). *The Philippine Economy in the 1960's*. Institute of Economic Development and Research, University of Philippines, Quezon City.

—— (1966). 'The Import Dependence of Import Substitution'. School of Economics, University of the Philippines, Quezon City (mimeo.).

—— *et al.* (1965). *Economics and Development*. University of the Philippines, Quezon City.

Taylor, George E. (1964). *The Philippines and the United States: Problem of Partnership*. Praeger, New York.

Treadgold, Malcolm and Hooley, Richard (1967). 'Decontrol and the Redirection of Income Flows: A Second Look', *The Philippine Economic Journal* (in press).

Wernstedt, F. L. and Spencer, J. Earle (1967). *The Philippine Island World: A Physical, Cultural and Regional Geography*. University of California Press, Berkeley.

Chapter 8: Indonesia

Bailey, K. V. (1961-2). 'Rural Nutrition Studies in Indonesia', *Tropical and Geographical Medicine*, 13 and 14, Amsterdam.

Bauer, P. T. (1948). *The Rubber Industry*. Harvard U.P., Cambridge, Mass.

Biro Pusat Statistik (Central Statistical Bureau) (1961). *Census of Population, 1961*. Djakarta.

—— (1966a). *Pendapatan Nasional Indonesia 1958-62* [The National Income of Indonesia]. Djakarta.

—— (1966b). *Sensus Pertanian, 1963 (Agricultural Census)*. Djakarta.

—— (1966c). *Sensus Pertanian, Sektor Perkebunan, 1963 (Agricultural Census: Estate Sector)*. Djakarta.

—— (1966d). *National Sample Surveys, 1963 and 1964*. Djakarta.

—— (1967). Bagian Laporan dan Analisa Ekonomi, 'Kemerosotan Daja Tukar Petani'. *Seminar Rentjana Pembangunan Pertanian Lima Tahun, 1969-73*. Jogjakarta.

Boeke, J. H. (1953). *Economics and Economic Policy of Dual Societies*. H. D. Tjeenk Willink and Zoon, Haarlem.

Department of Statistics (1967). *Monthly Statistical Bulletin of West Malaysia*. Kuala Lumpur.

Dewey, A. (1962). *Peasant Marketing in Java*. Free Press of Glencoe, New York.

ECAFE (1966). *Economic Survey of Asia and the Far East*. Bangkok.

Furnivall, J. S. (1944). *Netherlands India*. Cambridge U.P., Cambridge.

Geertz, C. (1963). *Agricultural Involution*. University of California Press, Berkeley and Los Angeles.

Gourou, P. (1953). *The Tropical World* (trans. E. D. Laborde). Longmans Green, London.

Iso, R. and Soedarsono, H. (1960). 'Perubahan Kepadatan Penduduk dan Penghasilan Bahan Makanan', *Agricultura*, Jogjakarta, (1): 3-107.

Jones, G. W. (1966). 'The Growth and Changing Structure of the Indonesian Labour Force, 1930-81', *Bulletin of Indonesian Economic Studies*, 4: 50-74.

Kampto Utomo (1967). 'Villages of Unplanned Resettlers in the Subdistrict Kaliredjo, Central Lampung', in Koentjaraningrat (ed.), *Villages in Indonesia*. Cornell U.P., Ithaca.

Kartono H. and Susilo (1967). 'Peranan Sensus dan Sample Survey Bagi Perkembangan Statistik Pertanian di Indonesia', *Seminar Rentjana Pembangunan Pertanian Lima Tahun, 1969-73*. Jogjakarta.

McVey, R. (ed.) (1963). *Indonesia*. Yale U.P. by arrangement with H.R.A.F. Press, New Haven.

Mears, L. A. (1961). *Rice Marketing in the Republic of Indonesia*. P. T. Pembangunan, Djakarta.

Mubyarto and Fletcher, L. (1966). *The Marketable Surplus of Rice in Indonesia. A Study in Java-Madura*. International Studies in Economics No. 4, Iowa State University, Ames.

Napitupulu, B. (1968). 'Hunger in Indonesia', *Bulletin of Indonesian Economic Studies*, (9): 60-70.

Nugroho (1967). *Indonesia: Facts and Figures*. Published privately, Djakarta.

Paauw, D. S. (1959). *Financing Economic Development*. Free Press of Glencoe, New York.

—— (1961). 'Some Frontiers of Empirical Research in Economic Development', *Economic Development and Cultural Change*, 9 (2): 180-91.

—— (1963a). 'Economic Progress in South East Asia', *Journal of Asian Studies*, 23 (1): 69-92.

—— (1963b). 'From Colonial to Guided Economy', in Ruth McVey (ed.), *Indonesia*. Yale U.P. by arrangement with H.R.A.F. Press, New Haven.

Pelzer, K. J. (1945). *Pioneer Settlement in the Asiatic Tropics*. American Geographical Society, Special Publication No. 29, New York.

Penny, D. H. (1964). 'The Transition from Subsistence to Commercial Family Farming in North Sumatra'. Ph.D. thesis, Cornell University, Ithaca.

—— (1966). 'The Economics of Peasant Agriculture: The Indonesian Case', *Bulletin of Indonesian Economic Studies*, (5): 22-44.

—— (1967). 'Development Opportunities in Indonesian Agriculture', *Bulletin of Indonesian Economic Studies*, (8): 35-64.

Polak, J. J. (1943). *The National Income of the Netherlands Indies, 1921-39*. Institute of Pacific Relations, New York.

Polanyi, K. *et al.* (eds.) (1957). *Trade and Markets in the Early Empires*. Free Press of Glencoe, New York.

Rangkuty, R. (1966). 'Peranan Penjuluhan dalam Perkembangan Pertanian Rakjat di Sumatera Utara'. Fakultas Pertanian, Medan (typescript).

Richter, H. V. (1966). 'Indonesia's Share in the Entrepot Trade of Malaya and Singapore prior to Confrontation', *Malayan Economic Review*, 11 (2): 43-4.

Roekasah, E. A. and Penny, D. H. (1967). 'Bimas: A New Approach to Agricultural Extension in Indonesia', *Bulletin of Indonesian Economic Studies*, (7): 60-9.

Selosoemardjan (1962). *Social Changes in Jogjakarta*. Cornell U.P., Ithaca.

Soedigdo, H. (1965). *Kebidjaksanaan Transmigrasi*. Bhratara, Djakarta.

Survey Agro-Ekonomi (in press a). *Produksi, Tataniaga, dan Export Djagung di Djawa Timur*. S.A.E.I. Departemen Pertanian, Djakarta.

—— (in press b). *Survey Report*. S.A.E.I. Departemen Pertanian, Djakarta.

Thomas, K. D. (1965). 'Shifting Cultivation and the Production of Smallholder Rubber in a South Sumatran Village', *Malayan Economic Review*, 10 (1): 100-15.

—— and Panglaykim, J. (1966-7). 'Indonesian Exports, Performance and Prospects, 1950-1970', *Bulletin of Indonesian Economic Studies*, (5): 71-102; (6): 66-88.

Timmer, M. (1961). *Child Mortality and Population Pressure in the D.I. Jogjakarta, Java, Indonesia*. Bronder Offset, Rotterdam.

Vries, E. de (1937). 'Rijstpolitik op Java in Vroeger Jaren', *Landbouw*, Dertiende Jaargang, 7 and 8.

Wertheim, W. F. (1956). *Indonesian Society in Transition*. Van Hoeve, The Hague.

—— (1964). *East-West Parallels*. Van Hoeve, The Hague.

Wharton, C. R. jun. (1962a). *Economic and Non-Economic Factors in the Agricultural Development of Southeast Asia: Some Research Priorities*. The Council on Economic and Cultural Affairs, New York.

—— (1962b). *The Inelasticity of South-East Asian Agriculture*. Paper of the Agricultural Economics Society of Thailand, Bangkok.

Zulkifli, M. (1962). 'Ongkos2 Perpasaran untuk Beberapa Hasil Bumi di Sumatera Utara', *Seri Penelitian*, Laporan Ke-1. Fakultas Pertanian, Medan.

Chapter 9: Papua-New Guinea

Anon. (1967). 'T.P.N.G.—the Economy in the Sixties', *Current Affairs Bulletin*, **39** (4). Department of Adult Education, University of Sydney.

Bureau of Agricultural Economics (1951). *An Economic and Cost Survey of the Copra Industry in the Territory of Papua and New Guinea*. Government Printer, Canberra.

Commonwealth of Australia (1941a). *Report to the Council of the League of Nations on the Administration of the Territory of New Guinea, 1st July 1939—30th June 1940*. Government Printer, Canberra.

—— (1941b). *Territory of Papua: Annual Report for the Year 1940-1941*. Government Printer, Canberra.

Crocombe, R. G. (1967). 'A Canberra View of Economic Development in New Guinea', *Australian Journal of Agricultural Economics*, **11** (2): 208-11.

Department of Territories (1964). *National Income Estimates for Papua and New Guinea, 1960-61—1962-3*. Canberra.

—— (1967). *Compendium of Statistics for Papua and New Guinea*. Government Printer, Canberra.

Epstein, T. Scarlett (1968). *Capitalism, Primitive and Modern*. A.N.U.P., Canberra.

Fisk, E. K. (1962). 'Planning in a Primitive Economy: Special Problems of Papua–New Guinea', *Economic Record*, **38**: 462-78.

—— (1964). 'Planning in a Primitive Economy: From Pure Subsistence to the Production of a Market Surplus', *Economic Record*, **40**: 156-74.

IBRD (International Bank for Reconstruction and Development) (1965). *The Economic Development of the Territory of Papua and New Guinea*. Johns Hopkins Press, Baltimore.

Klein, W. C. (1937). 'Economic Advancement of New Guinea', *Asiatic Review*, **33**: 568-75.

Mann, C. E. T. (1953). 'Investigation of the Rubber Industry in Papua and New Guinea'. Report to the Commonwealth of Australia (mimeo.).

Shand, R. T. (1963). 'The Development of Cash Cropping in Papua and New Guinea', *Australian Journal of Agricultural Economics*, **7** (1): 42-54.

—— (1965). 'The Development of Trade and Specialisation in a Primitive Economy', *Economic Record*, **41**: 193-206.

—— (1966). 'Trade Prospects for the Rural Sector', in E. K. Fisk (ed.), *New Guinea on the Threshold*. A.N.U.P., Canberra.

Territory of Papua and New Guinea (1959-60). *Production Bulletin*. Pt 1, Rural Industries, Nos. 1-8. Konedobu, Papua.

—— (1965-6). *Overseas Trade*. Bureau of Statistics, Konedobu.

—— (1967). *Economic Development of Papua and New Guinea*. Bureau of Statistics, Konedobu.

—— (1968). *Preliminary Bulletin* No. 19. Bureau of Statistics, Konedobu.

West, F. J. (1966). 'The Historical Background', in E. K. Fisk (ed.), *New Guinea on the Threshold*. A.N.U.P., Canberra.

White, R. C. (1964). *Social Accounts of the Monetary Sector of the Territory of Papua and New Guinea, 1956-57 to 1960-61*. New Guinea Research Unit Bulletin No. 3, New Guinea Research Unit, Australian National University, Canberra.

Zmudski, W. R. (1968). 'National Income of Papua and New Guinea', in *The Encyclopaedia of Papua and New Guinea*. Melbourne U.P., Melbourne.

Index

Index

Text set in 10 point Baskerville face, one point leaded, and printed on 85 gsm Burnie English Finish by Halstead Press Pty Ltd, Sydney